Boundary Collocation Techniques and their Application in Engineering

Boundary Collocation Techniques and their Application in Engineering

J.A. Kołodziej
Poznan University of Technology, Poland

&

A.P. Zieliński
Cracow University of Technology, Poland

WIT*PRESS* Southampton, Boston

J.A. Kołodziej
Poznan University of Technology, Poland

A.P. Zieliński
Cracow University of Technology, Poland

Published by

WIT Press
Ashurst Lodge, Ashurst, Southampton, SO40 7AA, UK
Tel: 44 (0) 238 029 3223; Fax: 44 (0) 238 029 2853
E-Mail: witpress@witpress.com
http://www.witpress.com

For USA, Canada and Mexico

WIT Press
25 Bridge Street, Billerica, MA 01821, USA
Tel: 978 667 5841; Fax: 978 667 7582
E-Mail: infousa@witpress.com
http://www.witpress.com

British Library Cataloguing-in-Publication Data

A Catalogue record for this book is available
from the British Library

ISBN: 978-1-84564-394-2

Library of Congress Catalog Card Number: 2009921879

The texts of the papers in this volume were set individually by the authors or under their supervision.

Printed in Great Britain by Athenaeum Press Ltd.

The first conference "Trefftz Method" organized in Cracow in 1996.
The persons in the photo are (from left):
S.Yu. Reutskiy (Ukraine), J.A. Teixeira de Freitas (Portugal), S. Abo Diab (Syria), J.A.Kołodziej (Poland), B. Tirozzi (Italy), H.A.Mang (Austria), E.A.W.Maunder (UK), E. Stein (Germany), Ch. Hochard (France), V. Kompis (Slovakia), A. Wróblewski (Poland), I. Herrera (Mexico), B. Szybiński (Poland), O.C. Zienkiewicz (UK), R. Piltner (USA), Mme Helen Zienkiewicz (UK), F. Ihlenburg (USA), J.M. Melenk (Switzerland) and A.P. Zieliński (Poland).

Contents

Chapter 6
2D harmonic problems – the Laplace-type equations 95

Chapter 7
Problems described by two-dimensional, biharmonic equations.............. 135

Chapter 8
Two-dimensional problem of elastostatics.. 175

Preface

This book deals with applications of the boundary collocation method – also called the generalized Trefftz approach – to solve problems in engineering and science. The method consists of the use of trial functions, which satisfy the differential equations governing the problem. The unknown coefficients of those functions are determined by the satisfaction of the boundary and connectivity conditions.

As the method involves applying analytical solutions, the approach can be seen as a mixed analytical–numerical method. Discretization is required only on the boundary of the domain under consideration, while the analytical series describes the behaviour of the solution inside.

Hence, the Trefftz type approach can be related to the well known Boundary Element Method, but while the classical BEM formulation applies singular fundamental solutions of the Green's type, the Trefftz trial functions are regular, usually based on series.

The main limitation of the Boundary Collocation or generalized Trefftz method is the necessity of dealing with complete analytical set of, so called, T- or F-functions, which are not always known. This book presents a thorough survey of those functions, which allow for the solution of a wide range of numerical problems. Another important issue is related to the conditioning of the final algebraic set of equations, which, additionally, do not always result in symmetric matrices. The book discusses these problems and presents practical suggestions of how to avoid any ill-conditioning. A detailed investigation of the accuracy and convergence of the numerical solutions is also presented.

The generalized Trefftz method can be widely and efficiently applied to solve numerous linear and non-linear boundary value problems in structural, solid and fluid mechanics. This book illustrates its behaviour through many examples of the type that will be useful to scientists and engineers involved in the development and application of numerical methods.

The book contains results obtained by the authors and some of their co-workers whose contributions are duly acknowledged throughout the text. We wish to express our gratitude for numerous valuable suggestions first and foremost to the late Professor O.C. Zienkiewicz (1921–2009), one of the originators of the finite element method and propagator of the Trefftz approach, as well as to Professors I. Herrera, J. Jirousek, M. Kleiber and M. ¯yczkowski (1930–2005). The very helpful technical assistance of T. Betleja, A. Garlicka, B. Rostworowska, M. Fryc and K. Forma are also acknowledged.

The authors.
Cracow, 2009

CHAPTER 1

Introduction

1 Boundary collocation versus other modelling methods

Methods of mathematical modelling applied in contemporary computational mechanics can be divided into purely numerical and analytical–numerical procedures. Purely analytical solutions lose their popularity because of strong limitations connected with simple regions and mostly linear equations to which they can be applied. Obviously, the fundamental monographs like [90] for elastic solids, [64] in fluid mechanics or [17] for heat exchange are always popular and often quoted, but rather as sources of comparative benchmarks confirming correctness and accuracy of computer solutions.

Observing the main branches of computational mechanics – analysis of solids and structures, fluid dynamics, thermodynamics and electromagnetism – one can recently notice considerable development of both, numerical and analytical–numerical techniques. In the former procedures, mainly the finite elements and the finite differences gained widespread popularity. The recent advances in meshless methods [3, 4, 31, 68] encouraged numerous researchers to be involved into these new techniques; however, they are still far from commercial applications.

The analytical–numerical formulations use sets of analytical solutions of governing equations to model a problem in a numerical sense. **This leads to an exchange of an inside-domain solution to a boundary procedure**. Obviously, this decrease of the problem dimension considerably diminishes the number of unknowns and reduces the time of the calculations. It is especially distinctly visible in case of large regions with homogeneous or slowly changing properties (e.g. infinite areas). On the contrary, when the properties of the modelled object often vary and require a dense mesh of subregions (elements), the boundary methods lose their advantages. Finally, it should be noted that the main limitation of this type of the modelling procedures is connected with the necessity of deriving sets of analytical solutions of the problem equations, which is not always possible. The most thorough survey and history of this family of methods can be found in monograph [67].

Up to now the most popular among the analytical–numerical techniques are the boundary element methods [2, 6, 9, 13, 14, 53]. This name appeared for the first time in 1977 [11] as an analogy to the finite element approach; however, already in the early 1960s Jaswon [37, 38] and Symm [88] proposed first numerical algorithms using the Somigliana boundary integral equations. These formulations were mainly singular, applying fundamental solutions of the governing equations, the singular points of which were located along the boundaries of the considered region. The methods became very popular in the 1970s and 1980s, mainly thanks to the strong scientific centre in Southampton organizing regular conferences (starting from [8, 10] up to [12, 85]) and editing numerous monographs.

It may also be noted that the boundary integral equations can be solved expanding the unknown boundary functions into the Fourier series instead of dividing the boundary of the considered regions into elements [106–108]. Sometimes, for example, in problems of shells with arbitrary contours, it can be even more effective [107].

The second large group of analytical–numerical methods is connected with the name of Trefftz [91]. He proposed a numerical solution of a boundary value problem using trial functions identically fulfilling the governing differential equations. In remembrance of the author of this idea, they are often called T-functions. The unknown coefficients of the T-functions are here calculated from the boundary and connectivity conditions. The boundary collocation method presented in this monograph can be considered as a basic version of this approach.

A large part (perhaps a majority) of the Trefftz-type solutions is formulated in an integral sense. However, the use of analytical integration is in contemporary engineering calculations practically excluded because of complexity of the integral functions and paths of integration. As it is shown in detail in Section 2 of Chapter 2, **the numerical boundary integral fit can be considered as a particular case of the boundary collocation**. This is a specific form of collocation in control points of the numerical integral, in which the function is weighted by weighting coefficients of this integral. The boundary collocation is a wider notion – it can be defined by a set of arbitrary collocation points along the boundary. Also, when the number of the control points exceeds the number of unknown solution coefficients, an arbitrary set of weights can be applied. Therefore, **the authors decided to use the name of this more general definition of the basic Trefftz approach**.

The main difference between the boundary collocation method and the standard boundary element formulation consists in the fact that the trial functions of the former approach are regular in the whole considered region Ω including its boundary Γ. This results in a specific form of the Trefftz solution. It has a character of a functional series defining the solution field in all the points of the region Ω.

On the contrary, in the first, main stage of the boundary element approach, we calculate only a system of approximate complementary boundary functions. Then, in the next step of the procedure, this system allows to calculate the solution inside the domain Ω, but only in discrete points. The calculation of the results in each internal point requires additional integration. The most difficult seems here to be accurate determination of the full solution of the problem in the vicinity of the boundary. For example, in elasticity problems we easily obtain accurate boundary tractions but have difficulties in calculating a full stress tensor near the boundary.

The functional series, which is the form of the solution in the Trefftz boundary collocation method, gives explicitly all the results in the whole region Ω. This advantage can, for example, be taken in case of elastic–plastic problems (Chapter 12), in which a dense mesh of inside results should be calculated. Additionally, **the lack of singularities on the boundary Γ makes all the operations of the solving procedures much easier**. Certain limitations of the global Trefftz approach are connected with conditioning of the solution matrices. This problem will be discussed in detail in Section 3 of Chapter 2.

As it is presented in detail in Chapter 3, the T-functions can in general be divided into two main groups. The first type, including the so-called Herrera functions, forms sets of general solutions of homogeneous differential equations governing the problem. The second group, connected with the name of V. Kupradze, consists of fundamental solutions of considered problem, singularities of which are situated outside region Ω.

It should also be noted that the Trefftz approach is very convenient in certain specific cases, like problems with physical singularities or large local gradients of solutions. In these cases, the so-called special T-functions are applied, which fulfill not only the governing differential equations but also the local requirements (e.g. local boundary conditions) – see Section 5 in Chapter 3 and also recent monograph ([67]; Chapter 11).

The Trefftz boundary collocation approach is now in a stage of intensive development [67]. However, mainly because of some difficulties with derivation of the complete sets of the T-functions, for many years the number of papers in this field was rather small in comparison to publications on the finite and boundary element methods. According to Mackerle [71], up to the year 2000, 467 books on the finite elements and 223 on the boundary element approach have been published. On the contrary, the Trefftz method for years did not have broader presentation except several review papers [41, 61, 109]. More complex variants of the method (so called T-element approach) were surveyed by Jirousek & Wroblewski in a large publication [40] and by Qin in monograph [81]. The basic versions of the boundary collocation approach were first gathered in two Polish monographs [59, 110]. The present book is a continuation and extension of those issues.

The relatively modest number of papers on the boundary collocation method is connected also with smaller universality of the Trefftz approach. Obviously, derivation of the T-functions for more complex partial differential equations, for example, with non-linear terms or variable coefficients, seems to be a difficult task. However, which will be presented in this monograph, there is a considerable number of equations modelling basic problems of modern engineering, for which the complete T-functional systems are known. Additionally, there are numerous cases in which the analytical–numerical solutions based on the boundary collocation approach are evidently more rational than any other methods. Let us mention:

- wave propagation problems based on Helmholtz equation [21, 22, 32, 44, 87],
- Stokes-type flow round a system of solid bodies [19, 34, 58, 65, 76, 99, 100, 102],
- effective properties of composite materials and porous media [18, 60, 62, 65, 69, 70, 82, 93],
- evaluation of stress intensity factors [5, 23, 35, 45, 92, 103],
- free surface flow, also including surface tension [30, 74, 94, 95] and
- boundary problems connected with hydroelectrochemistry [20, 50–52].

The investigations of the Trefftz-type problems, as well as applications of the method, are presented every 3 years in international workshops. They started in Cracow in 1996 with the opening lecture of Zienkiewicz [57], one of the animators of the Trefftz method. The year 1996 was considered as the 70th anniversary of the idea of Trefftz [91]. On this occasion also a special issue of *Advances in Engineering Software* [44], including recent developments of this approach, was edited. The second Trefftz workshop took place in Sintra near Lisbon in 1999 [54], the third one was organized in Exeter in September 2002 [56], the fourth one in Zilina (Slovakia, 2005 [55]) and the fifth one in Leuven (Belgium, 2008). The next workshop is planned in Taiwan in 2011.

2 Different names of presented formulation

The turbulent development of the *boundary collocation method* resulted in numerous different names of the formulation. The name accepted by the authors of this monograph was first proposed in the known book of Collatz [24]. It was also used, for example, in papers [22, 61, 63, 83].

However, there are also many different names of this method. In early applications, it was often called the *point matching method* [25–28] for the direct boundary collocation and the *point least square method* [36, 78] when the number of collocation points exceeded the number of unknowns. These names, however, are now becoming less popular. It should also be noted that the name of the point matching method was sometimes used for procedures very different from the boundary collocation [66, 89].

The next group of names is connected with the name of Trefftz, who first proposed this type of solution [91]. The formulation was called the *global Trefftz method* [109], the *modified Trefftz method* [79, 84, 86], the *collocation Trefftz method* [67] or simply the *Trefftz method* [80, 104]. Also the name of the *T-complete functions method* [105, 110] comes from the same origin. All of them used in fact the boundary collocation procedures, but in their wider sense, including the numerical integral fit.

The largest variety of names of the same procedure (the boundary collocation) is connected with application of the Kupradze functions, that are fundamental solutions with singularities outside the considered region Ω. This formulation was called the *superposition method* [15], the *boundary point method* [42], the *fundamental solutions method* [7, 33, 73], the *source functions method* [29], the *discrete singularity method* [101], the *fundamental collocation method* [16, 72] and the *charge simulation method* [1, 46–49, 75, 77]. As it is shown in Section 2 of Chapter 2, also the basic formulation of the *regular indirect boundary element method* [97, 98] is nothing else than the boundary collocation procedure with Kupradze functions.

The way of proceeding in the solution both with the Herrera and Kupradze functions is very similar. Therefore, the authors decided not to differ the names of these approaches and to use in general the name of the *Trefftz boundary collocation procedure* or simply the *boundary collocation method*.

3 Short review of present monograph

This monograph can be divided into two parts. In the first one (Chapters 1–5), a general presentation of the boundary collocation approach and its numerous variants is enclosed. In the second part, the method is applied to many different engineering problems, showing its properties, accuracy and convergence. Both evident advantages but also limitations of the approach are clearly presented. The observations are based mainly on investigations of the authors and their co-operators carried out in the last two decades. The monograph includes 113 figures and 40 tables, from which 57 figures (with 97 diagrams) and 26 tables present results of numerical examples. A considerable number (above 1000) of papers and monographs concerning the discussed approach are quoted. They are enclosed separately into each chapter, which makes the literature survey easier to use.

After the introduction, in Chapter 2, relations between particular versions of the method are presented and choice of the name 'boundary collocation approach' is explained, as the most general definition of the applied procedures. In Chapter 3, a survey of the so-called Trefftz functions (T-functions) – functional complete systems identically fulfilling respective governing differential equations – are gathered. They are divided into two groups – the Herrera sets, which are complete general solutions of homogeneous equations and the Kupradze functions – fundamental solutions of the equations with their singularities situated outside the investigated regions. The chapter also presents non-typical T-functions and special purpose systems fulfilling not only the governing equations but also, partly, boundary conditions of investigated problems.

In Chapter 4, a problem of substructuring and the Trefftz finite elements is signalized. This approach has been presented in detail in the monographic paper [40] and book [81]. The T-elements, mainly in a hybrid form, may be applied to more complex regions in which the standard boundary collocation approach can have limitations because of difficulties with conditioning of the final systems of algebraic equations. The hybrid T-elements were also investigated by one of the authors [41, 105], but they are not discussed in detail in the present monograph.

Basic numerical investigations of the boundary collocation approach are shown in Chapter 5. They include numerous variants of the collocational fit of the applied boundary conditions, present behaviour of the chosen error and conditioning measures and suggest ways of proceeding to obtain the most reliable results. The investigations were made for both, simply- and doubly-connected regions, showing specific properties of the latter solutions, in which the T-complete sets for both bounded and unbounded regions should be applied.

Starting from Chapter 6, the monograph introduces readers into several particular engineering problems both from solid and fluid mechanics and presents ways of solving them with help of the boundary collocation approach. Beginning from the typical bar torsion problems (Sections 1–3 in Chapter 6), it enters the specific Poiseuille flow, then thermal effects of composite structures and different singular problems described by the Laplace-type equations.

In Chapter 7, two problems governed by a 2D biharmonic equation are discussed – plane creeping flow and thin plate bending. The third one described by the same equation – 2D elasticity – has been moved to Chapter 8 and investigated in detail. This chapter contains also optimization of a helical spring cross-section, in which the boundary collocation approach has especially been convenient.

The boundary, analytical–numerical methods are also very efficient in case of inverse problems. Different variants of the boundary collocation approach are presented in Chapter 9 in numerical examples concerning boundary value recovery in 2D elastic structures, when some experimental measurement data are given. The results confirm efficiency of the investigated procedures.

Chapter 10 signalizes the possibility of application of the boundary collocation approach to 3D problems, giving a large spectrum of papers (above 150), in which the method was applied in this case. In Chapter 11, typical up-to-date procedures to solve non-homogenous equations and non-linear boundary conditions are presented. Some of the procedures (e.g. the radial basis functions method) belong to current investigation field of the authors.

Chapter 12 describes an algorithm of solving the non-linear, elastic–plastic problems. The procedure based on the initial stress approach is presented in detail in two numerical examples. This solution was first proposed by one of the authors in Computer Methods in Applied Mechanics and Engineering in 1988. Chapter 13 deals with boundary–initial problems presenting different variants of the boundary collocation approach. In the investigated numerical example, encouraging results have been obtained and compared with the finite element solutions. In particular, the comparative convergence is worth to observe.

The penultimate chapter (Chapter 14) signalizes ways of application of the boundary collocation method to determine eigenvalues regarding various mechanical and electrical problems, for example, free vibrations of membranes and plates and frequencies of waveguides. The last chapter (Chapter 15) summarizes the monograph underlining necessary precautions and ways of correct application of this easy, effective, analytical–numerical modelling method – the boundary collocation approach.

References

[1] Amano, K., A charge simulation method for numerical conformal mapping onto circular and radial slit domains. *SIAM Journal on Scientific Computing*, **19**, pp. 1169–1187, 1998.

[2] Banerjee, P.K. & Butterfield, R., *Boundary Element Method in Engineering Science*, McGraw-Hill Book Company: London, 1981.

[3] Belytschko, T., Krongauz, Y., Organ, D., Flemming, M. & Krysl, P., Meshless methods: An overview and recent development. *Computer Methods in Applied Mechanics and Engineering*, **139**, pp. 3–44, 1996.

[4] Belytschko, T., Lu, Y.Y. & Gu. L., Element-free Galerkin methods. *International Journal for Numerical Methods in Engineering*, **37**, pp. 229–256, 1994.

[5] Berger, J.R., Karageorghis, A. & Martin, P.A., Stress intensity factors computation using the method of fundamental solutions: Mixed-mode problems. *International Journal for Numerical Methods in Engineering*, **67**, pp. 469–483, 2007.

[6] Beskos, E.D., *Boundary Element Methods in Mechanics*, North-Holland, Amsterdam, 1987.

[7] Bogomolny, A., Fundamental solutions method for elliptic boundary value problems. *SIAM Journal of Numerical Analysis*, **22**, pp. 644–669, 1985.

[8] Brebbia, C.A. (ed.), *Recent Advances in Boundary Element Methods*, Pentech Press: London, 1978.

[9] Brebbia, C.A., *The Boundary Element Method for Engineers*, Pentech Press: London, 1978.

[10] Brebbia, C.A. (ed.), *New Developments in Boundary Element Methods*, CML Publ., Springer-Verlag: London, 1980.

[11] Brebbia, C.A. & Dominguez, J., Boundary elements for potential problems. *Applied Mathematical Modelling*, **1**, pp. 372–378, 1977.

[12] Brebbia, C.A. & Poljak, D., *Boundary Elements and Other Mesh Reduction Methods XXIX*, WITpress: Southampton, 2007.

[13] Brebbia, C.A., Telles, J.C.F. & Wrobel, L.C., *Boundary Element Techniques*, Springer-Verlag: New York, 1984.

[14] Burczyński T., Boundary Element Method, Part IV of *Handbook of Computational Solid Mechanics*, Kleiber, M. (ed.), Springer-Verlag: Berlin, 1998.

[15] Burgess, G. & Mahajerin, E., A comparison of the boundary element and superposition methods. *Computers & Structures*, **19**, pp. 697–705, 1984.

[16] Burgess, G. & Mahajerin, E., The fundamental collocation method applied to non-linear Poisson equation in two dimensions. *Computers & Structures*, **27**, pp. 763–767, 1987.

[17] Carslaw, H.S. & Jaeger, J.C., *Conduction of Heat in Solids*, Oxford University Press: Oxford, 1959.

[18] Chapman, A.M. & Higdom, J.J.L., Oscillatory Stokes flow in periodic porous media. *Physics of Fluids A*, **4**, pp. 2099–2116, 1992.

[19] Chen, S.B. & Keh. H.J., Axisymmetric electrophoresis of multiple colloidal spheres, *Journal of Fluid Mechanics*, **238**, pp. 251–276, 1992.

[20] Chen, S.B. & Ye, X., Boundary effect on slow motion of a composite sphere perpendicular to two parallel impermeable plates. *Chemical Engineering Science*, **55**, pp. 2441–2453, 2000.

[21] Chen, W. & Hon, Y.C., Numerical investigation on convergence of boundary knot method in the analysis of homogeneous Helmholtz, modified Helmholtz, and convection-diffusion problems. *Computer Methods in Applied Mechanics and Engineering*, **192**, pp. 1859–1875, 2003.

[22] Chen, W., Meshfree boundary particle method applied to Helmholtz problems. *Engineering Analysis with Boundary Elements*, **26**, pp. 577–581, 2002.

[23] Cheung, Y.K., Woo, C.W. & Wang, Y.H., The stress intensity factor for a double edge cracked plate by boundary collocation method. *International Journal of Fracture*, **37**, pp. 217–231, 1988.

[24] Collatz, L., *The Numerical Treatment of Differential Equations*, Springer-Verlag: Berlin, 1960.

[25] Conway, H.D. & Leissa, A.W., Application of the point-matching method to shallow-spherical-shell theory. *Journal of Applied Mechanics*, **29**, pp. 745–747, 1962.

[26] Conway, H.D., The approximate analysis of certain boundary value problems. *Journal of Applied Mechanics*, **27**, pp. 275–277, 1960.

[27] Conway, H.D., Torsion of prismatic rods with isosceles triangle and other cross sections. *Journal of Applied Mechanics*, **27**, pp. 209, 1960.

[28] Conway, H.D., Triangular plates analysed by point matching. *Journal of Applied Mechanics*, **29**, pp. 755–756, 1962.

[29] de Mey, G., Integral equation for potential problems with the source function not located on the boundary. *Computers & Structures*, **9**, pp. 113–115, 1978.

[30] Dias, F. & Tuck, E.O., A steady breaking wave. *Physics of Fluids A*, **5**, pp. 277–279, 1993.

[31] Duarte, C.A. & Oden, J.T., h-p clouds, an h-p meshless method. *Numerical Methods for Partial Differential Equations*, **12**, pp. 673–705, 1996.

[32] Fairweather G.F., Karageorghis A. & Martin P.A., The method of fundamental solutions for scattering and radiation problems. *Engineering Analysis with Boundary Elements*, **27**, pp. 759–769, 2003.

[33] Fairweather, G. & Johnston, R.L., The method of fundamental solutions for problems in potential theory, In: *Treatment of Integral Equations by Numerical Methods* eds C.T.H. Baker & G.F. Miller, Academic Press: London, pp. 349–359, 1982.

[34] Hassonjee, Q., Ganatos, P. & Pfeffer, R., A strong-interaction theory for the motion of arbitrary three-dimensional clusters of spherical particles at low Reynolds number. *Journal of Fluid Mechanics*, **197**, pp. 1–37, 1988.

[35] Heng, Z., McCammond, D. & Tabarrok, B., Stress determination in edge-cracked anisotropic plates by an extension of boundary – collocation method. *Computer Methods in Applied Mechanics and Engineering*, **54**, pp. 187–195, 1986.

[36] Hulbert, L.E. & Niedenfuhr, F.W., Accurate calculation of stress distribution in multi holed plates. *Journal of Engineering for Industry*, **87**, pp. 331–336, 1965.

[37] Jaswon, M.A. & Ponter, A.R., Integral equation solution of the torsion problem. *Proceedings Royal Society*, Ser. A, No. **273**, pp. 237–246, 1963.

[38] Jaswon, M.A. Integral equation methods in potential theory – I. *Proceedings Royal Society*, Ser. A, No. **275**, pp. 23–32, 1963.

[39] Ji, M. & Ishikawa, H., Analysis of an internal central crack with bridging fibers in a finite orhotropic plate. *International Journal of Engineering Science*, **35**, pp. 549–560, 1997.

[40] Jirousek, J. & Wróblewski, A., T-elements: State of the Art and Future Trends. *Archives of Computational Methods in Engineering*, **3-4**, pp. 323–434, 1996.

[41] Jirousek, J. & Zieliński, A.P., Survey of Trefftz-type element formulations. *Computers & Structures*, **63**, pp. 225–242, 1997.

[42] Johnson, D., Plate bending by a boundary point method. *Computers & Structures*, **26**, pp. 673–680, 1987.

[43] Kamiya, N. & Kita, E., (eds), "Trefftz method 70 years", *Advances in Engineering Software,* **24**, pp. 1–155, 1995.

[44] Kamiya, N. & Wu, S.T., Generalised eigenvalue formulation of the Helmholtz equation by the Trefftz method. *Engineering Computations*, **11**, pp. 177–186, 1994.

[45] Karageorghis, A., Poullikkas, A. & Berger, J.R., Stress intensity factor computation using the method of fundamental solutions. *Computational Mechanics*, **37**, pp. 445–454, 2006.

[46] Katsurada, M., A mathematical study of the charge simulation method I. *Journal of Faculty of Science, University of Tokyo*, Section IA, **35**, pp. 507–518, 1988.

[47] Katsurada, M., A mathematical study of the charge simulation method II. *Journal of Faculty of Science, University of Tokyo*, Section IA, **36**, pp. 135–162, 1989.

[48] Katsurada, M., Asymptotic error analysis of the charge simulation method in a Jardan region with an analytic boundary. *Journal of Faculty of Science, University of Tokyo*, Section IA, **37**, pp. 635–657, 1990.

[49] Katsurada, M., Charge simulation method using exterior mapping function. *Japan Journal of Industrial and Applied Mathematics*, **11**, pp. 47–61, 1994.

[50] Keh, H.J. & Jan, J.S., Boundary effects on diffusiophoresis and electrophoresis: motion of a colloidal sphere normal to plane wall. *Journal of Colloidal and Interface Science*, **183**, pp. 458–475, 1996.

[51] Keh, H.J. & Lin, Y.M., Particle interactions in diffusiophoresis: Axisymmetric motion of multiple spheres in nonelectrolyte gradients. *Langmuir*, **10**, pp. 3010–3017, 1994.

[52] Keh, H.J. & Yang, F.R., Particle interaction in electrophoresis. IV. Motion of arbitrary three-dimensional clusters of spheres. *Journal of Colloidal and Interface Science*, **145**, pp. 362–389, 1991.

[53] Kleiber, M., (ed.), *Handbook of Computational Solid Mechanics*, Springer-Verlag: Berlin, 1998.

[54] Kleiber, M. & Mang, H.A., (eds), Freitas, J.A.T. & Almeida, J.P.M., (guest eds), Special Issue 'Trefftz Method II' of *Computer Assisted Mechanics and Engineering Sciences*, **8**; 2/3, pp. 211–500, 2001.

[55] Kleiber, M. & Mang, H.A., (eds), Kompis, V., (guest ed.), Special Issue 'Trefftz Method IV' of *Computer Assisted Mechanics and Engineering Sciences*, **13**; 3, pp. 365–481, 2006.

[56] Kleiber, M., Mang, H.A. (eds) & Maunder, E., (guest ed.), Special Issue 'Trefftz Method III' of *Computer Assisted Mechanics and Engineering Sciences*, **10**; 4, pp. 373–684, 2003.

[57] Kleiber, M., Mang, H.A. (eds) & Zieliński, A.P., (guest ed.), Special Issue 'Trefftz Method I'of *Computer Assisted Mechanics and Engineering Sciences*, **4**; 3/4, pp. 299–668, 1997.

[58] Kołodziej, J.A. & Mendes, C.A., T-functions for 2-D creeping flow on domains with circular cylinders, corners, and possessing symmetry. *Computer Assisted Mechanics and Engineering Sciences*, **8**, pp. 359–384, 2001.

[59] Kołodziej, J.A., *Application of the Boundary Collocation Method In Applied Mechanics* (in Polish). Poznan University of Technology Press, Poznan, 2001.

[60] Kołodziej, J.A., Determination of the interaction coefficient in mixture equations for heat conduction in unidirectional composites (in Polish). *Archives of Thermodynamics*, **8**, pp. 87–107, 1987.

[61] Kołodziej, J.A., Review of application of boundary collocation methods in mechanics of continuous media. *Solid Mechanics Archives*, **12**, pp. 187–231, 1987.

[62] Kompis, V., Stiavnicky, M., Kompis, M. & Zmindak, M., Trefftz implementation of multi-domain boundary point method. *Engineering Analysis with Boundary Elements*, **29**, pp. 391–396, 2005.

[63] Kuang, J.S. & Wang, Y.H., Bimaterial interface cracks originating from holes. *Mechanics Research Communications*, **22**, pp. 577–582, 1995.

[64] Lamb, H., *Hydrodynamics*. Cambridge University Press, 6th Edition, Cambridge, 1945.

[65] Larson, R.E. & Higdon, J.J.L., A periodic grain consolidation model of porous media. *Physics of Fluids A*, **1**, pp. 38–46, 1989.

[66] Lean, M.H., Friedman, M. & Wexler, A., Application of the boundary element method in electrical engineering problems. In: *Developments in Boundary Element Method – 1*, eds. P.K. Banerjee & R. Butterfield, Applied Science Publishers, pp. 207–250, 1975.

[67] Li, Z.C., Lu T.T., Hu, H.Y. & Cheng, A.H.D., *Trefftz and Collocation Methods*, WIT Press: Southampton and Boston, 2008.

[68] Liu, G.R., *Mesh Free Methods: Moving Beyond the Finite Element Method*, CRC Press: Boca Raton, USA, 2002.

[69] Lu, S.Y & Kim, S., Effective thermal conductivity of composites containing spheroidal inclusions. *AIChE Journal*, **36**, pp. 927–938, 1990.

[70] Lu, S.Y., The effective thermal conductivitics of composites with 2-D arrays of circular and square cylinders. *Journal of Composite Materials*, **29**, pp. 483–506, 1995.

[71] Mackerle, J., FEM and BEM in the context of information retrieval – some remarks of the new millennium. In: *Proceedings of the VIIIth International Conference Numerical Methods in Continuum Mechanics, 19-24th September 2000,* Liptovsky Jan, Slovak Republic, 2000.

[72] Mahajerin, E. & Burgess, G., Fundamental collocation method applied to plane thermoelasticity problems. *Computers & Structures*, **57**, pp. 795–797, 1995.

[73] Mathon, R. & Johnston, R.L., The approximate solution of elliptic boundary-value problems by fundamental solutions. *SIAM Journal of Numerical Analysis*, **14**, pp. 638–650, 1977.

[74] Moore, D.W., Saffman, P.G. & Tanveer, S., The calculation of some Batchelor flow: The Sadovskii vortex and rational corner flow. *Physics of Fluids,* **31**, pp. 978–990, 1988.

[75] Murashima, S., Kuhara, S. & Nieda, H., Application of charge simulation method to the two-dimensional potential problem that contains curved arcs as its boundary. *Boundary elements, Proceedings of 5th International Conference,* Hiroshima 1983, eds C.A. Brebbia, T. Futagami & M. Tanaka, Springer-Verlag: New York, 1983.

[76] Ogata, H., Aname, K., Sugita, M. & Okano, D., A fundamental solution method for viscous flow problems with obstacles in a periodic array. *Journal of Computational and Applied Mathematics*, **152**, pp. 411–425, 2003.

[77] Okamoto, M. & Sekiya, T., Some practical solutions for elastic torsion of bars by charge simulation method. *Mechanics Research Communications*, **17**, pp. 47–53, 1990.

[78] Ondris, L., Solution of some symmetrical plane thermal problems by the boundary point least squares method. *Computer Methods in Applied Mechanics and Engineering*, **28**, pp. 309–325, 1981.

[79] Patterson, C. & Sheikh, M.A., A modified Trefftz method for three-dimensional elasticity. *Boundary elements, Proceedings of 5th International Conference,* Hiroshima 1983, eds C.A. Brebbia, T. Futagami & M. Tanaka, Springer-Verlag: New York, 1983.

[80] Patterson, C. & Sheikh, M.A., On the use of fundamental solutions in Trefftz method for potential and elasticity problems. *Boundary Element Methods in Engineering, Conference Boundary Elements IV – Southampton,* Brebbia, C.A. (ed.), Springer-Verlag: New York, 1982.

[81] Qin, Q.H. *The Trefftz Finite and Boundary Method.* WITpress: Southampton, 2000.

[82] Raamachandran, J. & Rajamohan, C., Analysis of composite plates using charge simulation method. *Engineering Analysis with Boundary Elements,* **28**, pp. 1271–1281, 2004.

[83] Sergeev, B., Madenci, E. & Ambur, D.R., Influence of bolt spacing and degree of anisotropy in single-lap joints. *Computers & Structures*, **76**, pp. 89–103, 1997.

[84] Sheikh, M.A. & Patterson, C., A modified Trefftz method for fluid flow. *IV International Symposium Finite Elements Methods in Flow Problems*, Tokyo, 1982.

[85] Skerget, L. & Brebbia, C.A., (eds), *Boundary Elements and Other Mesh Reduction Methods XXX*, WITpress: Southampton, 2008.

[86] Stein, E., Die Kombination des modifizierten Trefftzschen Verfahrens mit der Metode der finiten Elemente. In: '*Finite Elemente in der Static*', Stuttgart, 1973.

[87] Stojek, M., Least-squares Trefftz-type elements for the Helmholtz equation. *International Journal for Numerical Methods in Engineering*, **41**, pp. 831–849, 1998.

[88] Symm, G.T., Integral equation methods in potential theory – II. *Proceedings Royal Society*, Ser. A, No. **275**, pp. 33–46, 1963.

[89] Theocaris, P.S. & Tsamasphyros, G.J., A photoelastic point-matching method for the solution of integral equations in contact problems. *Mechanics Research Communications*, **9**, pp. 31–38, 1982.

[90] Timoshenko, S. & Goodier, J.N., *Theory of Elasticity*, 3rd edn, McGraw-Hill Book Company: New York, 1982.

[91] Trefftz, E., Ein Gegenstück zum Ritzschen Verfahren. *Proceedings 2nd International Congres of Applied Mechanics*, Zurich, pp. 131–137, 1926.

[92] Ugandaoker, A. & Murali, B., Stress intensity factors for finite plate by boundary collocation. *International Journal of Fracture*, **52**, pp. R17–R24, 1991.

[93] Uściłowska, A. & Kołodziej, J.A., Solution of the nonlinear equation for isothermal gas flows in porous medium by Trefftz method, *Computer Assisted Mechanics and Engineering Sciences*, **13**, No. 3, pp. 445–456, 2006.

[94] Vanden-Broeck, J.-M. & Keller, J.B., Pouring flows. *Physics of Fluids*, **29**, pp. 3958–3961, 1986.

[95] Vanden-Broeck, J.-M., Joukowskii's model for a rising bubble. *Physics of Fluids,* **31**, pp. 974–977, 1988.

[96] Wang, C.Y., Settling of discs inside a vertical fluid-filled tube. *Applied Scientific Research,* **56**, pp. 43–51, 1996.

[97] Wearing, J.L. & Sheikh, M.A., A regular indirect BEM for thermal analysis. *International Journal for Numerical Methods in Engineering*, **25**, pp. 495–515, 1988.

[98] Wearing, J.L., Abdul Rahman, A.G., Patterson, C. & Sheikh, M.A., A regular indirect boundary element method for stress analysis. *Proc. Conf. Boundary Elements IX – W. Germany*, eds C.A. Brebbia, W.L. Wendland & G. Kuhn, Springer-Verlag: New York, 1987.

[99] Wu, W.Y. & He, X.Y., The Stokes flow of the rotating double spheres and multiple Spheres. *Applied Mathematics and Mechanics*, **8**, pp. 123–136, 1987.

[100] Yan, Z., Weinbaum, S., Ganatos, S. & Pfeffer, R., The three-dimensional hydrodynamic interaction of a finite sphere with a circular orifice at low Reynolds number. *Journal of Fluid Mechanics*, **174**, pp. 39–68, 1987.

[101] Yano, H., Fukutani, S., Watanabe, T., Nakajima, C. & Kieda, A., Application of discrete singularity method in engineering. *Proc. Conf. Boundary Elements V – Japan*, eds C.A. Brebbia, T. Futagami & M. Tanaka, Springer-Verlag: New York, 1983.

[102] Young D.L., Chen C.W., Fan C.M., Murugesan K. & Tsai C.C., The method of fundamental solutions for Stokes flow in rectangular cavity with cylinders. *European Journal of Mechanics B/Fluids*, **24**, pp. 703–716, 2005.

[103] Zhu, Z., Wang, L., Mohaty, B. & Huang, C., Stress intensity factor for a cracked specimen under compression. *Engineering Fracture Mechanics*, **73**, pp. 482–489, 2006.

[104] Zieliński, A.P. & Herrera, I., Trefftz method: fitting boundary conditions. *International Journal for Numerical Methods in Engineering*, **24**, pp. 871–891, 1987.

[105] Zieliński, A.P. & Zienkiewicz, O.C., Generalized finite element analysis with T-complete boundary solution functions. *International Journal for Numerical Methods in Engineering*, **21**, pp. 509–528, 1985.

[106] Zieliński, A.P. & Życzkowski, M., The trigonometric contour series method in application to clamped plates of arbitrary contour. *Bulletin of Polish Academy of Sciences /Applied Mechanics/,* **29**, pp. 159–167, 1981.

[107] Zieliński, A.P., A contour series method applied to shells. *Thin Walled Structures*, **3**, pp. 217–229, 1985.

[108] Zieliński, A.P., On curvilinear distributions expressed by double Fourier series. *Journal of Applied Mathematics and Physics /ZAMP/*, **31**, pp. 717–729, 1980.

[109] Zieliński, A.P., On trial functions applied in the generalized Trefftz method. *Advances in Engineering Software*, **24**, pp. 147–155, 1995.

[110] Zieliński, A.P., *The T-complete Functions Method and its Application to Elastic and Elasto-Plastic Problems* (in Polish), Monograph 86, Cracow University of Technology Press: Cracow, 1989.

CHAPTER 2

General information

1 Description of the method – basic versions of boundary collocation

Engineering problems of applied mechanics can often be modelled by a system of partial differential equations describing the phenomenon (process) in a certain region Ω with boundary conditions on its contour Γ. Let us consider, for example, an elliptic boundary value problem:

$$L\left[\phi(\mathbf{x})\right]=f(\mathbf{x}) \quad \mathbf{x}\in\Omega \tag{1a}$$

with

$$B\left[\phi(\mathbf{x})\right]=g(\mathbf{x}) \quad \mathbf{x}\in\Gamma, \tag{1b}$$

where $L[\ ]$ is a linear elliptic operator, $B[\ ]$ is a known differential operator (not necessarily linear), $f(\mathbf{x})$ and $g(\mathbf{x})$ are known functions and $\phi(\mathbf{x})$ is an unknown solution of the problem.

One of the most general approaches to numerical solving process of the above problem is the weighted residual formulation [9, 19]. Usually the approximate solution $\hat{\phi}(\mathbf{x})$ is there searched in the form:

$$\phi(\mathbf{x})\approx\hat{\phi}(\mathbf{x})=\sum_{i=1}^{M_f}a_i\Phi_i(\mathbf{x}) \quad \mathbf{x}\in\Omega, \tag{2}$$

where $\Phi_i(\mathbf{x})$ are assumed trial functions and a_i are unknown coefficients. To obtain a system of algebraic equations, it is necessary to orthgonalize the residuals:

$$\varepsilon_1(\mathbf{x})=L\left[\hat{\phi}(\mathbf{x})\right]-f(\mathbf{x}) \quad \mathbf{x}\in\Omega \tag{3a}$$

$$\varepsilon_2(\mathbf{x})=B\left[\hat{\phi}(\mathbf{x})\right]-g(\mathbf{x}) \quad \mathbf{x}\in\Gamma \tag{3b}$$

by a system of linearly independent weighting functions V_j, W_j. This results in the weak formulation of the problem (1):

$$\int_{\Omega} V_j \varepsilon_1(\mathbf{x}) d\Omega + \int_{\Gamma} W_j(\mathbf{x}) \varepsilon_2(\mathbf{x}) d\Gamma = 0 \qquad (4)$$
$$j = 1, 2, 3, \ldots N_f \quad (N_f = M_f).$$

There exist many different possibilities of the choice of the trial functions Φ_j and the weighting functions V_j, W_j, which lead to different known numerical solving procedures [2, 9]. In this monograph, we consider only the trial functions identically fulfilling the governing differential equation in a homogeneous form, that is:

$$L[\Phi_i(\mathbf{x})] = 0. \qquad (5)$$

For example, for the two-dimensional (2D) Laplace equation, such functions Φ_i have the form (see also Chapter 3):

$$\begin{aligned} &\Phi_1(\mathbf{x}) = 1.0 \\ &\Phi_{2n}(\mathbf{x}) = \mathrm{Re}(z^n) \\ &\Phi_{2n+1}(\mathbf{x}) = \mathrm{Im}(z^n) \\ &\quad z = x + iy; \quad \mathbf{x} \equiv \{x, y\} \quad n = 1, 2, 3, \ldots. \end{aligned} \qquad (6)$$

This concept was first signalized by Erich Trefftz [24] and, therefore, this type of the numerical procedures is often connected with his name. If, additionally, a particular integral $\phi^P(\mathbf{x})$ of the problem is known, then the approximate solution can be written as

$$\hat{\phi}(\mathbf{x}) = \phi^P(\mathbf{x}) + \sum_{i=1}^{M_f} a_i \Phi_i(\mathbf{x}) \quad \mathbf{x} \in \Omega \qquad (7)$$

and eqn (4) changes to

$$\int_{\Gamma} W_j \left\{ B[\hat{\phi}(\mathbf{x})] - g(\mathbf{x}) \right\} d\Gamma = 0 \qquad (8)$$
$$j = 1, 2, 3, \ldots (N_f = M_f).$$

Observing the above condition, one can notice an important feature of system (6) and all the Trefftz functions: they are defined in non-dimensional coordinate systems. Hence, the coordinates {x,y} should be understood as $\{x, y\} \equiv \{\overline{x}/a, \overline{y}/a\}$, where

$\{\overline{x},\overline{y}\}$ is a dimensional system and a is a characteristic dimension of a considered problem (in particular, a can be a unit value, e.g. $a = 1[\mathrm{m}]$).

The boundary collocation method can be treated as a specific case of the boundary weighted residual approach (8) with weighting functions in the form of the Dirac deltas:

$$W_k(\mathbf{x}) = \delta(\mathbf{x} - \mathbf{x}_k) \quad \mathbf{x}, \mathbf{x}_k \in \Gamma \quad k = 1, 2, 3, \ldots K_p, \tag{9}$$

where $\mathbf{x}_k$ are singular points of these distributions. The choosing property of $\delta(\mathbf{x} - \mathbf{x}_k)$ changes eqn (8) into the system

$$\begin{aligned} &B\big[\,\Phi_i(\mathbf{x})\big]_{\mathbf{x}=\mathbf{x}_k} a_i = g(\mathbf{x}_k) - B\big[\,\phi^p(\mathbf{x})\big]_{\mathbf{x}=\mathbf{x}_k} \qquad \mathbf{x}_k \in \Gamma \\ &i = 1, 2, \ldots M_f; \quad k = 1, 2, \ldots K_p, \end{aligned} \tag{10}$$

which can shortly be written as

$$\begin{aligned} &\sum_{i=1}^{M_f} A_{ki} a_i = b_k \quad k = 1, 2, \ldots K_p \quad A_{ki} = B\big[\Phi_i(\mathbf{x})\big]_{\mathbf{x}=\mathbf{x}_k} \\ &b_k = g(\mathbf{x}_k) - B\big[\phi^p(\mathbf{x})\big]_{\mathbf{x}=\mathbf{x}_k} \end{aligned} \tag{11a}$$

or

$$\mathbf{Aa} = \mathbf{b}. \tag{11b}$$

If number K_p of the collocation points is equal to number M_f of the unknown coefficients a_i, then the procedure can be called a direct boundary collocation. This form is generally rather not recommended (especially for equidistant collocation) because of possible oscillations of the approximate solution $\hat{\phi}$ between the collocation points. This is analogical to the so-called 'Runge phenomenon', which appears in interpolation procedures [14]. To avoid it, it is suggested to apply the orthogonal collocation in the Gauss or Lobatto points [25], which will be presented in detail in Chapters 5 and 8.

However, the most natural and often used is the overdetermined boundary collocation method in which $K_p > M_f$. In this case, to calculate the unknown coefficients a_i, we can apply the standard least square procedure:

$$\mathbf{A}^{\mathrm{T}}\mathbf{Aa} = \mathbf{A}^{\mathrm{T}}\mathbf{b}. \tag{12}$$

If this classical form leads to the ill-conditional matrix $\mathbf{A}^{\mathrm{T}}\mathbf{A}$ (see detailed analysis in Section 3 and Chapter 5), it is recommended to apply the direct Legras method of transformation of $\mathbf{A}$ into the lower triangular matrix [15], which is a convenient possibility of solving the overdetermined problem. However, a considerable advantage of form (12) is symmetry of the main matrix. Therefore, it is sometimes convenient to apply this form even to direct collocation procedures.

2 Relation between boundary collocation, boundary integral Trefftz approach and regular boundary element method

Observing the system of eqns (11) for $K_p > M_f$, we notice that, in formulation of the least square expression, different weights α_k can be applied to particular equations:

$$J_{ls} = \sum_{k=1}^{K_p} \alpha_k \left(\sum_{i=1}^{M_f} A_{ki} a_i - b_k \right)^2 = \min. \tag{13}$$

Hence, eqn (12) can in general be written as

$$\tilde{\mathbf{A}}^{\mathrm{T}} \mathbf{A} \mathbf{a} = \tilde{\mathbf{A}}^{\mathrm{T}} \mathbf{b} \tag{14a}$$

or

$$\sum_{i=1}^{M_f} \left(\sum_{k=1}^{K_p} \tilde{A}_{jk} A_{ki} \right) a_i = \sum_{k=1}^{K_p} \tilde{A}_{jk} b_k \qquad j = 1, 2, \ldots M_f, \tag{14b}$$

where

$$\tilde{A}_{kj} = \alpha_k A_{kj} \quad \text{(no summation with respect to } k\text{).}$$

However, if we now repeat the same consideration for the least square integral fit:

$$J_{\text{int}} = \int_\Gamma \left\{ B\left[\hat{\phi}(\mathbf{x}) \right] - g(\mathbf{x}) \right\}^2 d\Gamma = \min, \tag{15a}$$

which means

$$\int_\Gamma B\left[\Phi_j(\mathbf{x}) \right] \left\{ B\left[\hat{\phi}(\mathbf{x}) - g(\mathbf{x}) \right] \right\} d\Gamma = 0, \qquad j = 1, 2, \ldots \left(N_f = M_f \right) \tag{15b}$$

and calculate the integral in a numerical way, we obtain exactly the same form (14) only with specific weights α_j and specific points X_j. In particular, for the Gaussian integration:

$$\alpha_j^G = \frac{l_\Gamma}{2} W_j^G, \tag{16}$$

where l_Γ is the length of boundary Γ (range of integration), W_j^G are Gaussian weights and $\mathbf{x}_j$ are Gaussian points (zeros of the Legendre polynomials). Obviously, sometimes the boundary Γ is divided into segments; the procedure is applied to each of them and eqn (15b) changes into the sum of the integrals.

A great majority of boundary integrals in real engineering applications are calculated in the numerical way. Therefore, it can be stated that **the numerical boundary integral fit is a specific weighted boundary collocation in the control points of the numerical integral**. This observation draws attention to the relation of the number of unknown approximation coefficients a_i and the number of the control points K_p of the integral. For example, if $M_f = K_p$, the numerical integral fit changes into the direct collocation in the control points of the integration and weights do not play any role.

The consideration presented above leads to the conclusion that the **notion of boundary collocation is more general than the numerical boundary integral fit (points and weights can be arbitrary)**. Therefore, sometimes in this monograph we shall discuss also the integral formulations of the presented method.

In certain applications, it is more convenient to use a different form of the functional (15):

$$\bar{J}_{\text{int}} = \int_{\Gamma} \left\{ B\left[\hat{\phi}(\mathbf{x})\right] - g(\mathbf{x}) \right\} \left\{ D\left[\hat{\phi}(\mathbf{x})\right] - h(\mathbf{x}) \right\} d\Gamma = \min, \tag{17}$$

where

$$D\left[\phi(\mathbf{x})\right] = h(\mathbf{x}) \tag{18}$$

is a complementary boundary condition and $h(\mathbf{x})$ is an unknown function. For example, in case of the potential problem with the Dirichlet boundary conditions, $B = 1$ and $D = \partial / \partial n$ is an operator – a derivative normal to Γ. In this case, the condition

$$\frac{\partial \bar{J}_{\text{int}}}{\partial a_i} = 0 \tag{19}$$

leads to

$$\int_{\Gamma} \frac{\partial \Phi_j}{\partial n} \left[\hat{\phi}(\mathbf{x}) - g(\mathbf{x}) \right] d\Gamma + \int_{\Gamma} \Phi_j(\mathbf{x}) \left[\frac{\partial \hat{\phi}(\mathbf{x})}{\partial n} - h(\mathbf{x}) \right] d\Gamma = 0 \tag{20}$$
$$j = 1, 2, \ldots (N_f = M_f).$$

Now assuming

$$h(\mathbf{x}) \equiv \frac{\partial \hat{\phi}(\mathbf{x})}{\partial n}, \tag{21}$$

we obtain the often used form

$$\int_{\Gamma} \frac{\partial \Phi_j}{\partial n} \left[\hat{\phi}(\mathbf{x}) - g(\mathbf{x}) \right] d\Gamma = 0 \tag{22}$$
$$j = 1, 2, \ldots (N_f = M_f)$$

and the equivalent boundary collocation changes to

$$\tilde{\mathbf{C}}^{\mathrm{T}}\mathbf{A} = \tilde{\mathbf{C}}^{\mathrm{T}}\mathbf{b} \tag{23a}$$

or

$$\sum_{i=1}^{M_f}\left(\sum_{k=1}^{K_p}\tilde{C}_{jk}A_{ki}\right)a_i = \sum_{k=1}^{K_p}\tilde{C}_{jk}b_k \qquad j = 1,\ 2,\ldots M_f, \tag{23b}$$

where

$$\tilde{C}_{kj} = a_k^G\left.\frac{\partial\Phi_j(\mathbf{x})}{\partial n}\right|_{\mathbf{x}=\mathbf{x}_k}.$$

Note that in this case, one additional equation should be added because the first T-function is constant and its derivative is equal to zero, which would result in singularity of the solution matrix (see also Chapter 5).

If on the boundary Γ two different conditions are defined, namely, the Dirichlet type on Γ_1 and the Neumann type on Γ_2, then system (22) changes into:

$$\int_{\Gamma_1}\frac{\partial\Phi_j}{\partial n}\left[\hat{\phi}(\mathbf{x}) - g(\mathbf{x})\right]d\Gamma + \int_{\Gamma_2}\Phi_j(\mathbf{x})\left[\frac{\partial\hat{\phi}(\mathbf{x})}{\partial n} - h(\mathbf{x})\right]d\Gamma = 0 \qquad j = 1,\ 2,\ldots\left(N_f = M_f\right). \tag{24}$$

This formulation known as 'opposite weights' is often met in the French literature [12] in spite of the fact that it leads to non-symmetric matrices. Its considerable advantage in comparison to the least square approach (analogical to (15))

$$\int_{\Gamma_1}\Phi_j(\mathbf{x})\left[\hat{\phi}(\mathbf{x}) - g(\mathbf{x})\right]d\Gamma + W\int_{\Gamma_2}\frac{\partial\Phi_j}{\partial n}\left[\frac{\partial\hat{\phi}(\mathbf{x})}{\partial n} - h(\mathbf{x})\right]d\Gamma = 0 \qquad j = 1,\ 2,\ldots\left(N_f = M_f\right) \tag{25}$$

is lack of the weighting constant W, which in eqn (25) appears to be necessary to equalize influence of both integrals. Without this weight, one boundary condition could be fulfilled much less accurately than the other. Formulations (24) can be symmetrized by the change of the sign between the integrals [26]:

$$\int_{\Gamma_1}\frac{\partial\Phi_j}{\partial n}\left[\hat{\phi}(\mathbf{x}) - g(\mathbf{x})\right]d\Gamma - \int_{\Gamma_2}\Phi_j(\mathbf{x})\left[\frac{\partial\hat{\phi}(\mathbf{x})}{\partial n} - h(\mathbf{x})\right]d\Gamma = 0 \qquad j = 1,\ 2,\ldots\left(N_f = M_f\right). \tag{26}$$

However, this convenience is costly the problem can sometimes be negatively definite.

The main idea of the regular indirect boundary element method consists in enforcing the given boundary conditions on Γ by auxiliary source functions applied on Γ_s situated outside Ω. For the 2D potential problems, the solution $\phi(\mathbf{x})$ is represented by

$$\phi(\mathbf{x})=\phi^P(\mathbf{x})+\int_{\Gamma_s} a(\mathbf{y})G(\mathbf{x},\mathbf{y})\,d_y\Gamma_s \quad \mathbf{x}\in\Omega^+, \quad \mathbf{y}\in\Gamma_s\in\Omega^-, \tag{27}$$

where Ω^+ and Ω^- are bounded (considered) and unbounded regions, respectively; $a(\mathbf{y})$ is an unknown function and $G(\mathbf{x},\mathbf{y})$ is the fundamental solution (constant $-1/2\pi$ is here omitted):

$$\begin{aligned} G(\mathbf{x},\mathbf{y}) &= \ln r \\ r &= \sqrt{(x-\xi)^2+(y-\eta)^2}, \end{aligned} \tag{28}$$

where $\mathbf{x}\equiv\{x,y\}$ is the point in which solution is calculated, $\mathbf{y}\equiv\{\xi,\eta\}$ is the position of the source.

In numerical version of this formulation [16, 20, 22, 25], expression (27) changes to

$$\phi(\mathbf{x})=\phi^P(\mathbf{x})+\sum_{i=1}^{M_f} G(\mathbf{x},\mathbf{y}_i)a_i \quad \mathbf{x}\in\Omega^+, \ \mathbf{y}_i\in\Gamma_s\in\Omega^- \tag{29}$$

where $a_i = a(\mathbf{y}_i)\alpha_i$ are unknown coefficients and a_i are numerical weights of the integral. Points $\mathbf{y}_i$ are control points of the integral and simultaneously positions of sources in the Kupradze T-functions $G(\mathbf{x},\mathbf{y}_i)$ (also called F-functions, F from 'fundamental').

To calculate the unknown coefficients a_i, it is necessary to introduce this expression into the respective boundary conditions of the problem. For example, for the Dirichlet condition, we calculate the values of the fundamental solutions in points $\mathbf{x}_k$ along Γ and $G(\mathbf{x}_k,\mathbf{y}_i)$ play a role of $\Phi_i(\mathbf{x}_k)$ and in this case directly A_{ki} from eqn (11):

$$A_{ki}=\Phi_i(\mathbf{x}_k)=G(\mathbf{x}_k,\mathbf{y}_i) \quad \mathbf{x}_k\in\Gamma \ \ \mathbf{y}_i\in\Gamma_s\in\Omega^-. \tag{30}$$

For more complex boundary conditions, the form of A_{ki} suitably changes, but the further way of proceeding remains the same. Hence, it is visible that **the numerical version of the regular indirect boundary element method can be considered as the boundary collocation procedure with the Kupradze trial functions**. The visualization of the above idea is presented in Fig. 1. The boundary collocation procedure with the Kupradze trial functions is also known in literature as the method of fundamental solutions (MFS). Comprehensive reviews on this method for various applications can be found in Refs [7, 8, 10]. A new version of the MFS, being its modification in the weighted least square sense, has been proposed in [21].

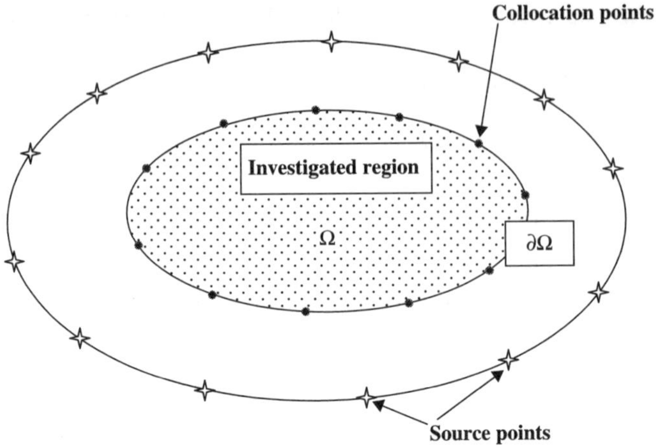

Figure 1: Distribution of collocation and source points.

An interesting comparison of different variants of the regular boundary methods can be found in [17]. Investigation of accuracy of collocational methods including the BCM has been demonstrated in [4].

3 Conditioning of the Trefftz-type formulations

Conditioning of a matrix **A** of algebraic equations

$$\mathbf{A}\mathbf{x} = \mathbf{b} \tag{31}$$

is a measure of its stability, that is sensitivity of roots on two kinds of perturbations:

1. initial errors resulting from approximation of the matrix **A** by a matrix **A** + Δ**A** and the vector **b** by a vector **b** + Δ**b** and
2. round-off errors occurring in the solving process of the system.

In case of weakly conditioned matrices, relatively small initial perturbations cause significant deviations Δ**x** of the roots **x**, which additionally increase in the solving process. Sometimes, in the whole numerical procedure, the above perturbations can result in total loss of reliability of the final results. In this case, we define the problem as ill conditioned.

The possibility of arising of weakly conditioned matrices in the solving systems is a considerable limitation of the basic, global formulation of the boundary collocation approach. It results from the global character of the T-functions, which in this case are non-zero in the whole investigated area Ω. If, for example, we consider the matrix **A** with the terms

$$A_{ki} = \Phi_i(\mathbf{x}_k) \quad i = 1, 2, \ldots M_f \quad k = 1, 2, \ldots K_p \tag{32}$$

and the T-functions $\Phi_i(\mathbf{x}_k)$ in the form (6), we observe that it is fully filled and its diagonal terms are not dominant over the others. A matrix of this type can be ill conditioned. Also, in case of the Kupradze functions $G(\mathbf{x},\mathbf{y}_i)$ for singular points, which are close to each other and relatively distant from Γ, respective matrix terms are similar:

$$A_{k(i-1)} \approx A_{ki} \approx A_{k(i+1)} \quad k = 1,\ 2,\ldots K_p \tag{33}$$

and the matrix diagonal loses its dominant character. Hence, for larger dimensions of the matrices (larger number of applied trial functions), a continuous testing of the conditioning becomes necessary.

The theory of the presented phenomenon is discussed in numerous monographs, for example [1, 16, 19]. They introduce certain measures of the conditioning allowing to estimate a relative error of the solution. In case of perturbing terms A_{ij} in the matrix $\mathbf{A}$ of system (31):

$$(\mathbf{A} + \Delta\mathbf{A})(\mathbf{x} + \Delta\mathbf{x}) = \mathbf{b}, \tag{34}$$

the estimation is expressed by:

$$\frac{\|\Delta\mathbf{x}\|}{\|\mathbf{x}\|} \le \frac{\kappa \dfrac{\|\Delta\mathbf{A}\|}{\|\mathbf{A}\|}}{1 - \kappa \dfrac{\|\Delta\mathbf{A}\|}{\|\mathbf{A}\|}}. \tag{35}$$

The norms present in the above formula have either the simpler (first) form

$$\|\mathbf{x}\| = \sum_{i=1}^{N} |x_i| \tag{36a}$$

$$\|\mathbf{A}\| = \max_{1 \le j \le N} \sum_{i=1}^{N} |A_{ij}| \tag{36b}$$

or more complex (second or Euclidean):

$$\|\mathbf{x}\|_E = \left(\sum_{i=1}^{N} |x_i|^2 \right)^{\frac{1}{2}} \tag{37a}$$

$$\|\mathbf{A}\|_E = \left(\sum_{i=1}^{N} \sum_{j=1}^{N} |A_{ij}|^2 \right)^{\frac{1}{2}} \tag{37b}$$

and the condition number κ is defined as

$$\kappa = \|\mathbf{A}\| \|\mathbf{A}^{-1}\|. \tag{38}$$

The inequality (35) is obviously valid only if its right-hand side is positive, that is

$$\left\|\mathbf{A}^{-1}\right\|\left\|\Delta\mathbf{A}\right\| < 1. \tag{39}$$

The perturbation of the free-term vector **b** has smaller influence on the solution. Assumption of the perturbed equation in the form

$$\mathbf{A}\left(\mathbf{x}+\Delta\mathbf{x}\right)=\mathbf{b}+\Delta\mathbf{b} \tag{40}$$

leads to estimation:

$$\frac{\left\|\Delta\mathbf{x}\right\|}{\left\|\mathbf{x}\right\|} \le \kappa \frac{\left\|\Delta\mathbf{b}\right\|}{\left\|\mathbf{b}\right\|}. \tag{41}$$

In case of simultaneous perturbation (both, the matrix **A** and the vector **b**), the following uniform estimation measure seems to be practical [1]:

$$\frac{\left\|\Delta\mathbf{x}\right\|}{\left\|\mathbf{x}\right\|} \le \frac{\left\|\Delta\overline{\mathbf{x}}\right\|}{\left\|\mathbf{x}\right\|} = \frac{\kappa}{1-\kappa\dfrac{\left\|\Delta\mathbf{A}\right\|}{\left\|\mathbf{A}\right\|}}\left(\frac{\left\|\Delta\mathbf{A}\right\|}{\left\|\mathbf{A}\right\|}+\frac{\left\|\Delta\mathbf{b}\right\|}{\left\|\mathbf{b}\right\|}\right). \tag{42}$$

In the above formula, the attention should be paid to two following doubts. First is the reliability of calculation of the number κ, which contains the norm of the inverse matrix $\mathbf{A}^{-1}$. In case of ill-conditioned matrix **A**, this norm can be calculated with such an error that the inequality (42) will loose its sense. The second problem is the question to what extent the left-hand side of this inequality is smaller than the right-hand-side. Both these problems will be discussed in detail on numerical examples presented in Chapter 5.

As it was already mentioned, the direct boundary collocation approach leads to eqn (31) with non-symmetric **A**. Therefore, in this case and, moreover, for the overdetermined formulation, we usually deal with equation

$$\mathbf{A}^{\mathrm{T}}\mathbf{A}\,\mathbf{x}=\mathbf{A}^{\mathrm{T}}\mathbf{b}. \tag{43}$$

The error estimation of this equation can be calculated in a simplified way by introducing $\mathbf{A}^{\mathrm{T}}\mathbf{A}$ as **A** and $\mathbf{A}^{\mathrm{T}}\mathbf{b}$ as **b** into the inequality (42).

There are also other, different estimators of the solution errors and the condition numbers for the system of algebraic equations (31). The quotient

$$\vartheta=\frac{\max_i\left|\lambda_i\right|}{\min_i\left|\lambda_i\right|} \tag{44}$$

of maximal and minimal absolute eigenvalues represents a rough condition number and

$$\zeta = \frac{1}{N}\left(\sum_{i=1}^{N}\left|\lambda_i\right|^2 \sum_{j=1}^{N}\left|\lambda_j\right|^{-2}\right)^{\frac{1}{2}} \tag{45}$$

is a little more informative [11]. However, the calculation of the eigenvalues is rather costly and therefore these condition numbers are less popular.

Besides the formal mathematical measures of the conditioning, one can also observe certain informal expressions and procedures giving, however, a very clear information about the matrix properties. For example, one of the often used conditioning measures is [22]:

$$\varepsilon = \frac{1}{N}\sum_{i=1}^{N}\left|D_{ii} - 1\right| \quad \mathbf{D} = \mathbf{A}\hat{\mathbf{A}}^{-1} \quad \hat{\mathbf{A}}^{-1} \approx \mathbf{A}^{-1} \tag{46}$$

giving information about the diagonal of the product **D**, which should be approximately equal to the identity matrix **I**. Also small artificial perturbations of the matrix **A** and observing the behaviour of the solution vector **x** is a good and relatively cheap way of error estimation [25]. Although less formal, both above measures are very clear for an engineer and therefore recommended in practical calculations.

The difficulties with the matrix conditioning in the global boundary collocation approach were investigated in numerous publications [6, 16, 25, 26]. They forced the authors to search for a remedy of this problem. One of them (the Legras direct solving algorithm [15]) was already mentioned in Section 1. Another possibility is the Tikhonov regularization procedure [23]. It consists in simultaneous minimization of the least square measure of both residuals **r** in the overdetermined equations

$$\mathbf{A}\mathbf{x} - \mathbf{b} = \mathbf{r} \tag{47}$$

and the solution vector **x**. Thus, finally we solve the regularized problem:

$$\left(\mathbf{A}^{\mathrm{T}}\mathbf{A} + \gamma\mathbf{I}\right)\mathbf{x} = \mathbf{A}^{\mathrm{T}}\mathbf{b}, \tag{48}$$

which can be much better conditioned than the original one. Of course, too large γ results in unacceptable deviations of **x**, and therefore a careful study of this parameter should be carried out (see Chapter 5).

Application of the Tikhonov procedure allowes to calculate successfully numerous ill-conditioned problems of applied mechanics solved by the boundary collocation and other techniques [5, 18, 27]. Different aspects of these problems and ways of dealing with them can be found in [3, 13, 20]. However, the most evident way to avoid the conditioning problem in case of more complex areas Ω is subtructuring of an investigated region, which will be discussed in detail in Chapter 4.

References

[1] Blum, E.K., *Numerical Analysis and Computation*, Addison-Wesley Publ. Comp: London, 1971.

[2] Brebbia, C.A., *The Boundary Element Method for Engineers*, Pentech Press: London, 1978.

[3] Chen, C.S., Cho, H.A. & Golberg, M.A., Some comments on ill-conditioning of the method of fundamental solutions. *Engineering Analysis with Boundary Elements*, **30**, pp. 405–410, 2006.

[4] Cheng, A.H.D., Golberg, M.A., Kansa, E.J. & Zammito, G., Exponential convergence and H-c multiquadrics collocation method for partial differential equations. *Numerical Methods for Partial Differential Equations*, **19(5)**, pp. 571–594, 2003.

[5] Duda, P. & Zieliński, A.P., Conditioning problems in Trefftz-type algorithms. *Proc. Int. Conf. 'Numerical Methods in Continuum Mechanics', Stara Lesna, Slovakia*, 1996.

[6] Duda, P. & Zieliński, A.P., Stability and convergence of solutions in Trefftz-type algorithms. *Strojnicky Casopis*, **48**, pp. 415–428, 1997.

[7] Fairweather, G. & Karageorghis, A., The metod of fundamental solutions for elliptic boundary value problems. *Advanced Computational Mathematics*, **9**, pp. 69–95, 1998.

[8] Fairweather, G., Karageorghis, A. & Martin, P.A., The method of fundamental solutions for scattering and radiation problems. *Engineering Analysis with Boundary Elements*, **27**, pp. 759–769, 2003.

[9] Finlayson, B.A., *The Method of Weighted Residuals and Variational Principles*, Ac. Press: London, 1972.

[10] Golberg, M.A. & Chen, C.S., The method of fundamental solutions for potential, Helmholtz and diffusion problems. In: ed. M.A. Golberg, *Boundary Integral Methods – Numerical and Mathematical Aspects*, Computational Mechanics Publications: Boston, pp. 103–176, 1998.

[11] Heise, U., Dependence of the round-off error in the solution of boundary integral equations on a geometric scale factor. *Computer Methods in Applied Mechanics and Engineering*, **62**, pp. 115–126, 1987.

[12] Hochard, Ch., Ladeveze, P. & Proslier, L., A simplified analysis of elastic structures. *European Journal of Mechanics, A/Solids*, **12**, pp. 509–535, 1993.

[13] Huang, H.T. & Li, Z.C., Effective condition number and superconvergence of the Trefftz method coupled with high order FEM singularity problems. *Engineering Analysis with Boundary Elements*, **30**, pp. 270–283, 2006.

[14] Lanczos, C., *Applied Analysis*, Prentice-Hall: Englewood Cliffs, N. J., 1956.

[15] Legras, J., *Methodes et Techniques de l'Analyse Numerique*, Dunod: Paris, 1971.

[16] Li, Z.C., Lu, T.T., Hu, H.Y. & Cheng, A.H.D., *Trefftz and Collocation Methods*, WITpress: Southampton and Boston, 2008.

[17] Li, Z.C., Lu, T.T., Huang, H.T. & Cheng, A.H.D., Trefftz, collocation and

other boundary methods – a comparison. *Numerical Methods for Partial Differential Equations*, **23**, pp. 93–144, 2007.

[18] Melenk, J.K., *Finite Element Methods with Harmonic Shape Functions for Solving Laplace's Equation* MSc. Thesis, Univ. Maryland, (supervised by I. Babuska), 1992.

[19] Philips, G.M. & Taylor, P.J., *Theory and Applications of Numerical Analysis*, Academic Press: London, 1973.

[20] Ramachandran, P.A., Method of fundamental solutions: singular value decomposition analysis. *Communications in Numerical Methods in Engineering*, **18**, pp. 789–801, 2002.

[21] Smyrlis, Y.S., The method of fundamental solutions: a weighted least-squares approach. *BIT Numerical Mathematics*, **46**, pp. 163–194, 2006.

[22] Szybiński, B. & Zieliński A.P., Alternative T-complete systems of shape functions applied in analytical Trefftz finite elements. *Numerical Methods for Partial Differential Equations*, **11**, pp. 375–378, 1995.

[23] Tikhonov, A., On certain problems of linear algebra and a stable method of their solutions (in Russian). *Doklady AN SSSR*, **163**, pp. 591–594, 1965.

[24] Trefftz, E., Ein Gegenstück zum Ritzschen Verfahren, *Proceedings 2nd International Congres of Applied Mechanics*, pp. 131–137, Zurich, 1926.

[25] Zieliński, A.P. & Herrera, I., Trefftz method: fitting boundary conditions. *International Journal for Numerical Methods in Engineering*, **24**, pp. 871–891, 1987.

[26] Zieliński, A.P. & Zienkiewicz, O.C., Generalized finite element analysis with T-complete boundary solution functions. *International Journal for Numerical Methods in Engineering*, **21**, pp. 509–528, 1985.

[27] Zieliński, A.P., Special Trefftz elements and improvement of their conditioning. *Communications in Numerical Methods in Engineering*, **13**, pp. 765–775, 1997.

CHAPTER 3

Analytical trial T-functions applied in the boundary collocation method

1 Complete systems of general solutions

As it was already mentioned, the boundary collocation method consists in identical fulfilment of a partial differential equation (or system of equations) governing a considered boundary value problem. The necessary condition of effective application of the method is thus the knowledge of trial functions, which are analytical solutions of this equation. The situation is here basically different from the finite element approach in which the shape functions can be identical for many different equations. In the boundary collocation techniques, the governing equation determines the system of applied functions. Hence, the conceptual simplicity of the method is paid by the necessary choice of the individual system of trial functions for each particular problem.

The most natural functional sets often applied in the Trefftz-type algorithms are complete general solutions of homogeneous partial differential equations. For certain basic types of equations, these solutions were derived more than a hundred years ago. To this type of equations belongs, for example, the two-dimensional (2D) Laplace equation:

$$\nabla^2 \phi(\mathbf{x}) = 0 \quad \mathbf{x} \in \Omega, \tag{1}$$

which in polar coordinates has a well-known solution

$$\begin{aligned} \phi(r,\theta) &= A_0 + A_1\theta + A_2\theta \ln r + A_3 \ln r \\ &\quad + \sum_{n=1}^{\infty} \left(B_n r^{\lambda_n} + C_n r^{-\lambda_n}\right)\cos(\lambda_n\theta) + \sum_{n=1}^{\infty} \left(D_n r^{\lambda_n} + E_n r^{-\lambda_n}\right)\sin(\lambda_n\theta), \end{aligned} \tag{2}$$

where A_0, A_1, A_2, A_3, B_n, C_n, D_n, E_n, and λ_n are constants which should be calculated from the boundary conditions.

This system was applied in numerous works concerning the boundary collocation approach, for example, by Barta [9], Fend, Barrody & Bell [35], Poritsky & Danforth [111], Sekiya & Takeuti [121], Conway [29–32], Sparrow & Loeffeler [127],

Balcerzak & Raynor [7, 8], Shuleshko [124], Beadle & Conway [10], Hocney [51], Leissa & Brann [87], Cheng [22–24], Schmidt [120], Axford [4], Bobone [15], Shih [122, 123], Hagan & Ratkowsky [43], Ratkowsky & Epstein [115], Takeuti [138–140], Takeuti & Sekiya [136], Takeuti & Sumi [137], Dwyer & Berry [34], Hoffman & Ariman [52], Hulbert [54], Symm [132], Kołodziej & Dudziak [73], Noda & Gryte [103], Sultanian & Sastrii [130, 131], Cieszko & Kołodziej [28], Han & Cosner [44], Isida & Noguchi [57], Ondris [107, 108], Kołodziej [80–82], Golik & Kołodziej [40], Marlewski & Kołodziej [99], Wang [146], Kołodziej & Stręk [76, 77], Kołodziej & Uściłowska [72], Kołodziej, Uściłowska & Ciałkowski [83], Kołodziej & Wachowska [78] and Kołodziej & Fraska [74].

Another typical equation often used in applied mathematical modelling is the 2D biharmonic equation:

$$\nabla^4 \phi(\mathbf{x}) = 0 \quad \mathbf{x} \in \Omega, \tag{3}$$

for which the solution was derived by J.H. Michell in 1899 [101]. In polar coordinates, it has a form:

$$\begin{aligned}
\phi(r,\theta) = {} & A_0 + A_1 \ln r + A_2 r^2 + A_3 r^2 \ln r + A_4 r^2 \theta + A_5 \theta \\
& + \frac{A_6 r}{2} \theta \cos\theta + \left(B_1 r + C_1 r^3 + B_1' r^{-1} + C_1' r \ln r\right) \cos\theta \\
& + \frac{A_7 r}{2} \theta \sin\theta + \left(D_1 r + E_1 r^3 + D_1' r^{-1} + E_1' r \ln r\right) \sin\theta \\
& + \sum_{n=2}^{\infty} \left(B_n r^n + C_n r^{-n+2} + B_n' r^{-n} + C_n' r^{-n+2}\right) \cos n\theta \\
& + \sum_{n=2}^{\infty} \left(D_n r^n + E_n r^{-n+2} + D_n' r^{-n} + E_n' r^{-n+2}\right) \sin n\theta,
\end{aligned} \tag{4}$$

with the unknown coefficients $A_0,\ldots A_7, B_1, C_1, \ldots D_1', E_1', B_n, C_n, \ldots D_n', E_n'$.

Similarly to the harmonic solution, the above form was applied in many engineering problems like bending of plates [2, 3, 5, 6, 9, 20, 30, 32, 39, 41, 42, 52, 53, 63, 88–91, 95, 102, 113, 141], 2-D elastostatics [10, 36, 37, 55, 63, 67, 93, 125], properties of composites [19, 45, 54, 80], thermal stresses [56, 107, 108, 114, 121, 136, 137, 140] and creeping flows [75, 100].

The authors considering axisymmetric creeping flow [17, 21, 38, 70, 104–106], which is governed by the equation:

$$DDu(\rho,\varphi) = 0, \tag{5}$$

where

$$D = \frac{\partial}{\partial \rho^2} + \frac{\sin\varphi}{\rho^2} \frac{\partial}{\partial \varphi}\left(\frac{1}{\sin\varphi} \frac{\partial}{\partial \varphi}\right), \tag{6}$$

often used the complete solution in the form

$$u(\rho,\varphi)=\frac{\rho^2}{2}\sin\varphi + \sum_{n=2}^{\infty}\left(A_n\rho^n + B_n\rho^{n+1} + C_n\rho^{-n+1} + D_n\rho^{-n+3}\right)I_n^{-\frac{1}{2}}(\cos\varphi), \quad (7)$$

where $I_n^{-1/2}$ is the Gegenbauer function (see [119] and [97]) expressed by the Legendre functions $P_n(t)$:

$$I_n^{-\frac{1}{2}}(t)=\frac{1}{2n-1}\left[P_{n-2}(t) - P_n(t)\right] \quad (8)$$

and A_n,..., D_n are unknown constants.

For a long time, certain systems of the Trefftz functions were used in numerical solutions of boundary value problems in an intuitive way. The real mathematically based development the T-functions owe to Ismael Herrera and his collaborators. Starting at the end of 1970s, the Mexican mathematician derived a series of such systems and proved their completeness [46, 48–50]. The systems (called Herrera functions) were presented both for bounded Ω^+ and unbounded Ω^- regions.

The difference between the two sets consists in their behaviour at the origin of the coordinates and in infinity. The functions applied in Ω^+ should be regular at the origin and can tend to large values (even to infinity) with the increase of the distance from the centre. The systems complete in Ω^- can be singular at the origin but should decay with the growing distance, fulfilling the Sommerfeld radiation condition [46]. In bounded but doubly connected regions, both systems ought to be superposed.

The basic sets of T-functions (Table 1) allow the formation of a large group of derivative T-complete systems. For example, using the T-sets for harmonic or biharmonic equations, one can derive T-systems for three-dimensional (3D) elasticity [110, 150]. The Laplace and Helmholtz equation solutions allow the formation of the Trefftz systems for Mindlin-type plates [61], etc. Also expansion of auxiliary unknown functions into a power series (as in Muskhelishvili-type solution for 2D [64] or 3D [109] elasticity) can be used in forming the T-complete system for certain more complex problems. The Herrera T-sets for a particular problem are not necessarily unique – one differential equation can have two or more such sets [155]. The selected, important T-complete systems are gathered in Table 1.

The Herrera functions can also be used in a-typical Trefftz algorithms. One of them is a solution of a boundary value problem with the help of the so-called boundary biorthogonal series [47, 98, 157]. These series are defined simultaneously in the bounded Ω^+ and unbounded Ω^- regions which are separated by a surface Γ:

$$\phi^+ = \sum_{n=0}^{\infty} a_n^+\Phi_n^+ \quad (9a)$$

$$\phi^- = \sum_{n=0}^{\infty} a_n^-\Phi_n^-, \quad (9b)$$

Table 1: Selected T-complete Herrera functions for typical engineering equations.

Equation	T-complete system
$\nabla^2 u(\mathrm{x}) = 0$ in 2D, Ω^+	$\left\{1, \mathrm{Re}\left(z^n\right), \mathrm{Im}\left(z^n\right); \quad n = 1, 2, \ldots\right\}$
$\nabla^2 u(\mathrm{x}) = 0$ in 2D, Ω^-	$\left\{\ln\left(x^2 + y^2\right), \mathrm{Re}\left(z^{-n}\right), \mathrm{Im}\left(z^{-n}\right); \quad n = 1, 2, \ldots\right\}$
$\nabla^2 u(\mathrm{x}) = 0$ in 3D, Ω^+	$\left\{\begin{matrix} r^n P_n^q(\cos\theta)\exp(iq\varphi); \\ n = 0, 1, 2, \ldots, \quad -n \le q \le n \end{matrix}\right\}$, where $P_n^q(\cos\theta)$ are associated Legendre functions
$\nabla^2 u(\mathrm{x}) = 0$ in 3D, Ω^-	$\left\{\begin{matrix} r^{-n-1} P_n^q(\cos\theta)\exp(iq\varphi); \\ n = 0, 1, 2, \ldots, \quad -n \le q \le n \end{matrix}\right\}$, where $P_n^q(\cos\theta)$ are associated Legendre functions
$\nabla^2 u(\mathrm{x}) + u(\mathrm{x}) = 0$ in 2D, Ω^+	$\left\{\begin{matrix} J_0(r), J_n(r)\cos(n\theta), J_n(r)\sin(n\theta); \\ n = 1, 2, 3, \ldots \end{matrix}\right\}$, where $J_n(r)$ are Bessel functions of the first kind
$\nabla^2 u(\mathrm{x}) + u(\mathrm{x}) = 0$ in 2D, Ω^-	$\left\{\begin{matrix} H_0^{(1)}(r), H_n^{(1)}(r)\cos(n\theta), H_n^{(1)}(r)\sin(n\theta); \\ n = 1, 2, 3, \ldots \end{matrix}\right\}$, where $H_n^{(1)}(r)$ are Hankel functions of the first kind
$\nabla^2 u(\mathrm{x}) + u(\mathrm{x}) = 0$ in 3D, Ω^+	$\left\{\begin{matrix} j_n(r) P_n^q(\cos\theta)\exp(iq\varphi); \\ n = 0, 1, 2, \ldots, \quad -n \le q \le n \end{matrix}\right\}$, where $j_n(r) = \sqrt{\pi/2r} J_{n+(1/2)}(r)$ are spherical Bessel functions of the first kind
$\nabla^2 u(\mathrm{x}) + u(\mathrm{x}) = 0$ in 3D, Ω^-	$\left\{\begin{matrix} h_n(r) P_n^q(\cos\theta)\exp(iq\varphi); \\ n = 0, 1, 2, \ldots, \quad -n \le q \le n \end{matrix}\right\}$, where $h_n(r) = \sqrt{\pi/2r} H_{n+(1/2)}(r)$ are spherical Hankel functions of the first kind

(*Continued*)

Table 1: Selected T-complete Herrera functions for typical engineering equations. (*Continued*)

$\nabla^4 u(x) = 0$ in 2D, Ω^+	$\{1, r^2, r^n\cos(n\theta), r^n\sin(n\theta),$ $r^{n+2}\cos(n\theta), r^{n+2}\sin(n\theta)\}$
$\nabla^4 u(x) = 0$ in 2D, Ω^-	$\{\ln(r), r^2\ln(r), r^{-n}\cos(n\theta), r^{-n}\sin(n\theta),$ $r^{2-n}\cos(n\theta), r^{2-n}\sin(n\theta)\}$

where Φ_n^+, Φ_n^- are respective T-complete sets and the coefficients a_n^+, a_n^- have a form of integrals defined only on the surface Γ. This last property allows the derivation of different boundary algorithms based on the biorthogonal series [94, 157].

Another a-typical Trefftz algorithm is based upon the generalized Betti's reciprocal theorem and is called the direct Trefftz approach. It has been proposed and investigated by O.C. Zienkiewicz and his collaborators [25, 26, 58, 59] as an analogy to the well-known direct BEM. It uses the regular Herrera systems instead of the singular fundamental solutions. Hence, for example, in an elasticity problem without body forces, the boundary equation arises [58, 155]:

$$\int_{\Gamma} \left(\mathbf{T}^{\mathrm{T}}\mathbf{u} - \mathbf{U}^{\mathrm{T}}\mathbf{t}\right) d\Gamma = 0, \tag{10}$$

in which **u** are displacements, **t** are tractions, **U** is a matrix containing a T-complete displacement system and **T** is a matrix of tractions derived analytically from **u**. The displacements **u** and the tractions **t** in eqn (10) are partly known as boundary conditions:

$$\begin{aligned} \mathbf{u} &= \bar{\mathbf{u}} \quad \text{on} \quad \Gamma_1 \\ \mathbf{t} &= \bar{\mathbf{t}} \quad \text{on} \quad \Gamma_2 \end{aligned} \tag{11}$$

and partly unknown, sought in the approximate form

$$\begin{aligned} \hat{\mathbf{u}} &= \Psi\mathbf{a} \quad \text{on} \quad \Gamma_2 \\ \hat{\mathbf{t}} &= \mathbf{X}\mathbf{b} \quad \text{on} \quad \Gamma_1, \end{aligned} \tag{12}$$

where Ψ, **X** are the interpolation functions, and **a, b** are their unknown coefficients. Hence, analogically to the direct BEM, the unknown variables are values of the displacements **u** and tractions **t** over the boundary and not directly the coefficients of the Trefftz functions. The algorithm presented above was used in solutions of harmonic problems [25], plane elasticity [58], Helmholtz equation [26] and plate-bending problems [59]. The above papers investigate also the standard Trefftz method (with the Herrera T-complete systems) which in these papers is called its indirect version.

2 Influence functions with singularities outside an investigated region

In the middle of 1960s, V. Kupradze proposed a regular boundary integral equation method in which he used, as trial functions, fundamental solutions with their singularities outside the investigated region [86]. For example, in a Dirichlet problem of elasticity defined in Ω^+, he obtained the regular integral relation [85]

$$\beta(\mathbf{x})\mathbf{u}(\mathbf{x})+\int_\Gamma \mathbf{G}^K(\mathbf{x},\mathbf{y})\mathbf{t}(\mathbf{y})d_y\Gamma=\mathbf{F}(\mathbf{x})$$
$$\beta(\mathbf{x})=\begin{cases}-1 & \mathbf{x}\in\Omega^+\\ 0 & \mathbf{x}\in\Omega^-\end{cases}, \tag{13}$$

where $\mathbf{G}^K(\mathbf{x},\mathbf{y})$ is the well-known fundamental Kelvin displacement solution. The right side of the equation is a known function (a particular integral here and further omitted):

$$\mathbf{F}(\mathbf{x})=\int_\Gamma \mathbf{T}(\mathbf{x},\mathbf{y})\bar{\mathbf{u}}(\mathbf{y})d_y\Gamma, \tag{14}$$

with $\mathbf{T}(\mathbf{x},\mathbf{y})$ as tractions analytically calculated from $\mathbf{G}^K(\mathbf{x},\mathbf{y})$. After introduction of discretization

$$\mathbf{G}^K(\mathbf{x},\mathbf{y})\rightarrow\mathbf{G}^K(\mathbf{x}_i,\mathbf{y}_j)\quad \mathbf{x}_i\in\Gamma_s\in\Omega^-,\ \mathbf{y}_j\in\Gamma$$
$$i,j=1,\ 2,\ 3,\ldots,(N_f=K_p), \tag{15}$$

where Γ_s is an arbitrary smooth surface enclosing Ω^+ and not having any common points with it, V. Kupradze obtained a system of algebraic equations for the unknown $\mathbf{t}(\mathbf{y}_j)$. He also proved completeness of the system of fundamental solutions $\mathbf{G}(\mathbf{x}_i,\mathbf{y})$ (called, because of this, Kupradze functions) when $\mathbf{x}_i$ is a countable set of points densely distributed on Γ_s. There exists a straightforward analogy between this Kupradze algorithm and Zienkiewicz's proposal presented at the end of the previous section. Both lead to the direct regular boundary integral equations for complementary functions (e.g. unknown displacements or tractions). Hence, both can be included into the Trefftz family although, as has been proposed earlier, as its a-typical members.

The Kupradze functions can be interpreted as fundamental solutions with pointwise sources (e.g. concentrated forces) applied to an infinite region on the surface Γ_s and modelling the unknown solution inside Ω^+. Therefore, as typical T-complete systems in Ω^+, they can be applied to the standard Trefftz algorithm. This simple idea was discovered independently by several researchers who, sometimes not quoting each other, gave it different names (see Section 2 in Chapter 1). The most often used Kupradze functions (F-functions) are presented in Table 2. For more complex governing equations, one can find fundamental solutions in the following papers: [11] for plane-layered elasticity, [68] for axisymmetric elasticity problems, [112] for 3D problems in elastostatics, [1] for 2D and 3D creeping flows.

Table 2: Selected examples of Kupradze functions in 2D (fundamental solutions of presented equations with singularities outside investigated regions – Fig. 1); $r_j = \sqrt{(x-\xi_j)^2 + (y-\zeta_j)^2}$, (ξ_j, ζ_j), positions of sources.

Equation	T-complete functions
$\nabla^2 u(\mathrm{x}) = 0$	$\{\ln r_j^2, \quad j = 1, 2, 3,...\}$
$\nabla^4 u(\mathrm{x}) = 0$	$\{r_j^2 \ln r_j^2, \quad \ln r_j^2, \quad j = 1, 2, 3,...\}$
$(\nabla^2 + \lambda^2)u(\mathrm{x}) = 0$	$\{iH_0^{(2)}(\lambda r_j), \quad j = 1, 2, 3,...\}$, where $H_0^{(2)}(\lambda r_j)$ are zero order Hankel functions of second kind
$(\nabla^2 - \lambda^2)u(\mathrm{x}) = 0$	$\{K_0(\lambda r_j), \quad j = 1, 2, 3,...\}$, where $K_0(\lambda r_j)$ are zero order modified Hankel functions

Considerations concerning equivalence of application of the Herrera (Trefftz) and Kupradze formulations can be found in [20].

An important feature of the Trefftz method with the Kupradze functions is the unknown position of the sources, that is the shape of the surface Γ_s and, arbitrary to a certain extent, its distance from Γ. Obviously, this 'free parameter' influences the convergence of the solutions and determines the effectiveness of the method.

Until now, the most often applied form of Γ_s is a homothetic contour (Fig. 1a). This choice opens a question about smoothness of Γ_s (mathematically formal requirement for completeness of the Kupradze functions) in case of Ω with corners (Fig. 1b). However, if limited numbers of functions are applied, this problem does not seem to be of considerable importance (observation confirmed by numerical studies of the authors [33, 133, 151, 156]). Also, many investigators situate the sources on straight lines perpendicular to surface Ω, omitting the zone of corners (Fig. 1c) or, as Wearing & Sheikh [147], propose their optimized position (Fig. 1d) in this zone.

An alternative choice has been suggested by Bogomolny [16], who proposed the position of the sources on a circle enclosing the area Ω^+ (Fig. 1e). Investigations carried out by authors of paper [133] have shown very similar behaviour of solutions for the sources situated either on a circle or on a homothetic contour, even in case of a triangular area Ω^+ (for concave areas Ω^+ the results were considerably worse). This suggests that the shape of Γ_s can be chosen relatively arbitrarily within certain acceptable forms.

The free parameter of Kupradze functions connected with the distance of Γ_s from Γ (e.g. δ or R_s) should be chosen possibly optimally for a particular problem.

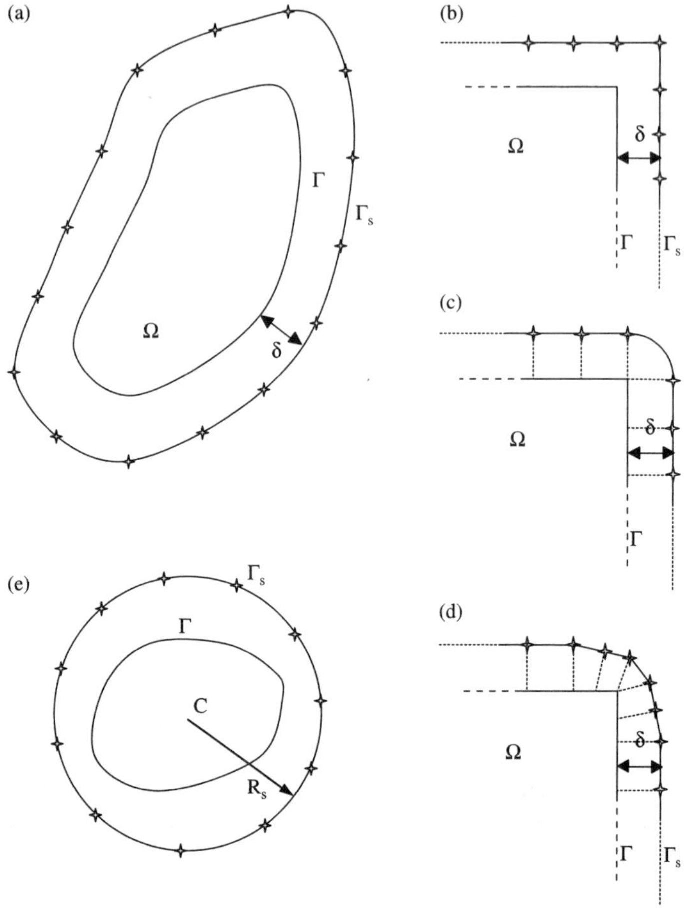

Figure 1: Kupradze functions; position of sources.

In all the investigated cases, the authors observe a certain relatively large range of acceptability of this parameter [133, 156]. Certain examples of such investigations are shown in Chapter 5. However, more careful and profound studies concerning the optimal choice of the distance parameter are still not completed.

Fundamental solutions are the influence functions most often used in the algorithm presented above, mainly because of their easy availability [12, 18, 71]. However, other kinds of the influence functions can also be applied, and sometimes even with better effectiveness. For example, Chmielewski [27] proposed for 2D elasticity the influence functions resulting from application of a concentrated force to a centre of a clamped circle with a radius R. He used the solution given by Lukasiewicz [96].

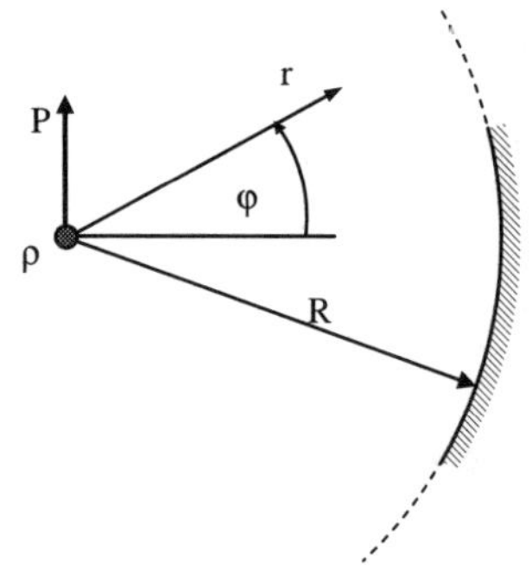

Figure 2: Chmielewski–Lukasiewicz influence function – source point.

$$u_r = -\frac{P}{8\pi G h}\left\{(3-\nu)\ln\frac{r}{R} - \frac{(1+\nu)(1-3\nu)}{2(3-\nu)}\left[1-\left(\frac{r}{R}\right)^2\right]\right\}\sin\theta,$$
$$u_\theta = -\frac{P}{8\pi G h}\left\{(3-\nu)\ln\frac{r}{R} + \frac{(1+\nu)(5+\nu)}{2(3-\nu)}\left[1-\left(\frac{r}{R}\right)^2\right]\right\}\cos\theta. \tag{16}$$

Chmielewski investigated also a variant in which the concentrated force was distributed on a small circle with a radius ρ (Fig. 2). The results obtained were encouraging. However, this kind of the Trefftz functions has two additional free parameters that obviously influence convergence of the solutions.

Summarizing, the F-functions presented above need further profound investigations. On one hand, they can give much better results than the Herrera functions [133]. On the other, their free parameters need sometimes optimization, which can result in final higher computational costs.

3 Influence T-functions represented by trigonometric series

The trigonometric T-functions can be applied to problems for which direct forms of influence functions are difficult to obtain (e.g. in case of shells). Therefore, in certain engineering solutions, it is convenient to use the influence T-functions expanded into double trigonometric series. They were first presented by A.P. Zieliński in several consecutive papers concerning plates and shells with arbitrary contours [152, 153, 158]. In these papers, a certain variant of the boundary integral equation method was proposed in which the unknown boundary functions were expanded into a single trigonometric series along the contour of the structure (this idea was followed by Birginie *et al.* [13, 14]).

The trigonometric influence functions are defined in rectangular areas containing considered structures with arbitrary contours. For example, for simply supported Kirchhoff plates, they have the form [152]:

$$w^P(x,y,\xi,\eta) = \frac{4P}{abD}\sum_{m,n}\gamma_{mn}\sin(\alpha_m\xi)\sin(\beta_n\eta)\sin(\alpha_m x)\sin(\beta_n y) \tag{17a}$$

for deflections caused by a concentrated force P situated in (ξ,η) and

$$w^{M_x}(x,y,\xi,\eta) = -\frac{4M_x}{abD}\sum_{m,n}\gamma_{mn}\beta_n \sin(a_m\xi)\cos(\beta_n\eta)\sin(a_m x)\sin(\beta_n y) \quad (17b)$$

$$w^{M_y}(x,y,\xi,\eta) = \frac{4M_y}{abD}\sum_{m,n}\gamma_{mn}a_m \cos(a_m\xi)\sin(\beta_n\eta)\sin(a_m x)\sin(\beta_n y) \quad (17c)$$

for deflections forced by concentrated moments M_x, M_y,
where

$$\gamma_{mn} = \frac{1}{\left(a_m^2+\beta_n^2\right)^2}, \quad a_m = \frac{m\pi}{a}, \; \beta_n = \frac{n\pi}{b}.$$

D is here the plate stiffness and $a \times b$ are the plate dimensions. It should be underlined that for correct application of such trial functions, it is necessary to know the rules of differentiation of double trigonometric series, which was first presented in paper [154]. This kind of series represents, in general, distributions (generalized functions) and, therefore, its differentiation is a little complex.

The trigonometric T-functions were successfully applied and presented in a series of papers by Lifits, Reutsky and Tirozzi [94, 116, 117]. They situated the characteristic 'singular' points of the functions outside the investigated region, chose suitable rectangular areas to define the T-functions, optimized the position of their characteristic points, etc. This form of the Trefftz functions requires, however, high-performance computers because of necessary investigation of convergence of the obtained solutions.

4 Heuristic T-functions

In certain cases, the direct derivation of the complete T-systems presented in the previous sections can be difficult. In such problems, some authors propose heuristic algorithms to derive the sets of T-functions which fulfil the governing differential equations, but the completeness of which is not proved. For example, Stein [128] and Ruoff [118] apply to shallow shells with constant curvatures the following system of Trefftz functions:

$$\left\{e^{\lambda_n x}\cos\beta_n y;\; e^{\lambda_n x}\sin\beta_n y;\; e^{\gamma_n y}\cos a_n x;\; e^{\gamma_n y}\sin a_n x\right\} \quad (18)$$

$$a_n = \frac{n\pi}{2a}, \; \beta_n = \frac{n\pi}{2b},$$

where $2a \times 2b$ are the dimensions of the panel and λ_n, γ_n are the complex roots of the characteristic equations.

This system is a superposition of two Levy-type solutions taken in two mutually perpendicular directions of principal curvatures (in the standard Levy method, the trigonometric functions in one direction fulfilled the given boundary or continuity conditions; this permitted decoupling of the variables). The system (18) was successfully applied in the works [118, 128] and then, as an internal field of the hybrid-Trefftz element, by Voros & Jirousek [143].

The standard identification process is another popular heuristic algorithm of derivation of the Trefftz functions. The system of polynominals

$$\Phi_k(x,y)=\sum_{j=0}^{k} a_{(k-j)j}x^{k-j}y^j \tag{19}$$

is introduced into the governing equation and the unknown coefficients $a_{(k-j)j}$ are calculated from the condition of fulfilment of the equation by these polynominals. This easy and natural procedure was used in the case of orthotropic plates [62] and elasticity [84], but obviously its range is more general.

The application of the T-functions presented in this section is not as obvious as in the case of the complete T-systems because of lack of the mathematical background. Therefore, the boundary collocation solutions with such trial functions should be carefully investigated (accuracy, convergence, error estimation, etc.).

5 Special purpose T-functions

The last group of the T-functions fulfil not only the governing differential equations but also certain given boundary conditions, for example, along the boundaries of a circle or an angle. Application of such functions is convenient especially in case of singularities or strongly concentrated effects which, in the standard FE solutions, require adaptive procedures.

Into this family, the Williams functions [148, 149] for angle singularities in 2D elasticity can be included. For polar coordinates $\{r,\theta\}$ with a pole situated in an angle corner point, they have a general form

$$\Phi_k(r,\theta)=r^{\lambda_k}f_k(\theta) \tag{20}$$

in which λ_k and $f_k(\theta)$ change depending on the problem. In a specific case, when a re-entrant angle θ_c of the corner tends to 2π, it leads to the crack effect [65, 142].

In general, if the investigated region Ω has straight lines, symmetry axes, cylindrical openings, etc., the special purpose T-functions fulfilling the boundary conditions along these contours can be formed. For example, using the general solutions (2) of the Laplace equation expressed in polar coordinates, we can derive the following special T-functions for a region with N_a symmetry axes [79]:

$$\Phi_1=1,\quad \Phi_k=r^{N_a(k-1)}\cos\left[N_a(k-1)\right]. \tag{21}$$

Such systems can also be found in [74, 78, 92]. The way of their derivation as well as other types of the special T-functions for the Laplace equation will be presented in Chapter 6.

For biharmonic equations and regions with corners, certain authors [69, 129, 144, 145] present analytical solutions as linear combinations of the following functions:

$$\left\{\begin{array}{l}\left(r^{1+\lambda},r^{1-\lambda}\right)\left[\cos\left(1+\lambda\right)\theta,\ \sin\left(1+\lambda\right)\theta,\ \cos\left(1-\lambda\right)\theta,\ \sin\left(1-\lambda\right)\theta\right];\\ \left[r\sin\left(n\ln r\right),r\cos\left(n\ln r\right)\right]\cdot\begin{bmatrix}\cos\theta\cosh n\theta,\sin\theta\sinh n\theta,\\ \cos\theta\sinh n\theta,\sin\theta\cosh n\theta\end{bmatrix};\\ \left(r^2,1\right)\left(\cos 2\theta,\sin 2\theta,1,\theta\right);\\ \left(r,r\ln r\right)\left(\cos\theta,\sin\theta,\cos\theta,\sin\theta\right);\\ \left(\sin\mu\theta,\cos\mu\theta\right)\left(r^{\mu},r^{-\mu},r^{2+\mu},r^{2-\mu}\right);\\ \left(\sinh v\theta,\cosh v\theta\right)\cdot\left[\sin\left(v\ln r\right),\cos\left(v\ln r\right),r^2\sin\left(v\ln r\right),r^2\cos\left(v\ln r\right)\right];\\ \left(\sin\theta,\cos\theta\right)\left(r,r^{-1},r\ln r,r^3\right);\\ \left(1,\theta\right)\left(1,\ln r,r^2,r^2\ln r\right)\end{array}\right\}, \tag{22}$$

where λ, n, v, μ are respective integer constants. In case of 2D creeping flow, an analytical solution for corners can be found in [75].

The special purpose functions representing analytical solutions in vicinity of circular holes in plates [61, 135] and discs (2D elasticity) [110] can in general be written as

$$\Phi_k\left(r,\theta\right)=g_k\left(r\right)\cos\left(k\theta\right)+h_k\left(r\right)\sin\left(k\theta\right). \tag{23}$$

Their explicit form is rather complex [64]. Additionally, their application can sometimes be troublesome because of great variety of solutions necessary to cover all the concentration cases. Hence, it is sometimes more efficient to divide the investigated area into the Trefftz finite elements (see Chapter 4) and apply a suitably refined mesh [65, 134].

References

[1] Alves, C.J.S. & Silvestre, A.L., Density results using Stokeslets and method of fundamental solutions for the Stokes equations. *Engineering Analysis with Boundary Elements*, **28**, pp. 1245–1252, 2004.

[2] Artemov, V.V. & Belkin, N.I., On bending of square plate with stiffening ring (in Russian). *Sbornik Trudov Vsesojuznovo Zaochnogo Politekhnicheskogo Instituta*, **81**, pp. 165–174, 1973.

[3] Artemov, V.V., Bending of square plates with central cylindrical holes supported by elastic rings (in Russian). *Sbornik Trudov Vsesojuznovo Zaochnogo Politekhnicheskogo Instituta*, **73**, pp. 87–91, 1972.

[4] Axford, R.A., Two-dimensional, multiregion analysis of temperature fields in reactor tube bundles. *Nuclear Engineering and Design*, **6**, pp. 25–42, 1967.

[5] Bailey, R. & Hicks, R., Behaviour of perforated plates under plane stress. *Journal of Mechanical Science*, **2**, pp. 143–167, 1960.

[6] Bailey, R.W. & Fidler, F., Stress analysis of plates and shells containing patterns of reinforced holes. *Nuclear Engineering and Design*, **3**, pp. 41–53, 1966.

[7] Balcerzak, E.M. & Raynor, S., Solution for several shapes of externally pressurised hydrostatic thrust bearings. *Applied Scientific Research*, Section A, **11**, pp. 189–217, 1962.

[8] Balcerzak, M.J. & Raynor, S., Steady state temperature distribution and heat flow in prismatic bars with isothermal boundary conditions. *International Journal of Heat and Mass Transfer*, **3**, pp. 113–125, 1961.

[9] Barta, J., Über die naherungsweise Lösung einiger Zwei-dimensionalier Elastizitatsaufgaben. *Zeitschrift fÜr Angewandte Mathematik und Mechanik*, **5**, pp. 187–237, 1934.

[10] Beadle, C.W. & Conway, H.D., Stress concentration in twisted structural beams. *Journal of Applied Mechanics*, **30**, pp. 138–141, 1963.

[11] Berger, J.R. & Karageorghis, A., The method of fundamental solutions for layered elastic materials. *Engineering Analysis with Boundary Elements*, **25**, pp. 877–886, 2001.

[12] Beskos, D.E. (ed.), *Boundary Element Methods in Mechanics*, North-Holland Publications: Amsterdam, 1987.

[13] Birginie, J.M., Allard F. & Kassab, A.J., Application of trigonometric boundary elements to heat and mass transfer problems. *Proc. Conf. Boundary Elements XVIII*, Wessex, 1996.

[14] Birginie, J.M., Dumargue, P. & Allard, F., Use of trigonometric boundary element method to characterize the boundary irregularity influence on a thermal field. *Proc. IX Conf. Numerical Methods for Thermal Problems*, Atlanta, 1995.

[15] Bobone, R., The method of solutions functions: a computer-oriented solution of boundary value problems as applied to nuclear reactors - Part 1, Cylindrical Reactors in R - θ geometry. *Nuclear Engineering Science*, **29**, pp. 337–353, 1967.

[16] Bogomolny, A., Fundamental solutions method for elliptic boundary value problems. *SIAM Journal of Numerical Analysis*, **22**, pp. 644–669, 1985.

[17] Bowen, B.D. & Masliyah, J.H., Drag force on isolated axisymmetric particles in Stokes flow. *Canadian Journal of Chemical Engineering*, **51**, pp. 8–15, 1973.

[18] Brebbia, C.A., Telles, J.C.F. & Wrobel, L.C., *Boundary Element Techniques*, Springer-Verlag: New York, 1984.

[19] Chen, C.H. & Cheng, S., Mechanical properties of fiber reinforced composites. *Journal of Composite Materials*, **1**, pp. 30–41, 1967.

[20] Chen, J.T., Wu, C.S., Chen, I.L. & Chen, K.H., On the equivalence of the Trefftz method of fundamental solutions for Laplace and biharmonic equations. *Comp. Math. Appl.*, **53**, pp. 851–879, 2007.

[21] Chen, S.B. & Keh, H.J., Axisymmetric electrophoresis of multiple colloidal spheres. *Journal of Fluid Mechanics*, **238**, pp. 251–276, 1992.

[22] Cheng, K.C., Analog solution of laminar heat transfer in non-circular ducts by Moire method and point matching. *Journal of Heat Transfer*, **88**, pp. 175–182, 1966.

[23] Cheng, K.C., Laminar flow and heat transfer characteristics in regular polygonal ducts. *Proceedings Third International Heat Transfer Conference AIChE*, **1**, pp. 64–76, 1966.

[24] Cheng, K.C., Laminar forced convection in regular polygonal ducts with uniform peripheral heat flux. *Journal of Heat Transfer*, **91**, pp. 159–167, 1969.

[25] Cheung, Y.K., Jin, W.G. & Zienkiewicz, O.C., Direct solution procedure for solution of harmonic problems using complete, non-singular Trefftz functions. *Communications in Numerical Methods in Engineering*, **5**, pp. 159–169, 1989.

[26] Cheung, Y.K., Jin, W.G. & Zienkiewicz, O.C., Solution of Helmholtz equation by Trefftz method. *International Journal for Numerical Methods in Engineering*, **32**, pp. 63–78, 1991.

[27] Chmielewski, L., *The Use of Collocation Method in Solving Planar Elastostatics Problem with Arbitrary Boundary Conditions* (in Polish), PhD Thesis, Warsaw University of Technology Press: Warszawa, 1984.

[28] Cieszko, M. & Kołodziej, J.A., Determination of heat transfer between bundle of regular rods and fluid in longitudinal laminar flow by means of boundary collocation method (in Polish). *Theoretical and Applied Mechanics*, **19**, pp. 605–619, 1981.

[29] Conway, H.D., The approximate analysis of certain boundary value problems. *Journal of Applied Mechanics*, **27**, pp. 275–277, 1960.

[30] Conway, H.D., The bending, buckling and flexural vibration of simply supported polygonal plates by point matching. *Journal of Applied Mechanics*, **28**, pp. 288–291, 1961.

[31] Conway, H.D., Torsion of prismatical rods with isosceles triangle and other cross sections. *Journal of Applied Mechanics*, **27**, pp. 209, 1960.

[32] Conway, H.D., Triangular plates analysed by point matching. *Journal of Applied Mechanics*, **29**, pp. 755–756, 1962.

[33] Duda, P. & Zieliński, A.P., Stability and convergence of solutions in Trefftz-type algorithms. *Strojnicki Casopis (Journal of Mechanical Engineering)*, **48**, pp. 1–14, 1997.

[34] Dwyer, O.E. & Berry, H.C., Laminar-flow heat transfer in-line flow through unbaffled rod bundles. *Nuclear Science and Engineering*, **42**, pp. 81–88, 1970.

[35] Fend, F.A., Baroody, E.M. & Bell, J.C., An approximate of the temperature distribution surrounding coolings holes in heat-generating solid. *BMI Report No. T-42, Battelle Memarial Institute, Columbus, Ohio, October 25,* 1950.

[36] Fett, T., A compendium of T-stress solutions. *FZKA-Report* 6057, Forschungszentrum Karlsruhe, 1998.

[37] Fett, T., A semi-analytical study of the edge cracked circular disc by use of the boundary collocation method. *Engineering Fracture Mechanics*, **56**, pp. 331–346, 1997.

[38] Gluckman, M.J., Pfeffer, R. & Weinbaum, S., A new technique for treating multiparticle slow viscous flow: axisymmetric flow past sphere and spheroids. *Journal of Fluid Mechanics*, **50**, pp. 705–740, 1971.

[39] Goldberg, J.E. & Jabbour, K.N., Stresses and displacement in perforated plates. *Nuclear Engineering and Design*, **2**, pp. 360–381, 1965.

[40] Golik, W.L. & Kołodziej, J.A., An adaptive boundary collocation method for linear PDEs. *Numerical Methods for Partial Differential Equations*, **11**, pp. 555–560, 1995.

[41] Guseva, E.B. & Belkin, V., Stress and displacement at bending of rectangular plates with central hole. *Izvesti Vyshebnych Uchebnych Zavedenij Stroitelstvo i Arkhitektura*, **4**, pp. 44–47, 1973.

[42] Guseva, E.B., Bending of rectangular plates with circular hole by point force (in Russian). *Sbornik Trudov Vsesojuznovo Zaochnogo Politekhnicheskogo Instituta*, **81**, pp. 182–188, 1973.

[43] Hagan, S.L. & Ratkowsky, D.A., Laminar flow in cylindrical ducts having regular polygonal shaped cores. *Canadian Journal of Chemical Engineering*, **46**, pp. 387–388, 1968.

[44] Han, L.S. & Cosner, A.A., Effective thermal conductivities of fibrous composites. *Journal of Heat Transfer*, **103**, pp. 387–393, 1981.

[45] Heaton, M.D., Calculation of the elastic constants of unidirectional fiber-reinforced composite. *British Journal of Applied Physics, Journal of Physics*, sec. D, ser. 2, **1**, pp. 1039–1048, 1968.

[46] Herrera, I. & Sabina, F., Connectivity as an alternative to boundary integral equations: Construction of bases. *Proceedings of National Academy of Sciences U.S.A. (Appl. Phys. Sc.)*, **75 (5)**, pp. 2059–2063, 1978.

[47] Herrera, I. & Spence, F., Framework for biorthogonal Fourier series. *Proceedings og National Academy Sciences U.S.A. (Appl. Phys. Sc.)*, **78 (12)**, pp. 7240–7244, 1981.

[48] Herrera, I., *Boundary Methods: an Algebraic Theory*, Pitman Adv. Publ. Program: London, 1984.

[49] Herrera, I., Boundary methods; A criterion for completeness. *Proceedings of National Academy of Sciences U.S.A. (Appl. Phys. Sc.)*, **77 (8)**, pp. 4395–4398, 1980.

[50] Herrera, I., Theory of connectivity for formally symmetric operators, *Proceedings of National Academy Sciences U.S.A. (Appl. Phys. Sc.)*, **74 (11)**, pp. 4722–4725, 1977.

[51] Hocney, R.W., A solution of Laplace equation for a round hole in square peg. *Journal of Society of Industrial Applied Mathematics*, **12**, pp. 1–14, 1964.

[52] Hoffman, R.E. & Ariman, T., Thermal bending of plates with circular holes. *Nuclear Engineering and Design*, **12**, pp. 231–238, 1970.

[53] Hoffman, R.E. & Ariman, T., Thermal and mechanical stresses in nuclear reactor vessels. *Nuclear Engineering and Design*, **20**, pp. 31–55, 1972.

[54] Hulbert, L.E. & Rybicki, E.F., Boundary point least squares analysis of the free edge effects in some unidirectional fiber composites. *Journal of Composite Materials*, **5**, pp. 164–175, 1971.

[55] Hulbert, L.E. & Niedenfuhr, F.W., Accurate calculation of stress distribution in multi-holed plates. *Journal of Engineering for Industry*, **87**, pp. 331–336, 1965.

[56] Hulbert, L.E., Solution of thermal stress problems in tube sheets by the boundary point least square method. *Journal of Engineering for Industry*, **92**, pp. 339–349, 1970.

[57] Isida, M. & Noguchi, H., Tension of a plate containing an embedded elliptical crack. *Engineering Fracture Mechanics*, **20**, pp. 387–408, 1984.

[58] Jin, W.G., Cheung, Y.K. & Zienkiewicz, O.C., Application of the Trefftz method in plane elasticity problems. *International Journal for Numerical Methods in Engineering*, **30**, pp. 1147–1161, 1989.

[59] Jin, W.G., Cheung, Y.K. & Zienkiewicz, O.C., Trefftz method for Kirchhoff bending problems. *International Journal for Numerical Methods in Engineering*, **36**, pp. 765–781, 1993.

[60] Jirousek, J., Wróblewski, A., Qin, Q.H. & He, X.Q., A family of quadrilateral hybrid-Trefftz p-elements for thick plate analysis. *Computer Methods in Applied Mechanics and Engineering*, **127**, pp. 314–344, 1995.

[61] Jirousek, J. & Lan Guex, The hybrid-Trefftz finite element model and its application to plate bending. *International Journal for Numerical Methods in Engineering*, **23**, pp. 651–693, 1986.

[62] Jirousek, J. & N'Diaye, M., Solution of orthotropic plates based on *p*-extension of the hybrid-Trefftz finite element model. *Computers & Structures*, **34**, pp. 51–62, 1990.

[63] Jirousek, J. & Teodorescu, P., Large finite elements method for the solution of problems in the theory of elasticity. *Computers & Structures*, **15**, pp. 575–587, 1982.

[64] Jirousek, J. & Venkatesh, A., Hybrid-Trefftz plane elasticity elements with p-method capabilities. *International Journal for Numerical Methods in Engineering*, **35**, pp. 1443–1472, 1992.

[65] Jirousek, J., Szybiński B. & Wróblewski, A., Mesh design and reliability assurance in hybrid-Trefftz p-element approach. *Finite Elements in Analysis and Design*, **22**, pp. 225–247, 1996.

[66] Joseph, D.D., A new separation variables theory for problems of Stokes flow and elasticity. In: *Trends in Applications of Pure Mathematics to Mechanics*, H. Zorski (ed.), Pitman: London, 1979.

[67] Kalinin, H.G. & Pankov, A.I., Stresses distribution in rectangular plate with central cylindrical hole (in Russian). *Sbornik Trudov RKIIGA*, **241**, Ryga, 1972.

[68] Karageorghis, A. & Fairweather, G., The method of fundamental solutions for axisymmetric elasticity problems. *Computational Mechanics*, **25**, pp. 524–532, 2000.

[69] Karageorghis, A., Modified method of fundamental solutions for harmonic and biharmonic problems with boundary singularities. *Numerical Methods for Partial Differential Equations*, **8**, pp. 1–19, 1992.

[70] Keh, H.J. & Lin, Y.M., Particle interactions in diffusiophoresis: Axisymmetric motion of multiple spheres in nonelectrolyte gradients. *Langmuir*, **10**, pp. 3010–3017, 1994.

[71] Kleiber, M. (ed.), *Handbook of Computational Solid Mechanics*, Springer Verlag: Berlin, 1998.

[72] Kołodziej, J.A. & Uściłowska, A., Regular polygonal shapes, and boundary collocation method. *The International Journal of Mechanical Engineering Education*, **27**, pp. 140–150, 2000.

[73] Kołodziej, J.A. & Dudziak, W., Determination of steady laminar viscous fluid flow in hexagonal array of cylindrical rods by mean of boundary collocation method (in Polish). *Archives of Mechanical Engineering*, **26**, pp. 103–114, 1979.

[74] Kołodziej, J.A. & Fraska, A., Elastic torsion of bars possessing regular polygon in cross-section using BCM. *Computers & Structures*, **84**, pp. 78–91, 2005.

[75] Kołodziej, J.A. & Mendes, C.A., T-functions for 2-D creeping flow on domains with circular cylinders, corners, and possesing symmetry. *Computer Assisted Mechanics and Engineering Sciences*, **8**, pp. 359–384, 2001.

[76] Kołodziej, J.A. & Stręk, T., Analytical approximations of the shape factors for conductive heat flow in circular and regular polygonal cross-sections. *International Journal of Heat and Mass Transfer*, **44**, pp. 999–1012, 2001.

[77] Kołodziej, J.A. & Stręk, T., Regular polygonal shapes, boundary collocation method and regression by exponential function. *Zeitschrift für Angewandte Mathematik und Mechanik /ZAMM/*, **80**, pp. S805–S806, 2000.

[78] Kołodziej, J.A. & Wachowska, A., Special purpose Trefftz functions for the torsion of bars with regular polygonal cross-section. *Proceedings of International Workshop on Mesh Free Methods*, Institute Seperior Tecnica, Lisboa, pp. 157–162, July 21–23, 2003.

[79] Kołodziej, J.A., *Application of the Boundary Collocation Method in Applied Mechanics* (in Polish), Poznan University of Technology Press: Poznan, 2001.

[80] Kołodziej, J.A., Determination of the interaction coefficient in mixture equations for heat conduction in unidirectional composites (in Polish). *Archives of Thermodynamic*, **8**, pp. 87–107, 1987.

[81] Kołodziej, J.A., Determination of the transverse effective thermal conductivity of unidirectionally fibre arranged composites by means of boundary collocation method (in Polish). *Theoretical and Applied Mechanics*, **23**, pp. 933–958, 1985.

[82] Kołodziej, J.A., Influence of the porosity of porous medium on the effective viscosity in Brinkman's filtration equation. *Acta Mechanika*, **75**, pp. 241–254, 1988.

[83] Kołodziej, J.A., Uściłowska, A. & Ciałkowski, M., Semi-analytical approximations of the laminar friction coefficients for flow in conduits with polygonal cross-section. *Acta Mechanica*, **158**, pp. 127–144, 2002.

[84] Kompis, V., Finite elements satisfying all governing equations inside the element. *Computers & Structures*, **30**, pp. 273–278, 1994.

[85] Kupradze, V.D. (ed.), *Three-dimensionnal Problems of the Mathematical Theory of Elasticity and Theomoelasticity*, North-Holland Publ. Comp.: Amsterdam, 1968.

[86] Kupradze, V.D., On a certain method for approximate solution of boundary problems of mathematical physics (in Russian). *Zadachi Vichislitielnoj Matiematiki i Matiematichieskoj Fiziki*, **4**, pp. 1118, 1964.

[87] Leissa, A.W. & Brann, J.H., On the solution of bars having symmetry axes. *Journal of Mechanical Science*, **6**, pp. 45–60, 1964.

[88] Leissa, A.W. & Clausen, W.E., Deflection of a circular plate having mixed boundary conditions. *AIAA Journal*, **5**, pp. 2287–2288, 1967.

[89] Leissa, A.W. & Neidenfuhr, F.W., Bending of square plate with two adjacent edges free and others clamped or simply supported. *AIAA Journal*, **1**, pp. 116–120, 1963.

[90] Leissa, A.W. & Niedenfuhr, F.W., A study of the cantilevered square plate subjected to a uniform loading. *Journal of Aerospace Science*, **29**, pp. 162–169, 1962.

[91] Leissa, A.W., Lo, C.C. & Niedenfuhr, F.W., Uniformly loaded plates of regular polygonal shape. *AIAA Journal*, **3**, pp. 566–567, 1965.

[92] Li, Z.C., Lu, T.T., Hu, H.Y. & Cheng, A.H.D., *Trefftz and Collocation Methods*, WITpress: Southampton and Boston, 2008.

[93] Liao, W.H. & Krueger, D.A., Multiple expansion calculation of slow viscous flow about spheroids of different sizes. *Journal of Fluid Mechanics*, **96**, pp. 223–241, 1980.

[94] Lifits, S., Reutskiy, S. & Tirozzi, B., Trefftz Spectral Method for initial-boundary problems. *Computer Assisted Mechanics and Engineering Sciences*, **4**, pp. 549–566, 1997.

[95] Lo, C.C. & Leissa, A.W., Bending of plates with circular holes. *Acta Mechanica*, **4**, pp. 64–78, 1967.

[96] Lukasiewicz, S., *Local Loads in Plates and Shells*, PWN-Noordhoff: Warszawa and Leyden, 1979.

[97] Magnus, W. & Oberhettinger, F., *Formulas and Theorems for the Functions of Mathematical Physics*, Chelsea Publishing Company: New York, 1954.

[98] Markiewicz, M. & Mahrenholtz, O., Application of biorthogonal functional series to hydrodynamic analysis of offshore structures. *Proceedings of Asia-Pacific Vibration Conference '93*, Vol. I, pp. 51–56, 1993

[99] Marlewski, A. & Kołodziej, J.A., DERIVE assistance to the collocation method for solving potential problems. *International Journal on Mathematical Education in Science and Technology*, **27**, pp. 875–882, 1997.

[100] Mendes, A.C. & Kołodziej, J.A., An adaptive boundary collocation method for creeping flow between eccentric cylinders. *Advances in Fluid Mechanics Series*, **9**, pp. 85–91, *First International Conference on Advances in Fluid Mechanics*, AFM 96, 1996.

[101] Michell, J.H., On the direct determination of stress in an elastic solid, with application to the theory of plates. *Proceedings London Mathematical Society*, **18**, pp. 1–18, 1899.

[102] Nash, W.A., Several approximate analysis of the bending of a rectangular cantilever plate by uniform pressure. *Journal of Applied Mechanics*, **19**, pp. 33–36, 1952.

[103] Noda, I. & Gryte, C.C., Mass transfer in regular arrays of hollow fibers in countercurrent dialysis. *AIChE Journal*, **25**, pp. 113–122, 1979.

[104] O'Brien, V., Form factors for deformed spheroids in Stokes flow. *AIChE Journal*, **14**, pp. 870–875, 1968.

[105] O'Brien, V., Stokes flow about deformed spheroids. *International Journal of Engineering Science*, **4**, pp. 925–937, 1966.

[106] O'Brien, V., Deformed spheroids in Stokes flow. *Appl. Phys. Lab. Rep.*, TG-716, Johns Hopkins University, Silver Spring, 1965.

[107] Ondris, L., Remarks concerning the solution of stationary plane thermal problems by the boundary point least squares method. *Computer Methods in Applied Mechanics and Engineering*, **31**, pp. 331–336, 1982.

[108] Ondris, L., Solution of some symmetrical plane thermal problems by the boundary point least squares method. *Computer Methods in Applied Mechanics and Engineering*, **28**, pp. 309–325, 1981.

[109] Piltner, R., Special finite elements with holes and internal cracks. *International Journal for Numerical Methods in Engineering*, **21**, pp. 1471–1485, 1985.

[110] Piltner, R., The use of complex valued functions for the solution of three-dimensional elasticity problems. *Journal of Elasticity*, **18**, pp. 191–225, 1987.

[111] Poritsky, H. & Danforth, C.E., On the torsion problem. *Proceedings 3rd U.S. National Congress on Aplied Mechanics,* pp. 95–98, 1958.

[112] Poullikkas, A., Karageorghis, A. & Georgiou, G., The method of fundamental solutions for three-dimensional elastostatics problems. *Computer & Structures,* **80**, pp. 365–370, 2002.

[113] Rao, A.K. & Rajaiah, K., Polygon-circle paradox of simply supported thin plates under uniform pressure. *AIAA Journal*, **6**, pp. 155–156, 1968.

[114] Rao, K.S., Bapu Rao, M.N. & Ariman, T., Thermal stresses in plates with circular holes. *Nuclear Engineering and Design*, **15**, pp. 97–112, 1971.

[115] Ratkowsky, D.A. & Epstein, N., Laminar flow in regular polygonal shaped ducts with circular centred cores. *Canadian Journal of Chemical Engineering*, **46**, pp. 387–388, 1968.

[116] Reutskiy, S. & Tirozzi, B., Trefftz spectral method for elliptic equations of general type. *Computer Assisted Mechanics and Engineering Sciences*, **8**, pp. 629–644, 2001.

[117] Reutskiy, S., A boundary method of Trefftz type with approximate trial functions. *Engineering Analysis with Boundary Elements*, **37**, pp. 341–353, 2002.

[118] Ruoff, G., Die praktische Berechnung der Kopplungsmatrizen bei der Kombination der Trefftzschen Methode. In: *Finite Elemente in der Static*, eds. K.E. Buck, D.W. Scharpf, E. Stein & W. Wunderlich, Ernst: Berlin, 1973.

[119] Sampson, R.A., On Stokes current function. *Philosophical Transactions of Royal Society*, **182A**, pp. 449, 1891.

[120] Schmidt, J., Longitudinal laminar flow in an array of circular cylinders. *International Journal of Heat and Mass Transfer*, **9**, pp. 925–937, 1966.

[121] Sekiya, T. & Takeuti, Y., Approximate solution of thermal problems in plane elasticity. *Proceedings 8th Japan National Congress on Applied Mechanics*, pp. 119–122, 1958.

[122] Shih, F.S., Laminar flow in axisymmetric conduits by a rational approach. *Canadian Journal of Chemical Engineering*, **45**, pp. 287–284, 1967.

[123] Shih, F.S., On the temperature field of a square column embedding a heating cylinder. *AIChE Journal*, **16**, pp. 134–138, 1970.

[124] Shuleshko, P., Comparative analysis of different collocation methods on the basis of the torsion problem. *Australian Journal of Applied Science*, **12**, pp. 194–210, 1961.

[125] Slot, T. & Yalch, J.P., Stress analysis of plate perforated structures by point-wise matching of boundary conditions. *Nuclear Engineering and Design*, **4**, pp. 163–176, 1966.

[126] Smith, R.C.T., The bending of a semi-infinite strip. *Australian Journal Scientific Research*, No. 5, 1952.

[127] Sparrow, E.M. & Loeffeler, A.L., Longitudinal laminar flow between cylinders arranged in regular array. *AIChE Journal*, **5**, pp. 323–330, 1959.

[128] Stein, E., Die Kombination des modifizierten Trefftzschen Verfahrens mit der Metode der finiten Element. In: *Finite Elemente in der Static*, eds. K.E. Buck, D.W. Scharpf, E. Stein & W. Wunderlich, Ernst: Berlin, 1973.

[129] Subrahmanyan, M.B. & Wah, T., Vibration of quadrilateral plates. *Computer Methods in Applied Mechanics and Engineering*, **43**, pp. 315–323, 1984.

[130] Sultanian, B.K. & Sastri, V.M.K., Effect of geometry on heat conduction in coolant channels of a liquid rocket engine. *Warme – und StoffÜbertragung*, **14**, pp. 254–251, 1980.

[131] Sultanian, B.K. & Sastri, V.M.K., Steady heat conduction in a circular duct with circumferentially varying heat transfer coefficients. *Regional Journal Energy, Heat and Mass Transfer*, pp. 101–110, 1979.

[132] Symm, G.T., The longitudinal shear modulus of an unidirectional fibrous composite. *Journal of Composite Materials*, **4**, pp. 426–428, 1970.

[133] Szybiński, B. & Zieliński, A.P., Alternative T-complete systems of shape functions applied in analytical Trefftz finite elements. *Numerical Methods for Partial Differential Equations*, **11**, pp. 375–388, 1995.

[134] Szybiński, B., Wróblewski, A. & Zieliński, A.P., General strategy of h-p adaptive solutions in structural Trefftz-type element analysis. *Computer Assisted Mechanics and Engineering Sciences*, **8**, pp. 469–478, 2001.

[135] Szybiński, B., Zieliński, A.P. & Karaś, M., Folded-plate structures with openings – analysis and optimization. *Computer Assisted Mechanics and Engineering Sciences*, **10**, pp. 629–640, 2003.

[136] Takeuti, Y. & Sekiya, T., Thermal stresses in a polygonal cylinder with a circular hole under internal heat generation. *Zeitschrift für Angewandte Mathematik und Mechanik (ZAMM)*, **48**, pp. 237–246, 1968.

[137] Takeuti, Y. & Sumi, N., Steady-state thermoelastic problem for a triangular prism with a circular hole under heat generation. *International Journal of Engineering Science*, **5**, pp. 145, 1968.

[138] Takeuti, Y., Approximate formulae for temperature distribution in heat generating polygonal cylinder with central circular hole. *Nuclear Engineering and Design*, **8**, pp. 241–246, 1968.

[139] Takeuti, Y., Steady temperature distribution in a heat-generating multibore cylinder. *Nuclear Engineering and Design*, **11**, pp. 41–56, 1969.

[140] Takeuti, Y., Thermal stresses in heat-generating multibore square or circular region. *Nuclear Engineering and Design*, **14**, pp. 201–210, 1970.

[141] Tolke, F., Über Spannungszustande in dünnen Rechteckplatten. *Ingenieur-Archiv*, **5**, pp. 187–237, 1934.

[142] Tong, P., Pian P.H. & Lasry, S.L., A hybrid element approach to crack problems in plane elasticity. *International Journal for Numerical Methods in Engineering*, **7**, pp. 297–308, 1973.

[143] Voros, G.M. & Jirousek, J., Application of the hybrid-Trefftz finite element model to thin shell analysis. *Proceedings of European Conference on New Advances in Computational Structural Mechanics,* Giens, France, pp. 547–554, 1991.

[144] Wah, T., Elastic quadrilateral plates. *Computers & Structures*, **10**, pp. 457–466, 1979.

[145] Wah, T., In-plane stresses in polygonal plates due to thermal inclusions. *Mechanics Research Communications*, **9**, pp. 301–310, 1982.

[146] Wang, C.Y., Torsion of polygonal bar with core of different material. *Journal of Engineering Mechanics. Transactions of ASCE*, **125**, pp. 1218–1221, 1999.

[147] Wearing, J.L. & Sheikh, M.A., A regular indirect BEM for thermal analysis. *International Journal for Numerical Methods in Engineering*, **25**, pp. 495–515, 1988.

[148] Williams, M.L., Stress singularities resulting from various boundary conditions in angular corners of plates in extension. *Journal of Applied Mechanics*, **19**, pp. 526–527, 1952.

[149] Williams, M.L., The stress around a fault or crack in dissimilar media. *Bulletin of Seismological Society of America*, **49**, 1959.

[150] Wróblewski, A., Zieliński, A.P. & Jirousek, J., Hybrid-Trefftz p-element for 3-D axisymmetric problems of elasticity. *Proc. Conf. Numerical Methods in Engineering, Brussels*, Elsevier, 1992.

[151] Zieliński, A.P. & Herrera, I., Trefftz method: fitting boundary conditions. *International Journal for Numerical Methods in Engineering*, **24**, pp. 871–891, 1987.

[152] Zieliński, A.P. & Życzkowski, M., The trigonometric contour series method in application to clamped plates of arbitrary contour. *Bulletin of Polish Academy of Siences /Appl. Mech/*, **29**, pp. 159–167, 1981.

[153] Zieliński, A.P., A contour series method applied to shells. *Thin Walled Structures*, **3**, pp. 217–229, 1985.

[154] Zieliński, A.P., On curvilinear distributions expressed by double Fourier series. *Journal of Applied Mathematics and Physics /ZAMP/*, **31**, pp. 717–729, 1980.

[155] Zieliński, A.P., On trial functions applied in generalized Trefftz method. *Advances in Engineering Software*, **24**, pp. 147–155, 1995.

[156] Zieliński, A.P., Special Trefftz elements and improvement of their conditioning. *Communications in Numerical Methods in Engineering*, **13**, pp. 765–775, 1997.

[157] Zieliński, A.P., *T-complete Functions Method and its Application to Elastic and Elasto-Plastic Problems* (in Polish), Monograph 86, Cracow University of Technology Press: Cracow, 1989.

[158] Zieliński, A.P., *The Boundary Series Method Applied to Plates and Shells with Curvilinear Contours* (in Polish), Monograph 98, Cracow University of Technology Press: Cracow, 1990.

CHAPTER 4

Substructuring and T-elements

As it was analysed in Section 3 of Chapter 2, the boundary collocation method can sometimes lead to weak or even ill-conditioned matrices. As a rule it happens in case of more complicated areas, in which it is necessary to apply a large number of trial T-functions. The basic remedy for the problem of conditioning is substructuring – division of given area Ω into subareas Ω_e. In this case, the systems of T-functions are applied to each subregion separately and the full matrix $\mathbf{A}$ changes to be block type.

The typical boundary collocation approach with substructuring was applied in works of Schmidt [31], Yu & Sparrow [47], Hulbert & Rybicki [6], Mathews & Hooke [23], Bowie, Freese & Neal [2], Bert & Chang [1], Checumov & Simonov [3], Tolley [39–41], Janssens & Tolley [7], Wah [42–46], Subrahmanyan & Wah [32], Zieliński & Zienkiewicz [48], Kołodziej [19, 20], Reichel [30], Kołodziej & Musielak [18], Portela & Charafi [27, 28], Leitao [21, 22], Ho & Chau [5], Mokady, Mestel & Winlove [24], Partridge & Sensale [25], De Medeiros, Partridge & Brandao [4]. In these works, the region Ω was divided into several subregions and in each subregion a different form of solution:

$$\phi_e(\mathbf{x}) = \phi_e^p(\mathbf{x}) + \sum_{i=1}^{M_f} \Phi_{ei} a_{ei} \quad \mathbf{x} \in \Omega_e \tag{1}$$

was considered. The constants a_{ei} were calculated from the collocational fulfilment of the boundary conditions and connectivity along the subregion interfaces. It should be noted that usually the number of subregions was in the above solutions considerably limited. The possibility of division of Ω into a large number of subregions (elements) is presented below.

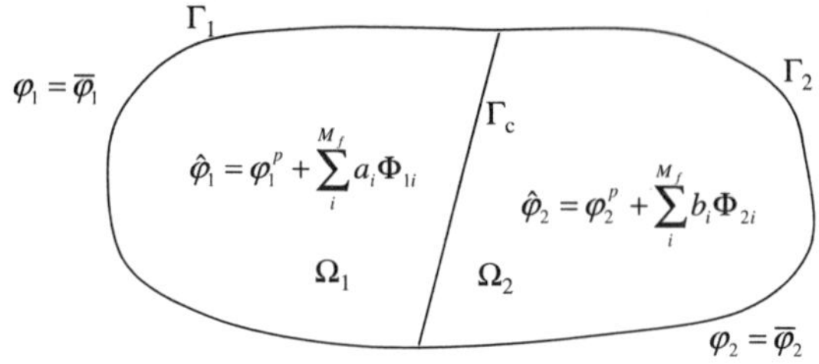

Figure 1: Coupling of two areas Ω_1 and Ω_2.

Let us first consider the division of Ω into two subregions (Fig. 1). It can be formulated in the integral weighted residual way as follows (Laplace equation with Dirichlet boundary conditions) [48]:

$$\int_{\Gamma_1} W_{1j} \left(\hat{\varphi}_1 - \overline{\varphi}_1\right) d\Gamma + \int_{\Gamma_c} \left\{ W_{1j} \left(\hat{\varphi}_1 - \hat{\varphi}_2\right) + V_{1j} \left[\left(\hat{\varphi}_1\right)'_n + \left(\hat{\varphi}_2\right)'_n \right] \right\} d\Gamma = 0$$

$$\int_{\Gamma_c} \left\{ W_{2j} \left(\hat{\varphi}_2 - \hat{\varphi}_1\right) + V_{2j} \left[\left(\hat{\varphi}_2\right)'_n + \left(\hat{\varphi}_1\right)'_n \right] \right\} d\Gamma + \int_{\Gamma_2} W_{2j} \left(\hat{\varphi}_2 - \overline{\varphi}_2\right) d\Gamma = 0 \tag{2}$$

$$j = 1,\ 2,\ 3, \ldots \left(N_f = M_f\right),$$

where $(\ )'_n$ means differentiation in the outward normal direction, and the weighting functions can be taken as:

a. direct weights:

$$W_{1j} = W\Phi_{1j} \qquad V_{1j} = \left(\Phi_{1j}\right)'_n$$
$$W_{2j} = W\Phi_{2j} \qquad V_{2j} = \left(\Phi_{2j}\right)'_n \tag{3a}$$

b. opposite weights:

$$W_{1j} = \left(\Phi_{1j}\right)'_n \qquad V_{1j} = -\Phi_{1j}$$
$$W_{2j} = \left(\Phi_{2j}\right)'_n \qquad V_{2j} = -\Phi_{2j}. \tag{3b}$$

W is here a positive weighting constant equalizing influence of the terms in eqn (2).

In case of many subareas (T-elements), the solution takes more complicated form. The following least square functional has to be minimized [10, 11, 13]:

$$J(\mathbf{a}) = \int_{\Gamma_\varphi} \left(\hat{\varphi} - \overline{\varphi}\right)^2 d\Gamma + W \int_{\Gamma_{\varphi n}} \left[\left(\hat{\varphi}\right)'_n + \left(\overline{\varphi}\right)'_n \right]^2 d\Gamma$$
$$+ \int_{\Gamma_I} \left(\hat{\varphi}^- - \hat{\varphi}^-\right)^2 d\Gamma + W \int_{\Gamma_I} \left[\left(\hat{\varphi}^+\right)'_n + \left(\overline{\varphi}^-\right)'_n \right]^2 d\Gamma = \min, \tag{4}$$

where Γ_ψ, $\Gamma_{\psi n}$ are the parts of the boundary on which the Dirichlet and Neumann boundary conditions are given, respectively, Γ_I are interelement segments of the boundary and the superscripts +, – denote the approximate solutions in both sides of Γ_I.

The minimization of the functional $J(\mathbf{a})$ leads to the variational equation:

$$\delta J(\mathbf{a}) = \delta \mathbf{a}^{\mathrm{T}} \frac{\partial J}{\partial \mathbf{a}} = \delta \mathbf{a}^{\mathrm{T}} (\mathbf{Ka} - \mathbf{R}) = 0, \tag{5}$$

where $\mathbf{a}^{\mathrm{T}} = \left\{ \mathbf{a}_1^{\mathrm{T}}, \mathbf{a}_2^{\mathrm{T}}, \ldots \mathbf{a}_{\mathrm{N_e}}^{\mathrm{T}} \right\}$ includes all the unknown coefficients a_{ie} from the N_e subareas, $\mathbf{K}$ can be called a global stiffness matrix and $\mathbf{R}$ is a known right-hand side vector. Indeed, taking into consideration that

$$\delta J(\mathbf{a}) = \sum_{e=1}^{N_e} \delta J_e(\mathbf{a}_e), \tag{6}$$

we observe the analogy to the finite element formulation and can easily build the local stiffness matrices $\mathbf{k}_e$ and the known vectors $\mathbf{r}_e$ on the level of one element, which appear in:

$$\delta J_e(\mathbf{a}_e) = \delta \mathbf{a}_e^{\mathrm{T}} \frac{\partial J_e(\mathbf{a}_e)}{\partial \mathbf{a}_e} = \delta \mathbf{a}_e^{\mathrm{T}} (\mathbf{r}_e^p + \mathbf{k}_e \mathbf{a}_e) = \delta \mathbf{a}_e^{\mathrm{T}} \mathbf{r}_e \tag{7}$$

and form the standard force–displacement relationship

$$\mathbf{r}_e = \mathbf{r}_e^p + \mathbf{k}_e \mathbf{a}_e. \tag{8}$$

Note, however, that in this formulation the parts of the boundary Γ and interelement borders Γ_I play a role of elements, and the coefficients of the T-functions are directly the degrees of freedom. The assembly means here building the integral fit along all the segments of subregion boundaries. Hence, for example, the problem defined in Fig. 1 consists of three border elements connected with boundary segments Γ_1, Γ_2, Γ_c, and the respective stiffness matrices have the form:

$$\mathbf{k}_1 = \int_{\Gamma_1} \Phi_1^{\mathrm{T}} \Phi_1 d\Gamma \tag{9a}$$

$$\mathbf{k}_2 = \int_{\Gamma_2} \Phi_2^{\mathrm{T}} \Phi_2 d\Gamma \tag{9b}$$

$$\mathbf{k}_c = \begin{bmatrix} \int_{\Gamma_c} \Phi_1^{\mathrm{T}} \Phi_1 d\Gamma + W \int_{\Gamma_c} (\Phi_1)_n'^{\mathrm{T}} (\Phi_1)_n' d\Gamma & -\int_{\Gamma_c} \Phi_1^{\mathrm{T}} \Phi_2 d\Gamma + W \int_{\Gamma_c} (\Phi_1)_n'^{\mathrm{T}} (\Phi_2)_n' d\Gamma \\ -\int_{\Gamma_c} \Phi_2^{\mathrm{T}} \Phi_1 d\Gamma + W \int_{\Gamma_c} (\Phi_2)_n'^{\mathrm{T}} (\Phi_1)_n' d\Gamma & \int_{\Gamma_c} \Phi_2^{\mathrm{T}} \Phi_2 d\Gamma + W \int_{\Gamma_c} (\Phi_2)_n'^{\mathrm{T}} (\Phi_2)_n' d\Gamma \end{bmatrix}. \tag{9c}$$

Here, Φ_1, Φ_2, $(\Phi_1)'_n$ and $(\Phi_2)'_n$ are vectors containing T-functions Φ_{1i}, Φ_{2i} and their normal derivatives, respectively. Observe that the element Γ_c has $2M_f$ degrees of freedom a_i, b_i, $i=1,...,M_f$ and the elements connected with Γ_1 and Γ_2 have M_f degrees of freedom each. Numerical examples showing effectiveness of the substructuring are given in Chapter 6.

Another possibility of defining the T-elements (coupling the subareas) is the introduction of the 'frame' function

$$\psi_e(\mathbf{x}) = \sum_{i=1}^{N} d_{ie} \Psi_{ie}(\mathbf{x}) \tag{10}$$

along element boundary Γ_e. This leads to the so-called hybrid Trefftz elements, which are now in the state of intensive development. The best known hybrid Trefftz displacement (HT-D) element was first proposed by Jirousek [8, 14] and then developed into different variants and investigated in numerous publications [9, 14, 26, 33–38]. A review of applications of this element was presented in the large monographic work [11] and in monograph [29].

The coupling between the internal field $\varphi_e(\mathbf{x})$ and the boundary function $\psi_e(\mathbf{x})$ is here stated in the weighted residual form (e.g. for the Laplace equation):

$$\int_{\Gamma_e} \delta(\varphi_e)'_n (\varphi_e - \psi_e) d\Gamma = 0, \tag{11}$$

which results in the relation between the Trefftz coefficients a_{ie} and the frame unknowns d_{ie} (element degrees of freedom). The shape functions Ψ_{ie} of the frame can have different forms [8, 12, 15, 35]; however, the most often used are hierarchic polynominals defined along each element side (Fig. 2)

$$\begin{aligned}
\Psi_1 &= \frac{1}{2}(1-\xi) \\
\Psi_2 &= \frac{1}{2}(1+\xi) \\
\Psi_3 &= 1-\xi^2 \\
\Psi_4 &= \xi(1-\xi^2) \\
\Psi_5 &= \xi^2(1-\xi^2) \\
&\ldots\ldots
\end{aligned} \tag{12}$$

Now, the customary finite element relationship of equivalence of the generalized virtual work can be formed as

$$\int_{\Gamma_e} \delta\psi_e (\varphi_e)'_n \, d\Gamma = \int_{\Gamma_{en}} \delta\psi_e (\overline{\varphi}_e)'_n \, d\Gamma + \sum_{i=1}^{N} \delta d_{ie} r_{ie}, \tag{13}$$

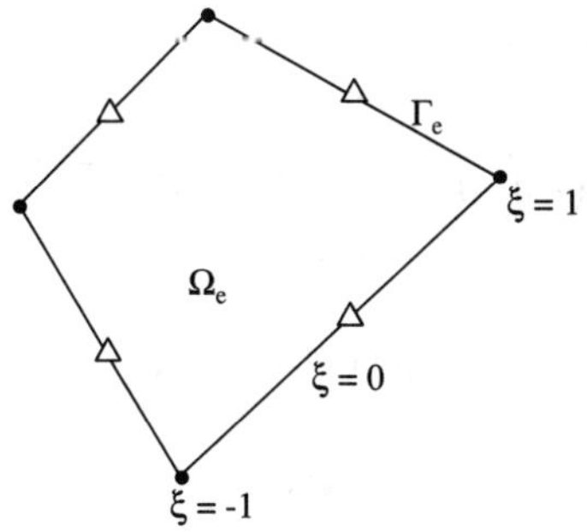

Figure 2: Scheme of the hybrid Trefftz finite element.

where Γ_{en} is nth part of Γ_e along which the normal derivative $(\varphi_e)'_n$ is defined as $(\overline{\varphi}_e)'_n$ and r_{ie} are generalized element nodal forces. From (11–13) we can easily define the standard finite element force–displacement relation:

$$\mathbf{r}_e = \mathbf{k}_e \mathbf{d}_e + \overline{\mathbf{r}}_e, \tag{14}$$

where $\mathbf{k}_e$ is a symmetric element stiffness matrix which can be introduced into any standard finite element code [11, 13, 16] and $\overline{\mathbf{r}}_e$ is a known vector representing the particular integral and the traction boundary conditions. The stiffness matrix has a characteristic symmetric, positive definite form

$$\mathbf{k}_e = \mathbf{G}_e^{\mathrm{T}} \mathbf{H}_e^{-1} \mathbf{G}_e, \tag{15a}$$

where (for the Laplace equation):

$$\mathbf{G}_e = \int_{\Gamma_e} (\Phi_e)_n'^{\mathrm{T}} \Phi_e d\Gamma \text{ and } \mathbf{H}_e = \int_{\Gamma_e} (\Phi_e)_n'^{\mathrm{T}} \Phi_e d\Gamma. \tag{15b}$$

Another type of the T-element is the least square hybrid structure HTLS [17, 26], in which instead of the residual equation (11), the fit of the Trefftz solution ϕ_e and the frame function ψ_e is stated as

$$\int_{\Gamma_e} (\varphi_e - \psi_e)^2 \, d\Gamma = \min, \tag{16}$$

which results in

$$\int_{\Gamma_e} \delta\varphi_e (\varphi_e - \psi_e)^2 \, d\Gamma = 0. \tag{17}$$

Supplementing this equation by

$$\int_{\Gamma_e} \delta\varphi_e (\varphi_e)'_n \, d\Gamma = \int_{\Gamma_{en}} \delta\varphi_e (\overline{\varphi}_e)'_n \, d\Gamma + \sum_{i=1}^{N} \delta d_{ie} r_{ie} \tag{18}$$

leads to the relation analogical to eqn (14) but with

$$\mathbf{k}_e = \mathbf{A}_e^{\mathrm{T}} \mathbf{H}_e \mathbf{A}_e. \tag{19a}$$

where

$$\mathbf{A}_e = \mathbf{M}_e^{-1} \mathbf{L}_e, \ \mathbf{M}_e = \int_{\Gamma_e} \Phi_e^{\mathrm{T}} \Phi_e d\Gamma, \ \mathbf{L}_e = \int_{\Gamma_e} \Phi_e^{\mathrm{T}} \Psi_e d\Gamma. \tag{19b}$$

The detailed comparison between HT-D and HTLS elements as well as numerical examples of both formulations can be found in [17, 48].

Obviously, in all the above formulations, the boundary integrals are calculated in a numerical way, namely using Gauss-Legendre quadrature. This can be interpreted as a specific boundary collocation in the Gaussian points and with appropriate weights (see Section 2 in Chapter 2). However, in one of the Trefftz element formulations, the boundary collocation is explicitly used [13]. In this formulation (called CT), the frame function ψ_e changes into a set of K_p discrete values d_k situated in points x_k along Γ_e (usually in the Lobatto points). If in this element the Trefftz solution ϕ_e is calculated in the same points

$$\phi_k = \phi_k^p + \sum_{i=1}^{M_f} \Phi_{ik} a_i, \quad k = 1, 2, \ldots K_p \tag{20a}$$

or, in vector/matrix notation:

$$\mathbf{f}_e = \mathbf{f}_e^p + \mathbf{F}_e \mathbf{a}_e \tag{20b}$$

where

$$\phi_k = \phi_e(\mathbf{x}_k), \ \phi_k^p = \phi_k^p(\mathbf{x}_k), \ \Phi_{ik} = \Phi_{ie}(\mathbf{x}_k), \tag{20c}$$

the minimization of the differences between ϕ_k and d_k

$$\sum_{k=1}^{K_p} (\phi_k - d_k)^2 = \min \tag{21}$$

leads to

$$\sum_{i=1}^{M_f}\sum_{k=1}^{K_p}\Phi_{jk}\left(\phi_k^p+\Phi_{ik}a_i-d_k\right)=0 \tag{22a}$$

or

$$\mathbf{F}_e^{\mathrm{T}}\left(\mathbf{f}_e^p+\mathbf{F}_e\mathbf{a}_e-\mathbf{d}_e\right)=\mathbf{0}. \tag{22b}$$

From this system of equations, the relation between the degrees of freedom d_{ie} and the Trefftz coefficients a_{ie} can be derived:

$$\mathbf{a}_e=\mathbf{B}_e\left(\mathbf{d}_e-\mathbf{f}_e^p\right),\ \mathbf{B}_e=\left(\mathbf{F}_e^{\mathrm{T}}\mathbf{F}_e\right)^{-1}\mathbf{F}_e. \tag{23}$$

Equivalency of virtual work (18) in this case results in the relation (14) with the stiffness matrix

$$\mathbf{k}_e=\mathbf{B}_e^{\mathrm{T}}\mathbf{H}_e\mathbf{B}_e, \tag{24}$$

where $\mathbf{H}_e$ is defined in (15b).

An alternative formulation designated as WCT (weighted collocation Trefftz approach) applies the least square boundary collocation (21) in the weighted form:

$$\sum_{k=1}^{K_p}W_k\left(\phi_k-d_k\right)^2=\min \tag{25}$$

and eqn (22b) changes to

$$\mathbf{F}_e^{\mathrm{T}}\mathbf{W}\left(\mathbf{f}_e^p+\mathbf{F}_e\mathbf{a}_e-\mathbf{d}_e\right)=\mathbf{0}, \tag{26}$$

where $\mathbf{W}$ is a diagonal matrix containing the weights W_k. This finally leads to the stiffness matrix in the weighted form

$$\mathbf{k}_e=\overline{\mathbf{B}}_e^{\mathrm{T}}\mathbf{H}_e\overline{\mathbf{B}}_e, \tag{27}$$

where

$$\overline{\mathbf{B}}_e=\left(\mathbf{F}_e^{\mathrm{T}}\mathbf{W}\mathbf{F}_e\right)^{-1}\mathbf{F}_e^{\mathrm{T}}\mathbf{W}.$$

Note that this approach makes sense only if $K_p > M_f$. If $K_p = M_f$ (the same number of the collocation points and the unknown Trefftz coefficients a_i), this variant yields the same results as the basic CT formulation. The numerical investigation of this approach can be found in [13].

As it was already mentioned, the T-elements presented above were discussed in detail in the monographic publications [11] and [29]. Therefore, the authors decided to put stronger accent on other different variants of the boundary collocation approach not discussed in the above studies.

References

[1] Bert, C.W. & Chang, S., Analysis of twisting stiffness for a multifiber composite layer. *Journal of Applied Mechanics*, **41**, pp. 658–662, 1974.

[2] Bowie, O.L., Freese, C.E. & Neal, D.M., Solution of plane problems of elasticity utilising partitioning concept. *Journal of Applied Mechanics*, **40**, pp. 767–772, 1973.

[3] Checumov, R.A. & Simonov, V.S., Bending of plates with circular holes of big diameter (in Russian). *Sbornik Trudov Belgorodskogo Tekhnicheskogo Instituta, Stroitelstvo Materialov*, **10**, pp. 10–16, 1975.

[4] De Medeiros, G.C., Partridge, P.W. & Brandao, J.O., The method of fundamental solutions with dual reciprocity for some problems in elasticity. *Engineering Analysis with Boundary Elements*, **28**, pp. 453–461, 2004.

[5] Ho, K.C. & Chau, K.T., A finite strip loaded by a bonded-rivet of a different material. *Computers & Structures*, **70**, pp. 203–218, 1999.

[6] Hulbert, L.E. & Rybicki, E.F., Boundary point least squares analysis of the free edge effects in some unidirectional fiber composites. *Journal of Composite Materials*, **5**, pp. 164–175, 1971.

[7] Janssens, P. & Tolley, M., On the deformation of elastic plates. *Zeitschrift für Angewandte Mathematic und Physik (ZAMP)*, **30**, pp. 234–242, 1979.

[8] Jirousek, J. & Leon, N., A powerful finite element for plate bending. *Computer Methods in Applied Mechanics and Engineering*, **12**, pp.77–96, 1977.

[9] Jirousek, J. & Teodorescu, P., Large finite elements method for the solution of problems in the theory of elasticity. *Computers & Structures*, **15**, pp. 575–587, 1982.

[10] Jirousek, J. & Wróblewski, A., Least-square T-elements: equivalent FE and BE forms of a substructure-oriented boundary solution approach. *Communications in Numerical Methods in Engineering*, **10**, pp. 21–32, 1994.

[11] Jirousek, J. & Wróblewski, A., T-elements: state of the art and future trends. *Archives of Computational Methods in Engineering*, **3**, pp. 323–434, 1996.

[12] Jirousek, J. & Zieliński, A.P., Dual hybrid-Trefftz element formulation based on independent boundary traction frame. *International Journal for Numerical Methods in Engineering*, **36**, pp. 2955–2980, 1993.

[13] Jirousek, J. & Zieliński, A.P., Survey of Trefftz-type element formulations. *Computers and Structures*, **63**, pp. 225–242, 1997.

[14] Jirousek, J., Basis for development of large finite elements locally satysfying all field equations. *Computer Methods in Applied Mechanics and Engineering*, **14**, pp. 65–92, 1978.

[15] Jirousek, J., Hybrid-Trefftz plate bending elements with p-method capabilities. *International Journal for Numerical Methods in Engineering*, **24**, pp. 1367–1393, 1993.

[16] Jirousek, J., Structural analysis program SAFE – special features and advanced finite element models, *Advanced Engineering Software*, **7**, pp. 68–76, 1985.

[17] Karaś, M. & Zieliński, A.P., Hybrid Trefftz least square element for plane elasticity. *Journal of Mechanical Engineering (Strojnicki Casopis)*, **53**, pp. 24–35, 2002.

[18] Kołodziej, J.A. & Musielak, G., Domain decomposition in boundary collocation method. *Parallel Algorithms for Partial Differential Equations, Proceedings of the Sixth GAMM-Seminar Kiel, January 19–21, 1990, Notes on Numerical Fluid Mechanics*, **31**, pp. 140–156, 1991.

[19] Kołodziej, J.A., Determination of the transverse effective thermal conductivity of unidirectionally fibre arranged composites by means of boundary collocation method (in Polish). *Theoretical and Applied Mechanics*, **23**, pp. 933–958, 1985.

[20] Kołodziej, J.A., Influence of the porosity of porous medium on the effective viscosity in Brinkman's filtration equation. *Acta Mechanica*, **75**, pp. 241–254, 1988.

[21] Leitao, V.M.A., Application of multi-region Trefftz to fracture mechanics. *Engineering with Boundary Elements*, **22**, pp. 251–256, 1998.

[22] Leitao, V.M.A., On the implementation of a multi-region Trefftz-collocation formulation for 2-D potential problems. *Engineering Analysis with Boundary Elements*, **20**, pp. 51–61, 1997.

[23] Matthews, G.J. & Hooke, C.J., Solution of axisymmetric torsion problem by point matching. *Journal of Strain Analysis*, **6**, pp. 217–222, 1971.

[24] Mokady, A.J., Mestel, A.J. & Winlove, C.P., Flow through a charged biopolymer layer. *Journal of Fluid Mechanics*, **383**, pp. 353–378, 1999.

[25] Partridge, P.W. & Sensale B., The method of fundamental solutions with dual reciprocity for diffusion and diffusion-convection using subdomains. *Engineering Analysis with Boundary Elements*, **24**, pp. 633–641, 2000.

[26] Piltner, R., Special finite elements with holes and internal cracks. *International Journal for Numerical Methods in Engineering*, **21**, pp. 1471–1485, 1985.

[27] Portela, A. & Charafi, A., Trefftz boundary element method for domains with slits. *Engineering Analysis with Boundary Elements*, **20**, pp. 299–304, 1997.

[28] Portela, A. & Charafi, A., Trefftz boundary elements-multi-region formulations. *International Journal for Numerical Methods in Engineering*, **45**, pp. 821–840, 1999.

[29] Qin, Q.H., *The Trefftz Finite and Boundary Element Method*, WITpress: Southampton, 2000.

[30] Reichel, L., On the numerical solution of some 2-D electromagnetic interface problems by the boundary collocation method. *Computer Methods in Applied Mechanics and Engineering*, **53**, pp. 1–11, 1985.

[31] Schmidt, J., Longitudinal laminar flow in an array of circular cylinders. *International Journal of Heat and Mass Transfer*, **9**, pp. 925–937, 1966.

[32] Subrahmanyan, M.B. & Wah, T., Vibration of quadrilateral plates. *Computer Methods in Applied Mechanics and Engineering*, **43**, pp. 315–323, 1984.

[33] Szybiński, B., *Trefftz Finite Elements for Analysis of Complex Thin-walled Structures* (in Polish), PhD Thesis (supervisor A.P. Zieliński), Cracow University of Technology, Cracow, 1997.

[34] Teixeira de Freitas, J.A. & Cismasiu, I., Adaptive p-refinement of hybrid-Trefftz finite element solutions. *Finite Elements in analysis and Design*, **39**, issue 11, pp. 1095–1121, 2003.

[35] Teixeira de Freitas, J.A. & Cismasiu, I., Developments with hybrid-Trefftz stress and displacement elements. *Computer Assisted Mechanics and Engineering Sciences*, **8**, pp. 289–311, 2001.

[36] Teixeira de Freitas, J.A., Formulation of elastostatic hybrid-Trefftz stress elements. *Computer Methods in Applied Mechanics and Engineering*, **153**, pp. 127–151, 1998.

[37] Teixeira de Freitas, J.A., Time integration and the Trefftz Method Part I-First-order and parabolic problems. *Computer Assisted Mechanics and Engineering Sciences*, **10**, pp. 453–463, 2003.

[38] Teixeira de Freitas, J.A., Time integration and the Trefftz Method Part II-Second-order and hyperbolic problems. *Computer Assisted Mechanics and Engineering Sciences*, **10**, pp. 465–477, 2003.

[39] Tolley, M.D., Application de la methode des grands elements finis singuliers a la resolution de equation biharmonique avec conditions aux limites discontinues. *C. R. Acad. Sc. Paris,* ser **A 287**, pp. 875–878, 1978.

[40] Tolley, M.D., La methode des grands elements finis singuliers: prinnciples et applications. *Bulletin de la Societe Mathematique de Belgique*, **32**, ser B, pp. 17–28, 1980.

[41] Tolley, M.D., Torsion des barres polygonales. *Bulletin de la classe des Sciences,* Academie Royal de Belgique, **LXIII**, ser. 5, pp. 1977–11, pp. 902–912, 1977.

[42] Wah, T., Elastic quadrilateral plates. *Computers & Structures*, **10**, pp. 457–466, 1979.

[43] Wah, T., In-plane thermal stresses in elastic plates. *Journal of Strain Analysis*, **19**, pp. 97–101, 1984.

[44] Wah, T., Partially loaded elastic plates. *International Journal of Mechanical Science*, **24**, pp. 89–93, 1982.

[45] Wah, T., Stress intensity factors determined by using Westergards's stress functions. *Engineering Fracture Mechanics*, **20**, pp. 65–73, 1984.

[46] Wah, T., In-plane thermal stresses in elastic plates. *Journal of Structural Mechanics*, **12**, pp. 79–85, 1984.

[47] Yu, H.S. & Sparrow, E.M., Stratified laminar flow in ducts of arbitrary shape. *AIChE Journal*, **13**, pp. 10–16, 1967.

[48] Zieliński, A.P. & Zienkiewicz, O.C., Generalized finite element analysis with T-complete boundary solution functions. *International Journal for Numerical Methods in Engineering*, **21**, pp. 509–528, 1985.

[49] Zieliński, A.P., Special Trefftz elements and improvement of their conditioning. *Communications in Numerical Methods in Engineering*, **13**, pp. 765–775, 1997.

CHAPTER 5

Basic numerical investigations of the boundary collocation method

1 Simply connected regions

The analytical systems of T-functions used in the boundary collocation approach result in certain specific features of the solutions. In this section, these properties are discussed on numerical examples of the Laplace-type equation; however, as it has been observed in numerous works [6, 7, 16], they are very similar in case of other governing equations. The non-homogeneous two-dimensional (2D) Laplace equation (called Poisson's equation) in the following form

$$\nabla^2 \phi(x,y) = -2 \quad (x,y) \in \Omega \tag{1a}$$

with the boundary condition

$$\phi(x_\Gamma, y_\Gamma) = 0 \quad (x_\Gamma, y_\Gamma) \in \Gamma \tag{1b}$$

will first be considered. It can be interpreted as a classical problem of torsion of a bar. In this case, the particular solution takes the form

$$\phi^P(x,y) = -\frac{1}{2}\left(x^2 + y^2\right) \quad (x,y) \in \Omega. \tag{2}$$

The simplest analytical solution of this problem can be derived for an elliptic area

$$\phi_{ell}^{EX} = \frac{a^2 b^2 - b^2 x^2 - a y^2}{a^2 + b^2}, \tag{3}$$

where a, b are the ellipse semi-axes. Hence, the system (6) (in Chapter 2) of T-functions together with the particular integral (2) solves the problem with its first terms:

$$\Phi_1 = 1, \quad \Phi_4 = x^2 - y^2. \tag{4}$$

The other terms can be considered as perturbations, which is visible in Table 1. The solution has been calculated using only a constant and even terms in the system ((6) in Chapter 2), which is equivalent to the assumption of symmetry with respect to x-axis. Formulation ((15) in Chapter 2) with the Gaussian quadrature along a half of elliptic boundary was applied. In Table 1, we can observe influence of the number of control points of the integral (collocation points) and the result of displacement x_0 of the T-function origin.

The number of control points ($N_G = 16$, or 32) considerably exceeded the number of T-functions (max $M_f = 9$), therefore their increase did not result in significant increase in the solution accuracy. It should be underlined that very similar results were obtained in this example for the opposite weighting ((22) in Chapter 2) in which constant a_0 was calculated by assuming the first weight equal to 1.0 or adding the equation

$$\phi\left[\left(a+x_0\right),0\right]=0. \tag{5}$$

A square area Fig. 1 defines a more informative numerical example. The series-type solution can here be written in the form [13]:

$$\phi(x,y)=\frac{32a^2}{\pi^3}\sum_{n=1,3,5}^{N}\frac{1}{n^3}(-1)^{(n-1)/2}\left[1-\frac{\cosh\left(n\pi y/2a\right)}{\cosh\left(n\pi b/2a\right)}\right]\cos\frac{n\pi x}{2a}, \tag{6}$$

which in this example was calculated for $N = 200$ to represent the exact solution. In the first variant, the Herrera functions ((6) in Chapter 2) were applied. Because of the specific symmetry of the square, only the coefficients of the fully symmetric T-functions took significant values (Table 2) – the other were merely disturbances. Table 3 presents local convergence of the approximate solution $\hat{\phi}$ and a certain measure of the derivative

$$\tau=\left[\left(\frac{\partial\phi}{\partial x}\right)^2+\left(\frac{\partial\phi}{\partial y}\right)^2\right]^{\frac{1}{2}}, \tag{7}$$

which can be interpreted as a resulting stress in a twisted bar (see also Figs 2 and 3).

The applied here Aitken procedure calculates a horizontal asymptote of an exponential function

$$y = A_A + B_A e^{-k_A x} \tag{8}$$

led through three equidistant terms of the sequence:

$$\phi_3=\phi\left(M_f=3\right),\qquad \phi_6=\phi\left(M_f=6\right),\qquad \phi_9=\phi\left(M_f=9\right), \tag{9}$$

which results in a simple formula [1, 11]:

$$A_A=\frac{\phi_3\phi_9-\phi_6^2}{\phi_3-2\phi_6+\phi_9} \tag{10}$$

Table 1: Coefficients of T-complete functions (ellipse a = 2.0, b = 1.0). M_f – number of T-functions applied, N_G – number of control points in Gauss–Legendre procedure, x_0 – coordinate of ellipse centre.

a_μ	$M_f = 3$ $N_G = 16$ $x_0 = 0.0$	$M_f = 6$ $N_G = 16$ $x_0 = 0.0$	$M_f = 9$ $N_G = 16$ $x_0 = 0.0$	$M_f = 9$ $N_G = 32$ $x_0 = 0.0$	$M_f = 6$ $N_G = 16$ $x_0 = 2.0$	$M_f = 9$ $N_G = 16$ $x_0 = 2.0$
a_0	0.8000	0.8000	0.7997	0.8000	$-0.268 \cdot 10^{-7}$	$0.1690 \cdot 10^{-2}$
a_1	$0.394 \cdot 10^{-11}$	$-0.101 \cdot 10^{-10}$	$0.936 \cdot 10^{-3}$	$0.382 \cdot 10^{-10}$	0.8000	0.7912
a_2	0.3000	0.3000	0.2997	0.3000	0.3000	0.3168
a_3		$0.212 \cdot 10^{-10}$	$-0.134 \cdot 10^{-2}$	$-0.584 \cdot 10^{-10}$	$0.958 \cdot 10^{-7}$	$-0.171 \cdot 10^{-1}$
a_4		$-0.484 \cdot 10^{-11}$	$0.154 \cdot 10^{-2}$	$0.386 \cdot 10^{-10}$	$-0.252 \cdot 10^{-7}$	$0.103 \cdot 10^{-1}$
a_5		$-0.617 \cdot 10^{-11}$	$-0.159 \cdot 10^{-3}$	$0.277 \cdot 10^{-10}$	$0.245 \cdot 10^{-8}$	$-0.387 \cdot 10^{-2}$
a_6			$-0.430 \cdot 10^{-3}$	$-0.185 \cdot 10\_^{10}$		$0.872 \cdot 10^{-3}$
a_7			$0.218 \cdot 10^{-3}$	$-0.443 \cdot 10^{-11}$		$-0.109 \cdot 10^{-3}$
a_8			$-0.322 \cdot 10^{-4}$	$0.268 \cdot 10^{-11}$		$0.573 \cdot 10^{-5}$

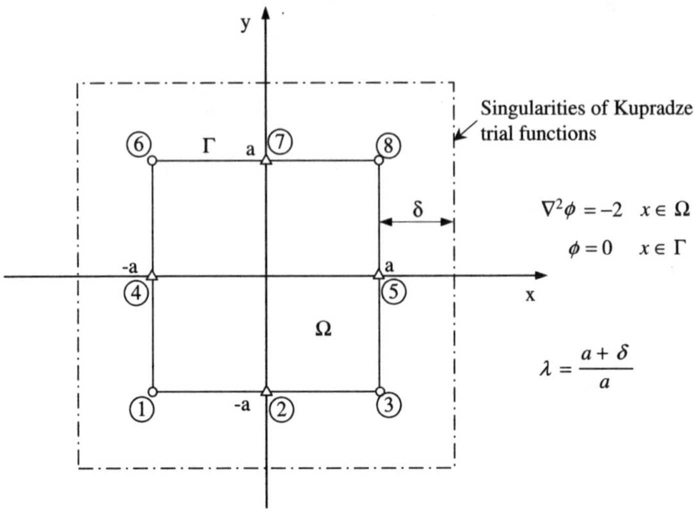

Figure 1: Square area – numerical example.

The coordinate A_A of the asymptote is considered to be much better approximation of the solution than the last term of the sequence (9). The detailed discussion on the Aitken and relative procedures can be found, for example, in [2, 15].

In all the examples presented above, the specific proportions of the investigated areas were chosen so that the boundary did not stray too much from the circle of the radius $r_0 = 1.0$. This was caused by the power form of series (6) (in Chapter 2). For example, in case where $a = 20.0$, $b = 10.0$ or $a = 0.2$, $b = 0.1$ for the ellipse, the integration along the boundaries produced very big or very small terms in the main matrices of equations. This can result in unacceptable value of the determinant and cause failures of procedures solving the equations. To avoid this problem, the following simple transformation can be introduced:

$$\xi = \frac{x}{a}, \qquad \eta = \frac{y}{a} \tag{11}$$

where

$$a = \left(l_{\max} l_{\min}\right)^{\frac{1}{2}}.$$

$l_{\max}$, $l_{\min}$ are the maximal and minimal dimension of Ω, and eqn (1a) becomes

$$\frac{\partial^2 \tilde{\phi}}{\partial \xi^2} + \frac{\partial^2 \tilde{\phi}}{\partial \eta^2} = -2 \tag{12}$$

where

$$\phi = \tilde{\phi} a^2$$

Table 2: Coefficients of T-complete functions – square area a = 1, Fig. 1, least square integral fit (15) (in Chapter 2), $N_G = 32$.

a_0	a_1	a_2	a_3	a_4	a_5	a_6	a_7	a_8
0.6667	$0.6366 \cdot 10^{-11}$	$0.6423 \cdot 10^{-10}$						
0.5891	$-0.9697 \cdot 10^{-11}$	$0.1005 \cdot 10^{-9}$	$-0.5565 \cdot 10^{-10}$	$-0.9698 \cdot 10^{-1}$	$0.3350 \cdot 10^{-10}$			
0.5894	$-0.7760 \cdot 10^{-11}$	$0.7000 \cdot 10^{-10}$	$-0.3249 \cdot 10^{-10}$	$-0.9077 \cdot 10^{-1}$	$0.3912 \cdot 10^{-10}$	$-0.3030 \cdot 10^{-10}$	$0.1436 \cdot 10^{-10}$	$0.2631 \cdot 10^{-2}$

Table 3: Values of function $\hat{\phi}$ and resulting stresses τ inside the square area, least square integral fit (15) (in Chapter 2), ($a = 1.0, N_G = 32$), A – Aitken's extrapolation.

Coordinates		Stress function ϕ					Resulting stresses τ				
x	y	$M_f = 3$	$M_f = 6$	$M_f = 9$	A	Exact result	$M_f = 3$	$M_f = 6$	$M_f = 9$	A	Exact result
0.0	0.0	0.666667	0.589080	0.589377	0.589376	0.589370	0.000000	0.000000	0.000000	0.000000	0.000000
0.0	0.5	0.541667	0.458019	0.458714	0.458708	0.458680	0.500000	0.548491	0.545219	0.545426	0.545475
0.0	1.0	0.166667	−0.007902	0.001242	0.000787	0.000000	1.000000	1.387931	1.342017	1.346876	1.350628
1.0	1.0	−0.333333	−0.022989	−0.005467	−0.004418	0.000000	1.414214	0.316979	0.149205	0.118921	0.000000

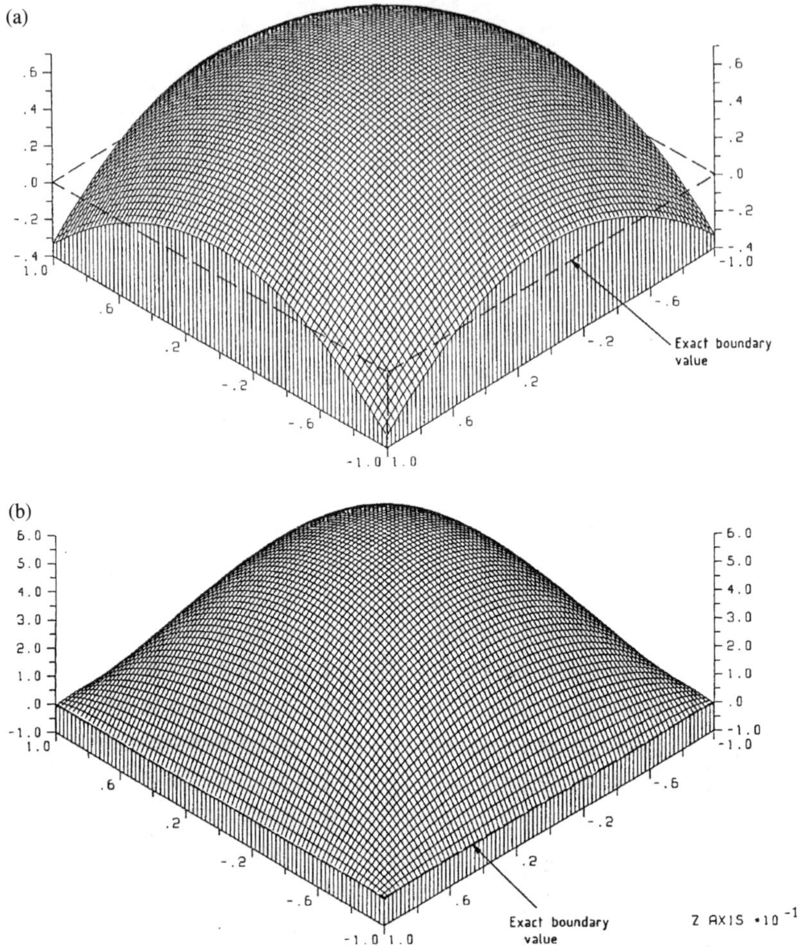

Figure 2: Function ϕ in square region; (a) $M_f = 3$; (b) $M_f = 9$.

The second variant of this example presents behaviour of the Kupradze functions. It was published in [18] as an investigation of the HTLS (hybrid Trefftz least square) formulation (17) (in Chapter 4) ((18) in Chapter 4); however, the majority of the results were calculated for a single element, which is analogical to the global approach with the boundary integral fit (15) (in Chapter 2). In the example $M_f = 44$, T-functions (28, 29 in Chapter 2) were applied. With increase of distance parameter λ, accuracy of the solution increased and conditioning worsened. To avoid the ill-conditioned matrices, the Tikhonov procedure (47) (in Chapter 2) was applied.

Figure 4 presents influence of the Tikhonov parameter γ on engineering measure (46) (in Chapter 2) of the conditioning. Without the application of this procedure, only relatively small distance δ of singularities of the Kupradze functions could be

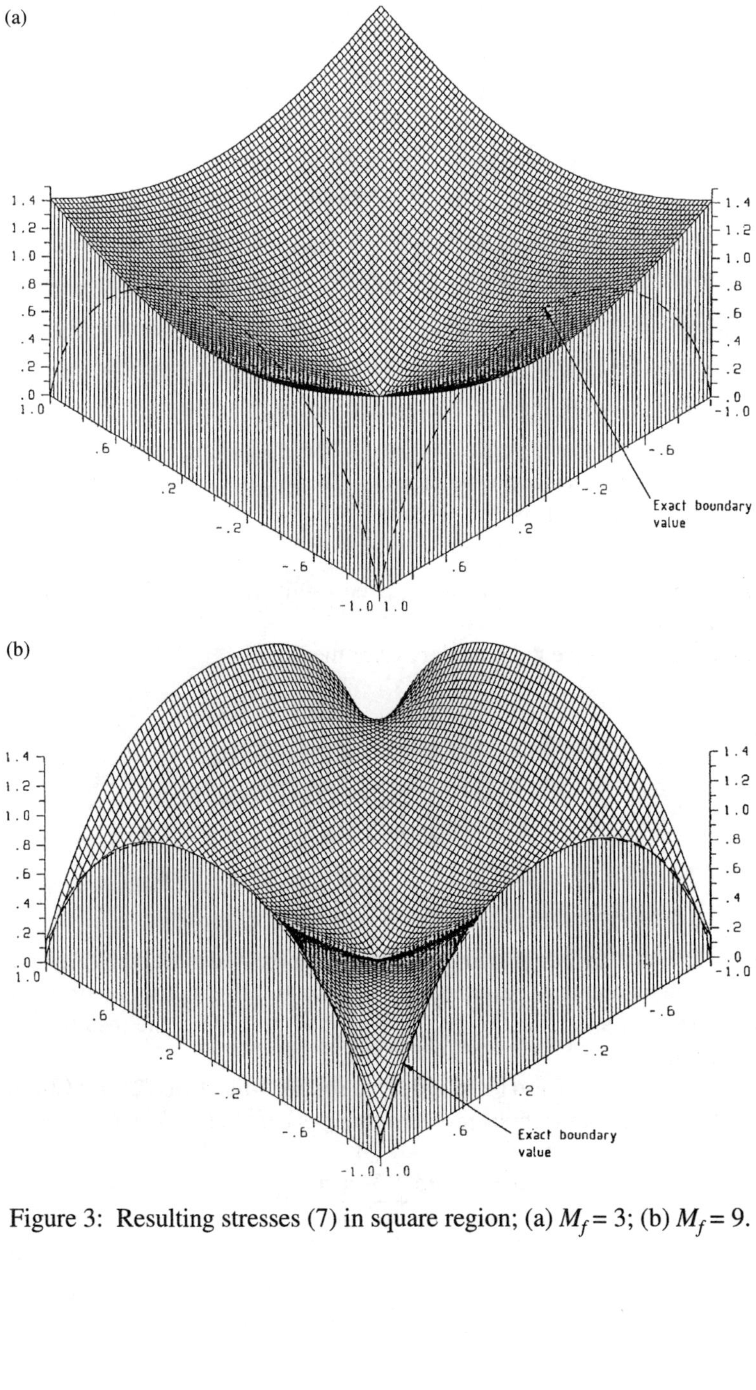

Figure 3: Resulting stresses (7) in square region; (a) $M_f = 3$; (b) $M_f = 9$.

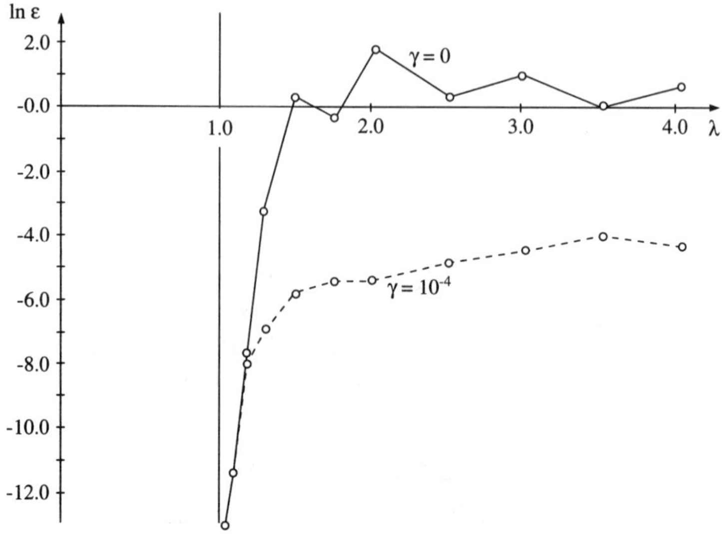

Figure 4: Influence of Tikhonov parameter γ (see (47) in Chapter 2) on engineering conditioning measure ε ((46) in Chapter 2).

chosen. With the non-zero corrector γ, even originally ill-conditioned problems can be solved (Fig. 5).

In Table 4, we observe the boundary error measures

$$\begin{aligned} \delta_{1-3} &= \frac{1}{K}\sum_{k=1}^{K}\left|\phi(\mathbf{x}_k)\right| \\ \delta_{1-3}' &= \frac{1}{K}\sum_{k=1}^{K}\left|\frac{\partial\phi(\mathbf{x})}{\partial y}\right|_{\mathbf{x}=\mathbf{x}_k} \end{aligned} \tag{13}$$

$$\mathbf{x} = \{x, y\}, \quad y = -1.0, \ -1.0 \le x \le 1.0,$$

where $\mathbf{x}_k$ ($K = 11$) are equidistant points between the nodes 1 and 3 (Fig. 1). The large distance parameter λ resulted here in strongly ill-conditioned problem for $\gamma = 0$; however, after application of the Tikhonov procedure, the results changed to be acceptable.

The same example can also be solved using special T-functions (21) (in Chapter 3) for $N_a = 4$ [6]. The homogenous problem changes here to (see Fig. 6):

$$\begin{aligned} &\frac{\partial^2\phi}{\partial r^2} + \frac{1}{r}\frac{\partial\phi}{\partial r} + \frac{1}{r^2}\frac{\partial^2\phi}{\partial r^2} = 0 \\ &0 \le \theta \le \frac{\pi}{4}, \quad 0 \le r \le \frac{1}{\cos\theta}, \end{aligned} \tag{14a}$$

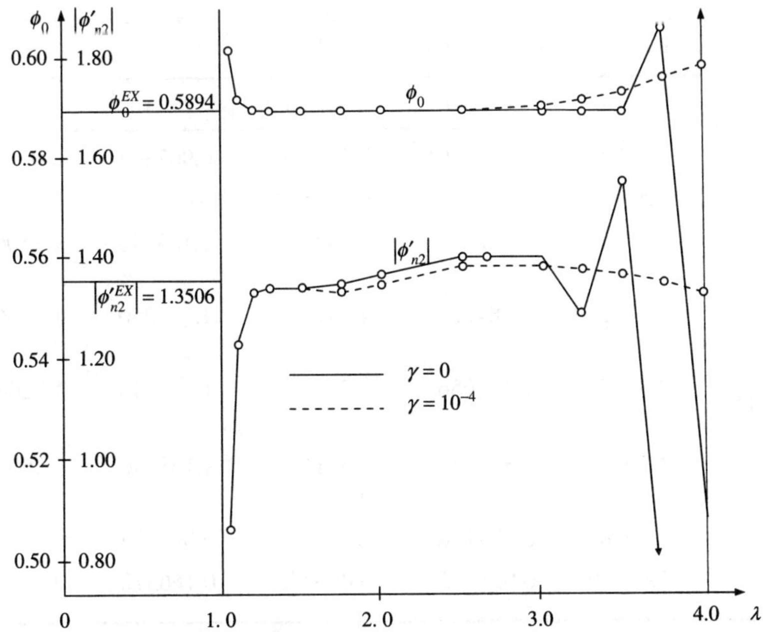

Figure 5: Solutions of problem (1) solved with the Kupradze functions (M_f = 44) versus distance parameter λ (square region); $\phi_0 = \phi(0,0)$, $|\phi'_{n2}| = |\partial\phi/\partial y(0,-a)|$ – extreme values.

with the boundary conditions:

$$\frac{\partial \phi}{\partial \theta} = 0 \quad \text{for}\, \theta = 0 \;\text{ and }\; \theta = \frac{\pi}{4} \tag{14b}$$

$$\phi = -\frac{1}{2} r^2 \quad \text{for}\;\; 0 \le \theta \le \frac{\pi}{4}, \quad r = \frac{1}{\cos\theta}. \tag{14c}$$

In this example, the direct variant of the boundary collocation approach was applied. Taking K_p equidistant control points along the boundary X = 1, the coordinates of which are given by the formulae:

$$\theta_k = \arctan\left(\frac{k-1}{K_p - 1}\right) \qquad r_k = \frac{1}{\cos\theta_k} \quad k = 1,\, 2, \ldots K_p \tag{15}$$

from the condition (14c) we obtain a system

$$\mathbf{Aa} = \mathbf{b} \tag{16a}$$

Table 4: Influence of different values of corrector γ (see (47) in Chapter 2) on results of numerical example for $\lambda = 4.0$.

$\lambda = 4.0$	$\gamma = 0$	$\gamma = 10^{-5}$	$\gamma = 10^{-4}$	$\gamma = 10^{-3}$	Exact
ε	3.618574	$0.1123 \cdot 10^{-1}$	$0.5921 \cdot 10^{-4}$	$0.9673 \cdot 10^{-6}$	0.0
ϕ_0	0.501822	0.590234	0.598514	0.634134	0.58937
$\left\|(\phi)'_{n2}\right\|$*	0.094048	1.382830	1.341936	1.163240	1.350628
$\left\|(\phi)'_{n4}\right\|$*	2.778862	1.384222	1.341926	1.163240	1.350628
$\left\|(\phi)'_{n5}\right\|$*	3.305426	1.383056	1.341924	1.163240	1.350628
$\left\|(\phi)'_{n7}\right\|$*	1.396614	1.383638	1.341928	1.163240	1.350628
δ_{1-3}	0.192896	0.004128	0.010475	0.043323	0.0
δ'_{1-3}	1.029698	0.025965	0.050445	0.166372	0.0

*Normal derivatives in points 2, 4, 5 and 7 (Fig. 1), respectively.

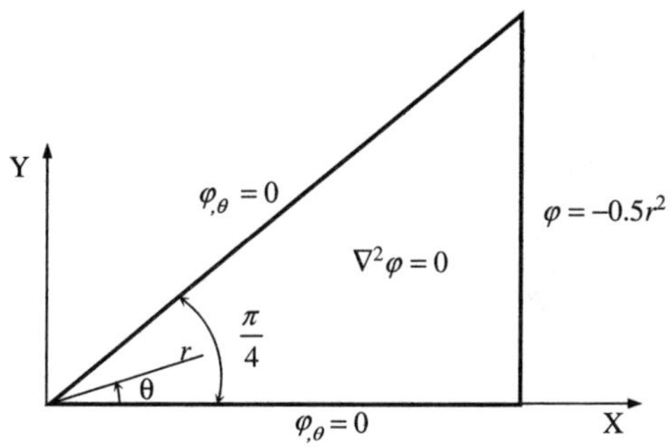

Figure 6: Example illustrating effectiveness of the special Trefftz functions.

where

$$A_{ki} = r_k^{4(i-1)} \cos\left[4(i-1)\theta_k\right]$$
$$b_k = -\frac{1}{2} r_k^2 \tag{16b}$$
$$i,k = 1,\ 2,\ldots\left(M_f = K_p\right).$$

Table 5: Coefficients of special T–functions (21) (in Chapter 3) for different number of functions and collocation points ($K_p = M_f$).

a	$M_f = 3$	$M_f = 5$	$M_f = 7$	$M_f = 10$
a_1	0.278926	0.2946985	0.2946878	0.2946859
a_2	$-3.0667 \cdot 10^{-2}$	$-4.5611 \cdot 10^{-2}$	$-4.5611 \cdot 10^{-2}$	$-4.5593 \cdot 10^{-2}$
a_3	$1.748 \cdot 10^{-3}$	$9.6799 \cdot 10^{-4}$	$9.6367 \cdot 10^{-4}$	$9.6238 \cdot 10^{-4}$
a_4		$-6.6703 \cdot 10^{-5}$	$-6.0923 \cdot 10^{-5}$	$-6.0397 \cdot 10^{-5}$
a_5		$1.2097 \cdot 10^{-5}$	$5.9227 \cdot 10^{-6}$	$5.9560 \cdot 10^{-6}$
a_6			$-7.5913 \cdot 10^{-7}$	$-7.3712 \cdot 10^{-7}$
a_7			$3.2135 \cdot 10^{-7}$	$-7.3712 \cdot 10^{-7}$
a_8				$-1.5408 \cdot 10^{-8}$
a_9				$-3.6930 \cdot 10^{-11}$
a_{10}				$-2.6862 \cdot 10^{-9}$

Table 5 contains the coefficients of the special T-functions (21) (in Chapter 3) for different number of the collocation points. In Table 6, a relative measure of the boundary error

$$\varepsilon_\Delta = \frac{\left|\hat{\phi}(x_\Gamma) - \overline{\phi}(x_\Gamma)\right|_{\max}}{\left|\overline{\phi}(x_\Gamma)\right|_{\max}} \qquad x_\Gamma \in \Gamma \tag{17}$$

was presented, where $\hat{\phi}(x_\Gamma)$ and $\overline{\phi}(x_\Gamma)$ are calculated and given boundary values of the solution, respectively. In the third column of Table 6, the condition numbers

$$\beta = \frac{1}{M_f} \|\mathbf{A}\| \|\mathbf{A}^{-1}\| \tag{18}$$

with

$$\|\mathbf{A}\| = \left(\sum_{i=1}^{M_f} \sum_{k=1}^{M_f} A_{ki}^2 \right)^{\frac{1}{2}} \tag{19}$$

versus the number of collocation points were enclosed. As it is observed, even the very large values of this parameter did not cause unacceptable results. This characteristic phenomenon was also noted in several other papers [12, 14, 16].

Table 6: Values of maximal relative error ε_Δ (17) and condition number β; $K_p = M_f$

M_f	ε_Δ	β
3	$0.62752 \cdot 10^{-2}$	$0.53319 \cdot 10^{1}$
6	$0.83438 \cdot 10^{-3}$	$0.15193 \cdot 10^{3}$
9	$0.29524 \cdot 10^{-3}$	$0.60685 \cdot 10^{4}$
12	$0.14687 \cdot 10^{-3}$	$0.31583 \cdot 10^{6}$
15	$0.86792 \cdot 10^{-4}$	$0.51692 \cdot 10^{8}$
18	$0.57066 \cdot 10^{-4}$	$0.19657 \cdot 10^{11}$
21	$0.39744 \cdot 10^{-4}$	$0.87759 \cdot 10^{13}$
24	$0.29588 \cdot 10^{-4}$	$0.43058 \cdot 10^{16}$
27	$0.22938 \cdot 10^{-4}$	$0.22685 \cdot 10^{19}$
30	$0.18000 \cdot 10^{-4}$	$0.12639 \cdot 10^{22}$
33	$0.13975 \cdot 10^{-4}$	$0.73676 \cdot 10^{24}$
36	$0.69074 \cdot 10^{-5}$	$0.44576 \cdot 10^{27}$
39	$0.92770 \cdot 10^{-5}$	$0.44576 \cdot 10^{30}$
42	$0.90482 \cdot 10^{-5}$	$0.17836 \cdot 10^{33}$
45	$0.21096 \cdot 10^{-5}$	$0.11697 \cdot 10^{36}$
48	$0.18280 \cdot 10^{-4}$	$0.78251 \cdot 10^{38}$
49	$0.22470 \cdot 10^{-4}$	$0.68708 \cdot 10^{39}$
50	$0.55810 \cdot 10^{-3}$	$0.60451 \cdot 10^{40}$

Investigations of the boundary collocation approach for convex and concave areas have been presented in [3]. Problem (1) was there solved with help of the Kupradze functions for a rectangle with one side modified (Fig. 7). In Fig. 8a, we can observe influence of the Tikhonov parameter on the global boundary error

$$\varepsilon_g = \frac{\sum_{j=1}^{N} \left| \hat{\phi}\left(x_{\Gamma_j}\right) - \bar{\phi}\left(x_{\Gamma_j}\right) \right|}{N \left| \bar{\phi}\left(x_\Gamma\right) \right|}, \quad x_\Gamma \in \Gamma, \; N >> K_p. \tag{20}$$

The introduction of non-zero γ widens the range of acceptable results because of improvement of conditioning; however, it disturbs the solution increasing error measure ε_g. Figure 8b compares the results for different convexity measure c. The negative effect of concavity can be to a large extent reduced by addition of one suitably situated singular point and two collocation points (Fig. 9). In Fig. 10, we can see distribution of the boundary error

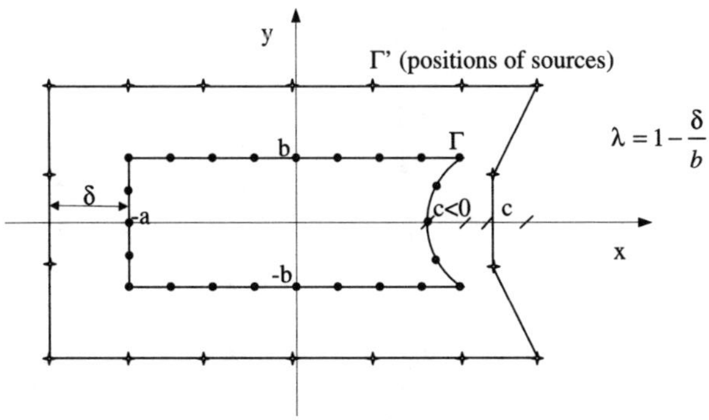

Figure 7: Example for eqn (1) and concave boundary segment ((a = 2b).

$$\varepsilon_b = \frac{\hat{\phi}(x_\Gamma) - \overline{\phi}(x_\Gamma)}{\left|\overline{\phi}(x_\Gamma)\right|_{\max}}, \quad x_\Gamma \in \Gamma \tag{21}$$

obtained without and with the additional source function.

Different types of collocation were investigated in detail in [14] also on the example (1). The orthogonal boundary collocation proposed in this paper meant collocation in the Gaussian points along the boundary Γ. Behaviour of the following direct boundary error measure (Chebyshev's norm)

$$\varepsilon_e = \left|\phi_\varepsilon(x_\Gamma)\right|_{\max} \quad x_\Gamma \in \Gamma \tag{22a}$$

and also

$$\phi_\varepsilon(x_\Gamma) = \hat{\phi}(x_\Gamma) - \overline{\phi}(x_\Gamma) \tag{22b}$$

was observed. Two families of the T-functions were applied. Because of the symmetry (see Fig. 11), the Herrera system (6) (in Chapter 2) was reduced to

$$\Phi_n^H = \mathrm{Re}\left(z^n\right) \quad n = 0,\ 2,\ 4,\ldots \quad z = x + iy \tag{23}$$

and the Kupradze functions (28, 29 in Chapter 2) were suitably coupled:

$$\begin{aligned} \Phi_1^K &= 1.0 \\ \Phi_2^K &= \sum_{i=1}^{2} \ln r_i \\ \Phi_3^K &= \sum_{i=1}^{6} \ln r_i \\ \Phi_4^K &= \sum_{i=1}^{10} \ln r_i . \end{aligned} \tag{24}$$

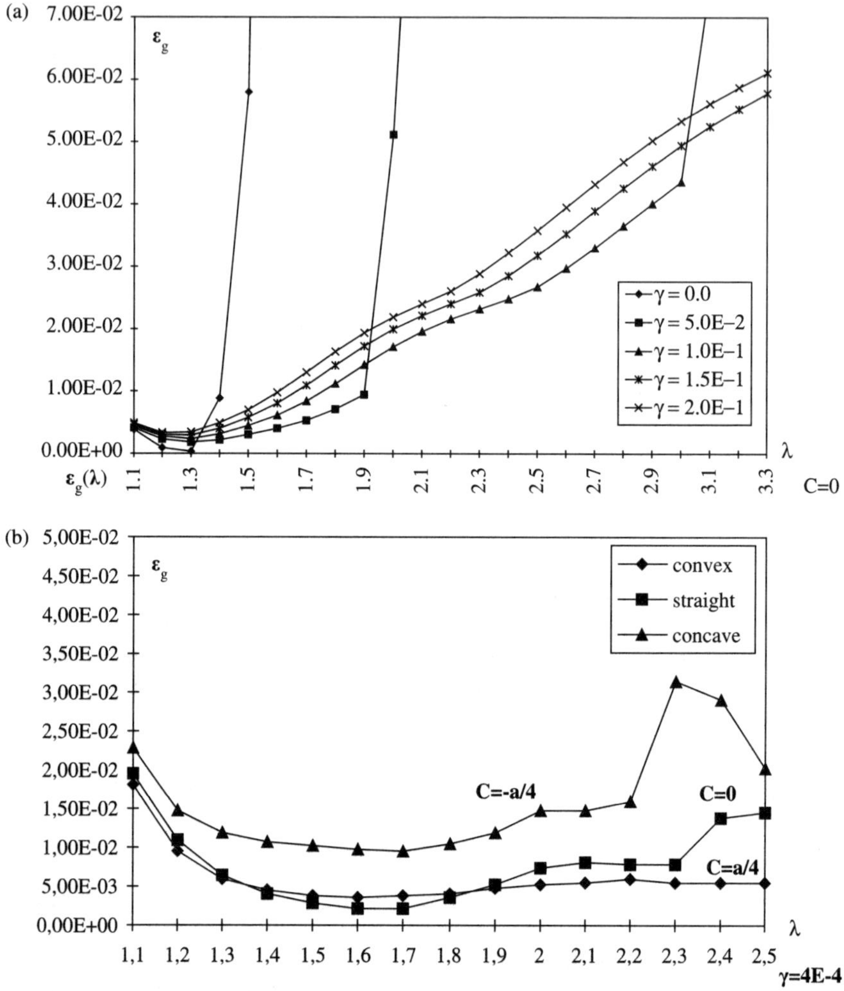

Figure 8: Global boundary error ε_g (20) versus distance parameter $\lambda = 1-(\delta / b)$; (a) $c = 0$, $M_f = 36$, $K_p = 48$, $N = 600$; (b) $c \neq 0$, $M_f = 18$, $K_{p.} = 24$, $N = 600$.

Note the additional constant function $\Phi_1^K = 1.0$ which represents the generalized rigid body mode. It improves the convergence of the solution and, in the T-elements, results in fulfilment of the patch test. However, in general, the addition of the rigid body terms does not introduce crucial effects on the Kupradze-type solutions [14].

In Fig. 12, we can see the results of calculations for the harmonic polynomials and a long rectangle (a = 2.0, b = 0.5). The number of trial functions applied is $M_f = 6$ $\left(\Phi_n^H \text{ for } n = 0, 2, 4 \ldots 10\right)$, thus the point $N_c = 3$ (N_c is a number of the control points on each side of the rectangle, Fig. 11) means the direct fit of the

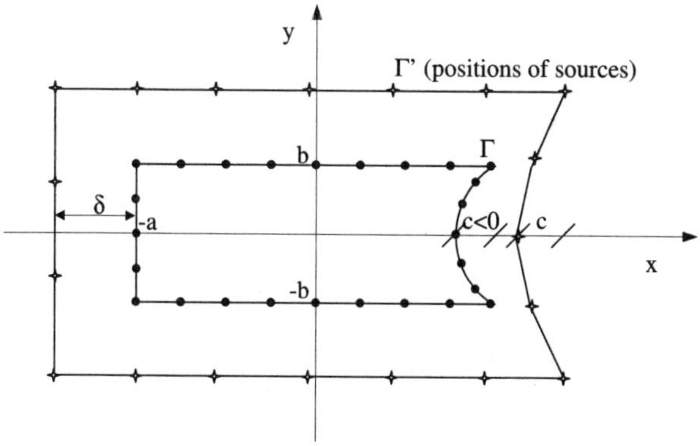

Figure 9: Additional source and collocation points minimizing boundary error.

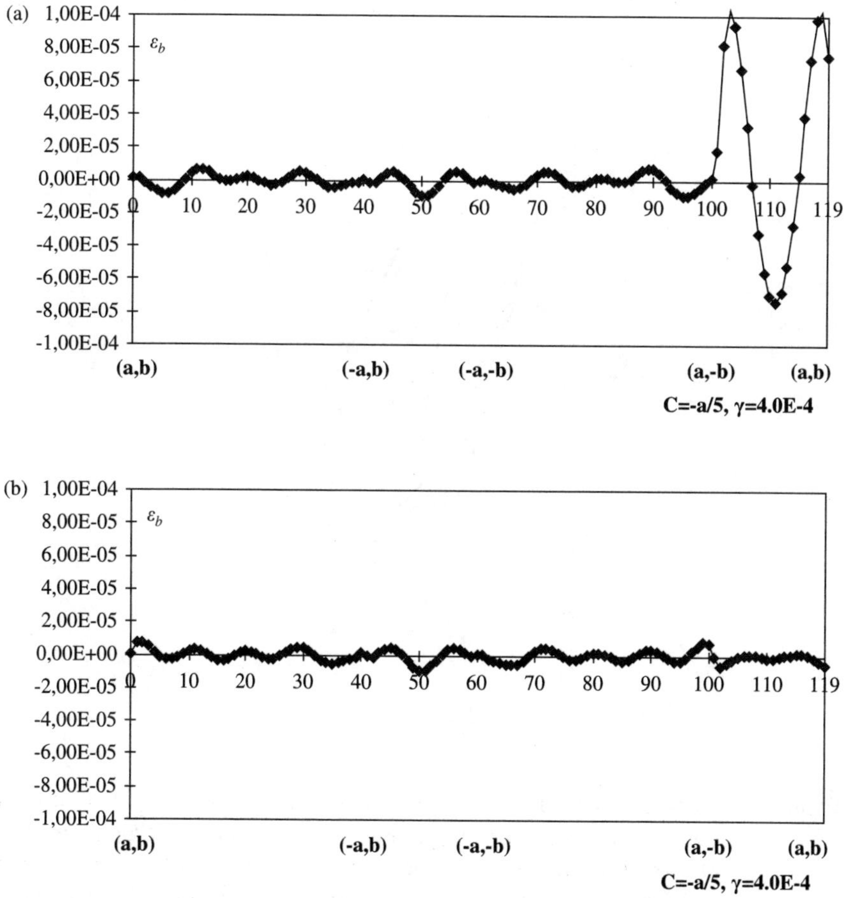

Figure 10: Boundary error ε_b (21) in case of source and collocation points according to (a) Fig. 7; (b) Fig. 9.

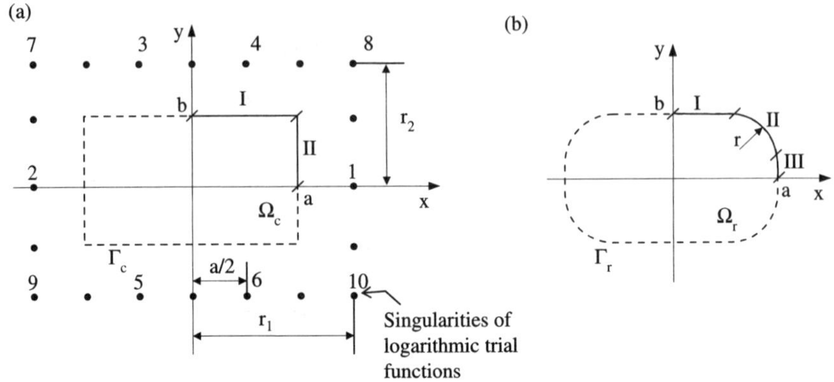

Figure 11: Influence of position and number of collocation points – definition of numerical examples.

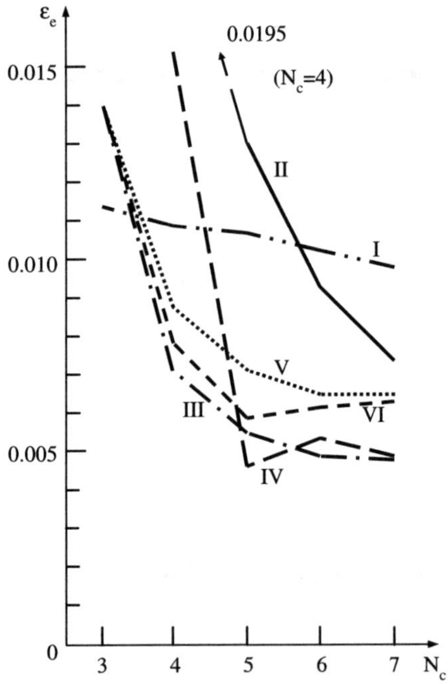

Figure 12: Extreme boundary error ε_e (22 a) versus number N_c of collocation points on one side of rectangle Fig. 11 (a) harmonic polynomials (23), $M_f = 6$; I – equidistant collocation without corners; II – equidistant collocation with corners; III – orthogonal collocation (Gauss points); IV – orthogonal collocation + corners; V – Gauss integral fit with N_c control points; VI – orthogonal collocation with Gaussian weights.

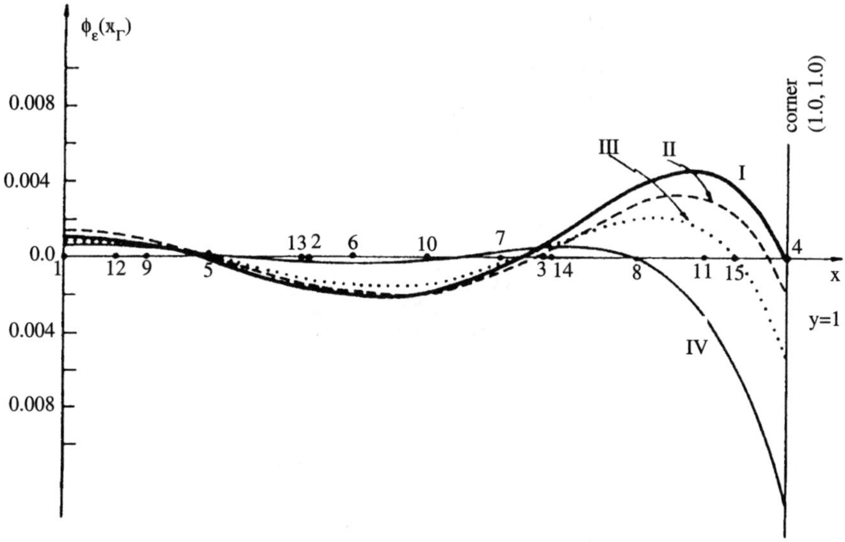

Figure 13: Influence of position of control points on error function ϕ_e – (22 b); $a = b = 1.0$, $M_f = 6$; I – equidistant points 1, 2, 3, 4; II – three Gaussian points 9, 10, 11 + p. 4; III – four Gaussian points 12, 14, 14, 15; IV – equidistant points 5, 6, 7, 8.

boundary conditions and $N_c > 3$ – the least square solution. In case of the control point located in the corner, the solution for $N_c = 3$ could not be calculated; the solution is singular.

Figure 13 presents influence of the choice of control points on the error function $\phi_\varepsilon(x_\Gamma)$. In case of the standard integral fit with the Gaussian numerical integration, it would be difficult to supplement the set of the control points by additional ones in places of special interest (e.g. vicinity of corners). However, applying the orthogonal collocation, we can add arbitrary control points without negative consequences. It evidently shows that **the boundary collocation is a notion broader than the numerical integral fit**.

In the results calculated for a rectangle with rounded corners (smooth contour), the observed tendency is very similar (Figs. 14 and 15). Also investigation of the solution conditioning confirms this similarity. Table 7 shows the results that were obtained for harmonic polynomials. The most conspicuous increase of κ ((38) in Chapter 2) and $\|\Delta x\|$ ((42) in Chapter 2) can be observed with changes of the contour shape (increase of the ratio a/b) and the number of trial functions M_f.

For the singular logarithmic functions, the problem of conditioning of the matrices is more complicated since the distance of the singularities from region Ω plays an additional important role. In Fig. 16, we can see the results of the calculations for the area Ω_r. It becomes obvious that the optimization of the distances r_1 and r_2 must include not only the error values (as was proposed by Mathon & Johnston [8]) but also the conditioning number κ ((38) in Chapter 2).

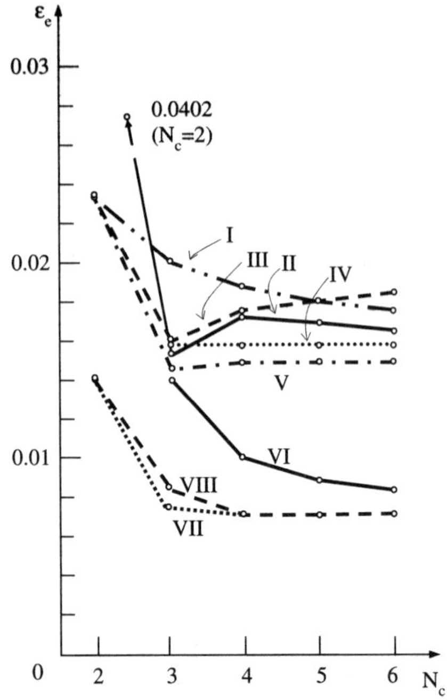

Figure 14: Extreme boundary error ε_e (22 a) versus number N_c of collocation points on one segment of Γ_r (Fig. 11 b); harmonic polynomials (23); I – equidistant collocation inside segments, $M_f = 4$; II – equidistant collocation including segment ends, $M_f = 4$; III – orthogonal collocation in Gauss points, $M_f = 4$; IV – integral fit with Gaussian integration, $M_f = 4$; V – orthogonal collocation with Gaussian weights, $M_f = 4$; VI – equidistant collocation including segment ends, $M_f = 6$; VII – orthogonal collocation, $M_f = 6$; VIII – orthogonal collocation with Gaussian weights, $M_f = 6$.

Since increasing the number of trial functions that are used in a given boundary value problem produces an increase in κ and $\|\Delta \mathbf{x}\|$, in general it is not possible to enlarge the system of shape functions that is used beyond certain limits. Thus, when such limits are reached, all that can be done to improve the solution is to use the multi-step fitting (discussed below) or to divide the region Ω into elements. However, it must be noticed that using $\|\Delta \overline{\mathbf{x}}\|$ and κ to determine such limits may be too conservative, since $\|\Delta \overline{\mathbf{x}}\|$ is only a bound of the actual error $\|\Delta \mathbf{x}\|$ (see (42) in Chapter 2). Table 8 displays some of the results for the condition number κ, which may be of interest in engineering calculations. One can also observe that, in spite of the large variations of the roots, the solution inside the investigated region does not change much.

To investigate the conditioning in a direct way, we introduced into the matrix **A** additional disturbances of order 10^{-8} situated incidentally inside the matrix. Reaction for these disturbances is compared with estimators of the result error (with the assumed data error of the same order $\delta_a = 10^{-8}$). In Table 8, we can see the considerable difference between this estimation and real errors of the roots.

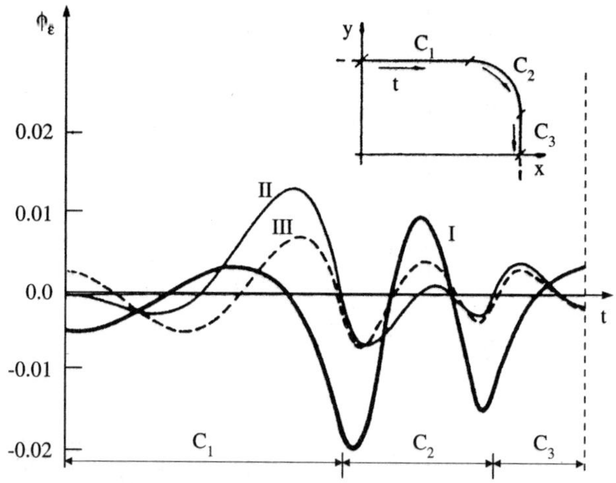

Figure 15: Error function ϕ_ε on boundary Γ_r (a = 2.0, b = 1.0, r = 1.0), N_c = 3, harmonic polynomials; I – equidistant collocation inside segments, M_f = 4; II – equidistant collocation inside segments, M_f = 6; III – orthogonal collocation inside segments, M_f = 6.

As it was mentioned above, even the great error of roots can result in relatively small changes of the field functions in the central region of Ω. All these facts must be taken into account while limiting the solution factors.

Additionally, it should be noted that order of the final solution error did not increase during the test when changing places of the matrix disturbances. Hence, the test described above can be considered as an image of real sensitivity of the solution on the matrix errors. Thus, the authors suggest it as a helpful tool in determination of limiting values of the solution factors a/b, M_f and r_2.

An important advantage of the boundary collocation method is the fact that the error of the approximate solution is introduced only through the boundary. **The boundary error is here directly given as a difference between the prescribed and calculated boundary functions (observe that the same is not possible while using the singular boundary integral equation procedures).** This feature of the boundary collocation approach allows to use the multi-step fitting of boundary conditions, that is applying procedures for successive correction of boundary errors.

Secondary collocation at the points of extreme boundary error in the first-step solution is one of the obvious methods for the fit improvement. In Fig. 17, we can see the result of application of three additional control points with weights proportional to the values of the error in the first-step solution. Repeating this procedure we could obtain the optimal fit for a given number of shape functions. This can especially be important when increase in the number of shape functions is difficult because of conditioning reasons.

Treating the boundary error $\phi_\varepsilon^{(1)}(x_\Gamma)$ from the first-step solution (which is known exactly) as a new boundary condition is another possibility for improving the result. Since the unknown exact solution is

Table 7: Conditioning measures of the matrices in numerical examples (trial functions–harmonic polynomials); assumed (introduced) errors of matrix terms $\delta_a = 10^{-8}$; accuracy of calculations – 11 digits.

Γ^*	a	b	r	M_f	N_c	Fit	$\lVert A \rVert$	$\lVert A^{-1} \rVert$	κ ((38) in Chapter 2)	$\lVert \Delta \bar{x} \rVert$
Γ_c	2	1		4	2	equidistant collocation	$3.140 \cdot 10^3$	1.426	$4.478 \cdot 10^3$	$3.818 \cdot 10^{-6}$
						orthogonal collocation	$9.562 \cdot 10^3$	0.782	$7.480 \cdot 10^3$	$2.661 \cdot 10^{-6}$
						integral fit	$5.838 \cdot 10^3$	0.777	$4.534 \cdot 10^3$	$2.295 \cdot 10^{-6}$
					6	equidistant collocation	$2.122 \cdot 10^4$	0.388	$7.177 \cdot 10^3$	$1.098 \cdot 10^{-6}$
						orthogonal collocation	$4.373 \cdot 10^4$	0.258	$1.127 \cdot 10^3$	$9.288 \cdot 10^{-6}$
						integral fit	$6.850 \cdot 10^4$	0.902	$6.180 \cdot 10^3$	$2.823 \cdot 10^{-6}$
Γ_c	3	1		4	2	equidistant collocation	$4.803 \cdot 10^5$	2.072	$9.954 \cdot 10^5$	$1.432 \cdot 10^{-5}$
						orthogonal collocation	$5.614 \cdot 10^5$	0.990	$5.558 \cdot 10^5$	$1.024 \cdot 10^{-5}$
					6	equidistant collocation	$1.584 \cdot 10^6$	0.465	$7.384 \cdot 10^5$	$4.034 \cdot 10^{-6}$
						orthogonal collocation	$1.838 \cdot 10^6$	0.337	$6.194 \cdot 10^5$	$4.484 \cdot 10^{-6}$

Γ_c	3	1		6	3	equidistant collocation	$5.965 \cdot 10^{9}$	2.975	$1.775 \cdot 10^{10}$	$9.510 \cdot 10^{-3}$
						orthogonal collocation	$1.262 \cdot 10^{10}$	0.719	$9.075 \cdot 10^{9}$	$2.359 \cdot 10^{-3}$
					6	equidistant collocation	$1.600 \cdot 10^{10}$	0.673	$1.076 \cdot 10^{10}$	$2.267 \cdot 10^{-3}$
						orthogonal collocation	$3.082 \cdot 10^{10}$	0.392	$1.014 \cdot 10^{10}$	$1.037 \cdot 10^{-3}$
Γ_r	2	1	0.5	4	4	equidistant collocation	$2.049 \cdot 10^{4}$	0.326	$6.681 \cdot 10^{3}$	$1.767 \cdot 10^{-6}$
						orthogonal collocation	$2.030 \cdot 10^{4}$	0.329	$6.671 \cdot 10^{3}$	$1.798 \cdot 10^{-6}$
				6		equidistant collocation	$6.828 \cdot 10^{6}$	0.317	$2.166 \cdot 10^{6}$	$5.376 \cdot 10^{-5}$
						orthogonal collocation	$7.129 \cdot 10^{6}$	0.315	$2.248 \cdot 10^{6}$	$5.024 \cdot 10^{-5}$
Γ_r	2	0.5	0.25	6	4	equidistant collocation	$4.468 \cdot 10^{6}$	10.10	$4.514 \cdot 10^{7}$	$3.120 \cdot 10^{-4}$
						orthogonal collocation	$4.351 \cdot 10^{6}$	9.095	$3.963 \cdot 10^{7}$	$2.693 \cdot 10^{-4}$

*See Fig. 11.

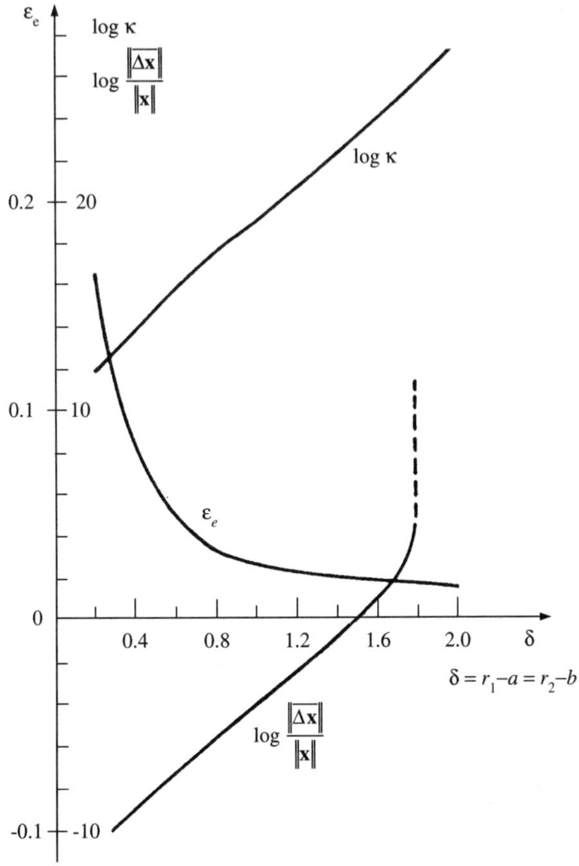

Figure 16: Influence of distance δ of sources on results; logarithmic functions, boundary Γ_r ($a = 2.0$, $b = 1.1$, $r = 0.5$), $M_f = 4$, $N_c = 4$ on each segment, orthogonal collocation.

$$\phi^{EX}(\mathbf{x}) = \hat{\phi}^{(1)}(\mathbf{x}) - \phi_\varepsilon^{(1)}(\mathbf{x}) \tag{25}$$

where

$$\phi_\varepsilon^{(1)}(\mathbf{x}) = \hat{\phi}(\mathbf{x}) - \phi^{EX}(\mathbf{x}) \;\rightarrow\; \mathbf{x} \in \Omega \qquad \text{(unknown)}$$

$$\phi_\varepsilon^{(1)}(\mathbf{x}_\Gamma) = \hat{\phi}(\mathbf{x}_\Gamma) - \phi^{EX}(\mathbf{x}_\Gamma) \quad \mathbf{x}_\Gamma \in \Gamma \qquad \text{(given)},$$

it is required to find possibly accurate approximation of $\phi_\varepsilon^{(1)}(\mathbf{x})$. The function $\hat{\phi}_\varepsilon^{(1)}(\mathbf{x})$ should, however, be calculated in a different way to the first-step solution. For example, one can use different T-complete systems of functions or a different division of Γ into segments. In Fig. 18, we can see the two-step procedure while

Table 8: Influence of the disturbances of order 10^{-8} introduced into matrix **A**. Logarithmic trial functions, $M_f = 4$, contour Γ_r (Fig. 11b) ($a = 2.0$, $b = 1.0$, $r = 0.5$), N_c in each segment.

Fit	r_1, r_2 Fig. 5.11a	$\|\mathbf{A}\|$ (for undisturbed data)		$\|\mathbf{A}^{-1}\|$	κ ((38) in Chapter 2)	$\|\overline{\Delta \mathbf{x}}\|$ Roots of equation $\mathbf{Ax} = \mathbf{b}$: 1	2	3	4	$\hat{u}(0,0)$	
Equidistant collocation	$r_1 = 3.0$	2.702×10^3	5.828×10^4	1.575×10^8	0.5900	32.601693	−0.174134	−0.013509	−3.005841	0.910178†	U
	$r_2 = 2.0$					32.602116	−0.174145	−0.013521	−3.005871	0.910178†	D
Orthogonal collocation	$r_1 = 3.0$	2.702×10^3	5.633×10^4	1.530×10^8	0.5690	32.285155	−0.166791	−0.005362	−2.983287	0.909756†	U
	$r_2 = 2.0$					32.285621	−0.166803	−0.005374	−2.883320	0.909756†	D
Equidistant collocation	$r_1 = 4.0$	4.439×10^3	1.673×10^8	7.425×10^{11}	–	−247.5385	6.308819	5.448441	12.681568	0.908618‡	U
	$r_2 = 3.0$				$-2.790\text{x}10^3$*	−260.7607	6.523186	5.642669	13.477234	0.908614‡	D
Orthogonal collocation	$r_1 = 4.0$	4.439×10^3	1.669×10^8	7.409×10^{11}	–	−276.2433	6.772079	5.869465	14.410267	0.908292‡	U
	$r_2 = 3.0$				$-3.121\text{x}10^3$*	−288.8782	6.976907	6.055041	15.170618	0.908309‡	D

U – undisturbed results; D – disturbed results.

*Formal application of formula (42) in Chapter 2 for these values of κ (see Fig. 16).

†Extreme difference between disturbed and undisturbed $\hat{u}$ was of the order 10^{-6}.

‡Extreme difference between disturbed and undisturbed $\hat{u}$ was of the order 10^{-4}.

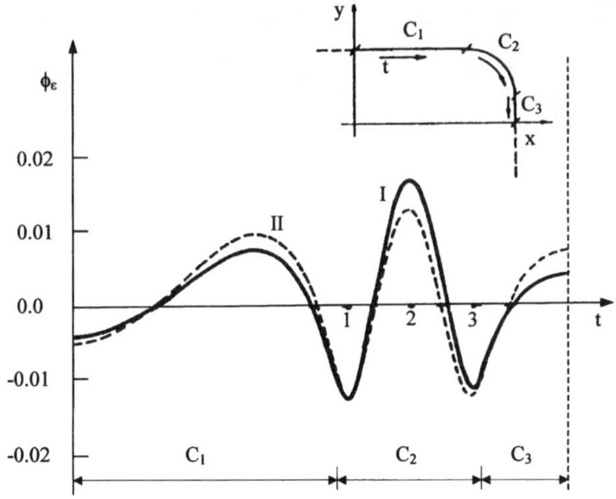

Figure 17: Two-step fitting on Γ_r, harmonic polynomials, orthogonal collocation, $M_f = 4$, $N_c = 4$, Γ_r ($a = 2.0$, $b = 1.0$, $r = 0.5$); I – first-step solution; II – improved solution with three additional control points.

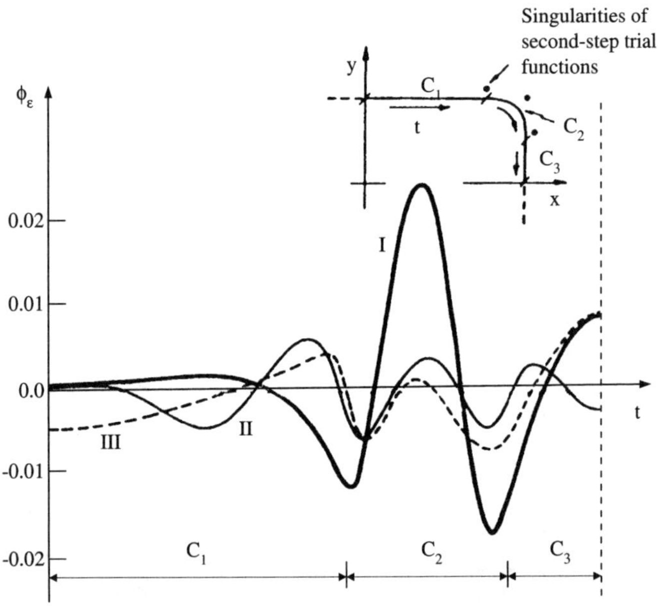

Figure 18: Improvement of fitting boundary conditions on Γ_r, logarithmic functions, orthogonal collocation, Γ_r ($a = 2.0$, $b = 1.0$, $r = 0.5$); I – one-step solution, $M_f = 4$, $N_c = 4$, $\kappa = 1.53 \cdot 10^8$; II – one-step solution, $M_f = 7$, $N_c = 4$, $\kappa = 4.7 \cdot 10^{11}$; III – two-step solution, $M_{f1} = 4$, $N_{c1} = 4$, $\kappa_1 = 1.53 \cdot 10^8$; $M_{f2} = 3$, $N_{c2} = 6$, $\kappa_2 = 4.89 \cdot 10^2$.

applying the singular logarithmic trial functions. Diminishing of the error can also be done in the one-step solution with larger number of shape functions (curve II), but in this case the condition number of the solution matrix increases.

The multi-step fitting can also be useful in typical engineering calculation which can be considerably improved by application of additional control points in regions where the largest values of investigated field functions occur.

2 Doubly connected regions

Doubly connected regions are, in general, more difficult for application of the boundary collocation approach. This does not concern regions with circular holes and inclusions, for which numerous investigations were carried out [4, 10, 16, 17]. The investigators used the special T-functions, identically fulfilling boundary conditions on the circular contour. In certain cases, the idea of conformal mapping of other contours to a circle was applied [9]. However, in general, applications of the method to regions Ω with arbitrary holes were rather rare.

The monograph [5] contains detailed investigations of that case. The numerical examples have here been calculated for a biharmonic equation with T-functions (4) (in Chapter 3). Because of the symmetry of the objects, only cosine terms have been taken into account. The majority of calculations have been performed for a rectangular region with circular and elliptic holes; however, along both contours (also along the edge of the hole), the boundary collocation fit has been applied.

The results can be interpreted as a 2D problem of elasticity solved on the level of stresses. Relation between the biharmonic stress function ϕ and the stresses has been taken in the customary form:

$$\begin{aligned} \sigma_r &= \frac{1}{r}\frac{\partial \phi}{\partial r} + \frac{1}{r^2}\frac{\partial^2 \phi}{\partial \theta^2} \\ \sigma_\theta &= \frac{\partial^2 \phi}{\partial r^2} \\ \tau_{r\theta} &= -\frac{\partial}{\partial r}\left(\frac{1}{r}\frac{\partial \phi}{\partial \theta}\right) \end{aligned} \tag{26a}$$

or

$$\begin{aligned} \sigma_x &= \sigma_{11} = \frac{\partial^2 \phi}{\partial y^2} \\ \sigma_y &= \sigma_{22} = \frac{\partial^2 \phi}{\partial x^2} \\ \tau_{xy} &= \sigma_{12} = -\frac{\partial^2 \phi}{\partial x \partial y} \end{aligned} \tag{26b}$$

which, after introducing the T-functions from (4) (in Chapter 3) but only with cosine terms and omission of certain unfitting expressions, results in the set of Trefftz functions for stresses:

$$\Psi_0 = \begin{Bmatrix} \sigma_r^0 \\ \sigma_\theta^0 \\ \tau_{r\theta}^0 \end{Bmatrix} = \begin{Bmatrix} 2B_0 + C_0 r^{-2} + D_0(1 + 2\ln r) \\ 2B_0 - C_0 r^{-2} + D_0(3 + 2\ln r) \\ 0 \end{Bmatrix} \tag{27a}$$

$$\Psi_1 = \begin{Bmatrix} \sigma_r^1 \\ \sigma_\theta^1 \\ \tau_{r\theta}^1 \end{Bmatrix} = \begin{Bmatrix} \left[2B_1 r - 2C_1 r^{-3} + D_1 r^{-1}\right]\cos\theta \\ \left[6B_1 r + 2C_1 r^{-3} + D_1 r^{-1}\right]\cos\theta \\ \left[2B_1 r - 2C_1 r^{-3} + D_1 r^{-1}\right]\sin\theta \end{Bmatrix} \tag{27b}$$

$$\Psi_n = \begin{Bmatrix} \sigma_r^n \\ \sigma_\theta^n \\ \tau_{r\theta}^n \end{Bmatrix} = \begin{Bmatrix} \left[(1-n)nA_n r^{n-2} + (2-n)(n+1)B_n r^n\right]\cos n\theta \\ \left[(1-n)nA_n r^{n-2} + (n+1)(n+2)B_n r^n\right]\cos n\theta \\ \left[(n-1)nA_n r^{n-2} + n(n+1)B_n r^n\right]\sin n\theta \end{Bmatrix}$$
$$+ \begin{Bmatrix} \left[-n(n+1)C_n r^{-n-2} + (1-n)(n+2)D_n r^{-n}\right]\cos n\theta \\ \left[(n+1)nC_n r^{-n-2} + (n-2)(n-1)D_n r^{-n}\right]\cos n\theta \\ \left[-n(n+1)C_n r^{-n-2} + (1-n)nD_n r^{-n}\right]\sin n\theta \end{Bmatrix}, \quad n = 2,3\ldots. \tag{27c}$$

The tractions are understood in the standard way:

$$t_i = \sigma_{ij} n_j \quad i = 1,\ 2 \quad \text{on} \quad \Gamma \tag{28}$$

where n_j is a cosine of an angle of the outward normal to the contour Γ.

In the first numerical example, an extended square plate with a circular hole (Fig. 19) is considered. As an error measure, the Chebyshev norm (17) is here assumed; however, to observe better behaviour of the boundary fit, two independent measures are introduced, separately for external and internal contours:

$$\varepsilon_{ext} = \frac{\left|\hat{\mathbf{t}}\left(\mathbf{x}_{\Gamma_{ext}}\right) - \bar{\mathbf{t}}\left(\mathbf{x}_{\Gamma_{ext}}\right)\right|_{\max}}{p} \cdot 100\% \tag{29a}$$

$$\varepsilon_{\text{int}} = \frac{\left|\hat{\mathbf{t}}\left(\mathbf{x}_{\Gamma_{\text{int}}}\right) - \bar{\mathbf{t}}\left(\mathbf{x}_{\Gamma_{\text{int}}}\right)\right|_{\max}}{p} \cdot 100\% \tag{29b}$$

where p is the intensity of the extending force, $\hat{\mathbf{t}}(\mathbf{x}_\Gamma)$ and $\bar{\mathbf{t}}(\mathbf{x}_\Gamma)$ are vectors of calculated and exact tractions, respectively, and Γ_{ext} and Γ_{int} are external and internal contours. Exemplary distribution of the fit error for the external contour is presented in Fig. 19 (ε_{int} appeared to be much smaller in this case). Because of

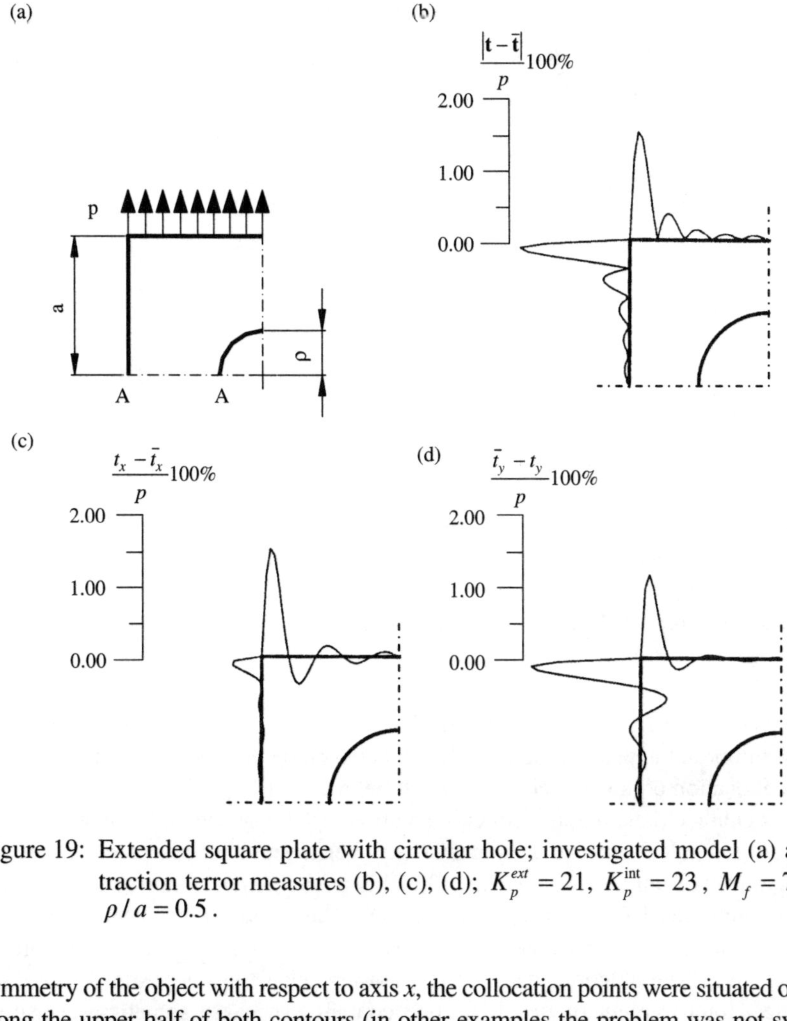

Figure 19: Extended square plate with circular hole; investigated model (a) and traction terror measures (b), (c), (d); $K_p^{ext} = 21$, $K_p^{\text{int}} = 23$, $M_f = 78$, $\rho / a = 0.5$.

symmetry of the object with respect to axis x, the collocation points were situated only along the upper half of both contours (in other examples the problem was not symmetric with respect to y). It should be noted that in the corner point, three independent equations appeared (instead of two). Two conditions were required for normal stresses and one for the tangent stress τ_{xy}, identical for both sides meeting in the corner.

Table 9 contains typical results for increasing number of the trial functions (see [5] for more detailed investigations). An additional measure of accuracy of the solutions is here the integral (see Fig. 19a)

$$I_{AA} = \frac{1}{pa} \int_{A-A} \sigma_y dx, \tag{30}$$

the exact value of which was 1.0. Observation of the tables leads to the following conclusions. In general, errors on the external contour are much larger than on the internal (smooth) boundary and they are situated in vicinity of the corners (typical behaviour of the Trefftz-type solutions). Next, we observe that for a

Table 9: Approximation errors (29) of boundary conditions for $\rho / a = 0.25$, $N_{\text{int}}^{eq} = N_{ext}^{eq} = 36$ – numbers of equations resulting from external and internal control points, respectively.

K_p^{ext}	N_{ext}^{eq}	K_p^{inp}	N_{int}^{eq}	M_f	$\varepsilon_{ext}\,[\%]$	$\varepsilon_{\text{int}}\,[\%]$	$I_{\text{A–A}}$
25	52	14	28	14	6.30	0.71	1.0282
				18	5.86	0.84	1.0297
				26	3.47	0.23	1.0013
				34	3.41	0.23	1.0009
				42	1.52	0.03	1.0027
				50	1.50	0.03	1.0029
				58*	0.46	1.40	1.0011
				62**	0.46	1.95	1.0013

*M_f with negative powers of r: $M_f^- = 29 > N_{\text{int}}^{eq}$.
**$M_f^- = 31 > N_{\text{int}}^{eq}$.
Remark: only upper half of the plate was considered (see explanation in the text).

relatively larger hole, more modelling T-functions should be applied. Finally, the most important appears to be the relation between the number of T-functions and the collocation points, which is discussed below.

It is characteristic that the effect of solution instability, the rapid increase of ε_{int} or ε_{ext} (see Table 9), occurs even in the case when the number of equations exceeds the number of unknowns (see Fig. 20). This phenomenon is caused by different behaviour of the T-functions on both contours. The functions with negative powers take on the internal boundary relatively large values and on the external contour rather small. The functions with positive powers behave just opposite. As a consequence, improvement of the fit on one contour can cause sometimes negative effects on the other. Additionally, the sinusoidal form of solution (27) requires certain necessary number of collocation points both on the internal and external contours to stop possible oscillations of the solution along these boundaries. Hence, if we introduce notion of a number of waves N_{v} (maximal n in sets (4) in Chapter 3, (27) in this chapter):

$$N_V = \frac{M_f - 2}{4} \quad \text{(because } M_f = 4N_V + 2,\text{)} \tag{31}$$

we obtain the basic condition of solution stability for non-simply connected regions:

$$K_p^{\text{int}} > N_V \quad \text{and} \quad K_p^{ext} > N_V \tag{32}$$

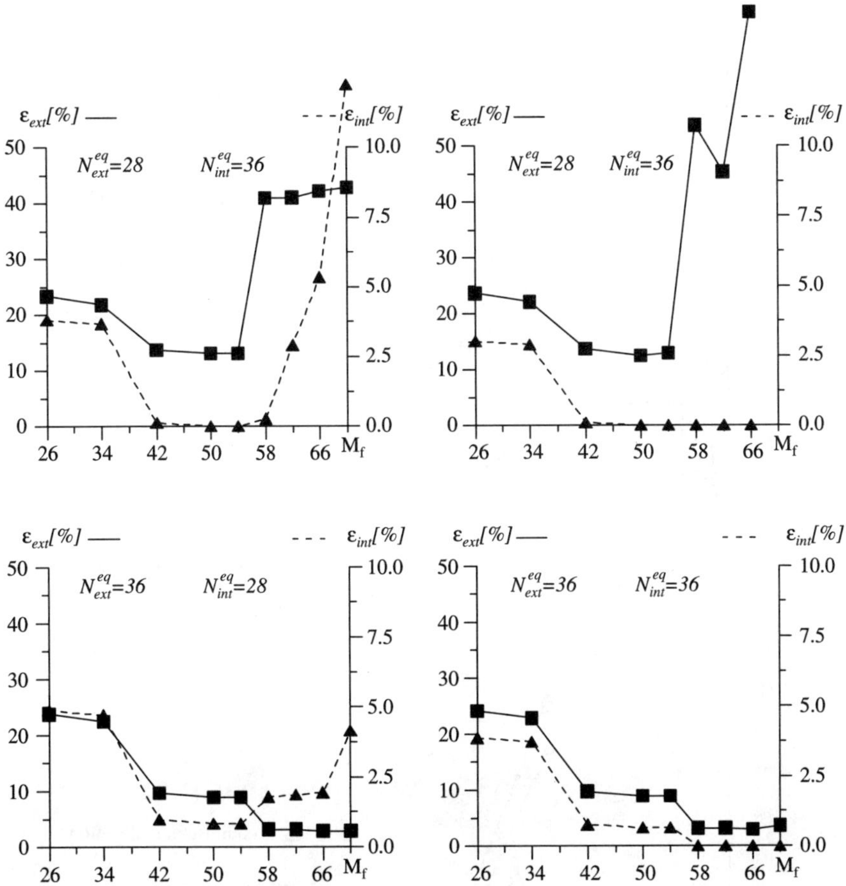

Figure 20: Different behaviour of ε_{ext} and ε_{int} (29) for $\rho / a = 0.5$, variable M_f and different relations of parameters N^{eq}_{ext} and N^{eq}_{int} (stability of solutions).

where $K_p^{\text{int}}, K_p^{ext}$ are numbers of collocation points on the internal and external boundaries, respectively. The discussed influence of the relation $K_p^{\text{int}}, K_p^{ext}$ and M_f is visible in Fig. 20; note the following relations of numbers of equations:

$$N^{eq}_{\text{int}} = 2K_p^{\text{int}}; \quad N^{eq}_{ext} = 2K_p^{ext} + 2. \tag{33}$$

In the above example, the equidistant boundary collocation was applied. Investigations of different variants of the position of control points were for doubly connected regions also investigated. In Fig. 21, we observe the behaviour of the error measure ε_{ext} versus the number of T-functions (along Γ_{int} the equidistant

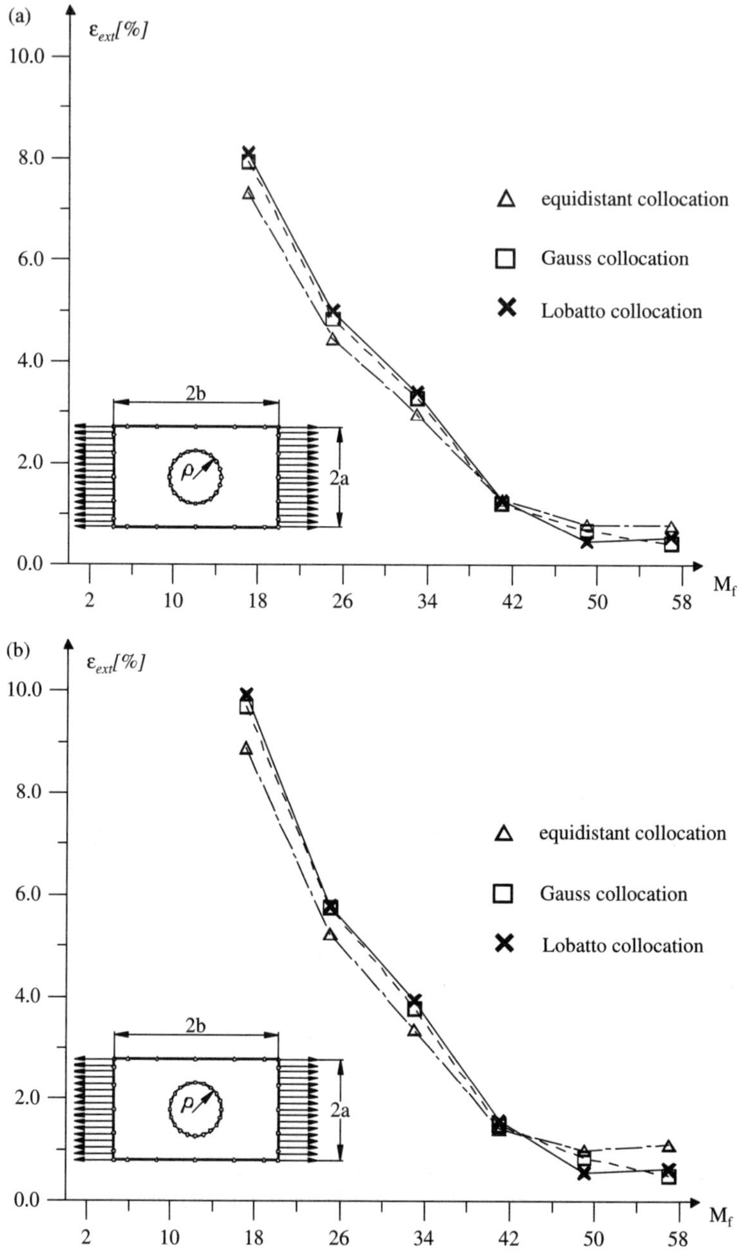

Figure 21: Error ε_{ext} in extended plate with circular hole $\rho / a = 0.25$, $N_{\text{int}}^{eq} = N_{ext}^{eq} = 36$; (a) $b / a = 2$; (b) $b / a = 0.5$.

collocation was applied and ε_{int} was much smaller). As we can see, the change of the collocation form to orthogonal (Gaussian or Lobatto-type) does not introduce significant improvement of the solution in this case.

References

[1] Aitken, A.C., On Bernoulli's numerical solution of algebraic equations. *Proceedings of Royal Society*, Edinburgh, **46**, pp. 283–305, 1926.

[2] Brezinski, C., Acceleration de la Convergence en Analyse Numerique. *Lecture Notes in Mathemetics*, **584**, pp. 313, 1977.

[3] Duda, P. & Zieliński, A.P., Stability and convergence of solutions in Trefftz-type algorithms. *Strojnicki Casopis (Journal of Mechanical Engineering)*, **48**, pp. 1–14, 1997.

[4] Jirousek, J. & Venkatesh, A., Hybrid-Trefftz plane elasticity elements with p-method capabilities. *International Journal for Numerical Methods in Engineering*, **35**, pp. 1443–1472, 1992.

[5] Karas´, M., *Generalized Trefftz Method Applied to Optimization of Elastic Struc*tures (in Polish), PhD Thesis (supervisor A.P. Zieliński), Cracow University of Technology, Cracow, 2001.

[6] Kołodziej, J.A., *Application of the Boundary Collocation Method in Applied Mechanics* (in Polish), Poznan University of Technology Press: Poznan, 2001.

[7] Li, Z.C., Lu, T.T., Hu, H.Y. & Cheng, A.H.D., *Trefftz and Collocation Methods*, WITpress: Southampton and Boston, 2008.

[8] Mathon, R. & Johnston, R.L., The approximate solution of elliptic boundary-value problems by fundamental solutions. *SIAM Journal of Numerical Analysis*, **14**, pp. 638–650, 1977.

[9] Piltner, R., Special finite elements with holes and internal cracks. *International Journal for Numerical Methods in Engineering*, **21**, pp. 1471–1485, 1985.

[10] Sanecki, H. & Zieliński, A.P., Special analytical finite elements and their application to optimal design. *Proceedings of Conference 'Numerical Methods in Continuum Mechanics'*, Slovakia, pp. 114–119, 1996.

[11] Shanks, D., Non-linear transformation of divergent and slowly convergent sequences. *Journal of Mathematics and Physics*, **34**, pp. 1–42, 1955.

[12] Szybiński, B. & Zieliński, A.P., Alternative T-complete systems of shape functions applied in analytical Trefftz finite elements. *Numerical Methods for Partial Differential Equations*, **11**, pp. 375–388, 1995.

[13] Timoshenko, S. & Goodier, J.N., *Theory of elasticity*, 3rd ed., McGraw-Hill Book Company: New York, 1982.

[14] Zieliński, A.P. & Herrera, I., Trefftz method: fitting boundary conditions. *International Journal for Numerical Methods in Engineering*, **24**, pp. 871–891, 1987.

[15] Zieliński, A.P. & Życzkowski, M., Certain modifications of Aitken's accelerator for slowly convergent sequences. *Periodica Polytechnica, Ser. Civil Engineering*, **37**, pp. 57–67, 1993.

[16] Zieliński, A.P. *T-complete Functions Metod and its Application to Elastic and Elasto-Plastic Problems*, (in Polish), Monograph 86, Cracow University of Technology Press: Cracow, 1989.

[17] Zieliński, A.P., Sanecki, H. & Karaś, M., Effectiveness of the Trefftz method in different engineering optimization procedures. *Computer Assisted Mechanics and Engineering Sciences*, **8**, pp. 479–493, 2001.

[18] Zieliński, A.P., Special Trefftz elements and improvement of their conditioning. *Communications in Numerical Methods in Engineering*, **13**, pp. 765–775, 1997.

CHAPTER 6

2D harmonic problems – the Laplace-type equations

1 Torsion of prismatic bars. Formulation of the boundary value problem

Torsion of homogenous, isotropic elastic bars is governed by a partial differential equation in the form [1]:

$$\frac{\partial^2 \psi}{\partial x^2} + \frac{\partial^2 \psi}{\partial y^2} = -2G\beta \tag{1}$$

where ψ is the Prandtl stress function, G is the Kirchhoff shear modulus and β is an angle of twist per unit length of the rod. Stresses τ_{xz} and τ_{yz} (others are equal to zero) can be obtained from:

$$\tau_{xz} = \frac{\partial \psi}{\partial y}, \quad \tau_{yz} = -\frac{\partial \psi}{\partial x}. \tag{2}$$

A solution of eqn (1) should fulfil the following boundary conditions:

- simply connected bar cross-section (Fig. 1a):

$$\psi = 0 \quad \text{on } \Gamma_0 \tag{3}$$

- multiply connected bar cross-section (Fig. 1b):

$$\psi = 0 \quad \text{on } \Gamma_0 \tag{4a}$$

$$\psi = c_k \text{ on } \Gamma_k, \quad k = 1, 2, \ldots, K \tag{4b}$$

where c_k are constants chosen to fulfil following relations:

$$\oint_{\Gamma_k} \frac{\partial \psi}{\partial n} ds = -2G\beta S_k \quad \text{for } k = 1, 2, \ldots, K \tag{4c}$$

Here S_k is the area closed by Γ_k and $\partial\psi/\partial n$ is the normal derivative outside to the contour Γ_k.

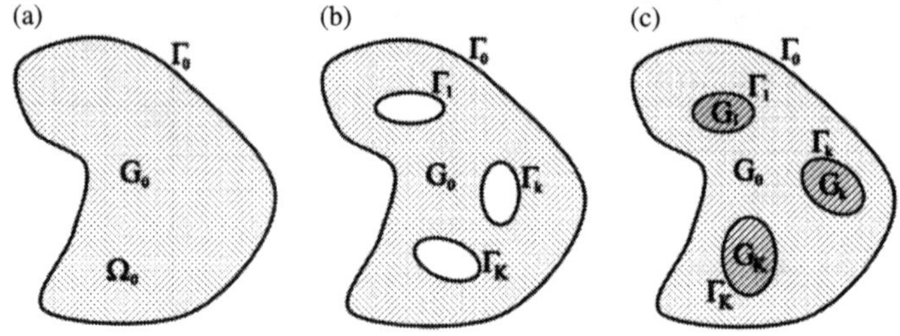

Figure 1: Cross-sections of different bars.

In case of a bar made of several different materials (Fig. 1c), the boundary value problem is more complex. Let us assume that the total bar consists of a matrix with shear modulus G_0 and internal K bars with shear moduli $G_1, G_2,..., G_K$ and contours $\Gamma_1, \Gamma_2,..., \Gamma_K$. Then in the region bounded outside by Γ_0 and inside by $\Gamma_1, \Gamma_2,..., \Gamma_K$ (matrix area), the stress function ψ_0 fulfils the following equation [1]:

$$\frac{\partial^2 \psi_0}{\partial x^2} + \frac{\partial^2 \psi_0}{\partial y^2} = -2G_0\beta. \tag{5}$$

Analogically, inside each contour Γ_k we have:

$$\frac{\partial^2 \psi_k}{\partial x^2} + \frac{\partial^2 \psi_k}{\partial y^2} = -2G_k\beta, \quad k = 1,\ 2,...,K. \tag{6}$$

The boundary conditions have then the form:

$$\psi = 0 \quad \text{on } \Gamma_0 \tag{7a}$$

$$\left.\begin{aligned} &\psi_0 = \psi_k \\ &\frac{1}{G_0}\frac{\partial \psi_0}{\partial n} = -\frac{1}{G_k}\frac{\partial \psi_k}{\partial n} \end{aligned}\right\} \text{on } \Gamma_k, \quad k = 1,\ 2,...,K. \tag{7b}$$

The last conditions mean specific relations of tangent stresses and conformity of displacements along boundaries of the matrix internal bars.

2 Application of the boundary collocation method to twisted bars – review

The method was widely applied to analysis of bar torsion. It was used for problems of simply and multi-connected bar cross-sections, as well as to composite bars. A review of related papers is given in Table 1.

In the most often cited applications of the method in the literature [4, 5, 16, 57, 60, 76, 77, 79, 82, 85, 90, 93, 95], the authors assume one approximate solution in

Table 1: Papers in which the boundary collocation method was applied to problems of twisted bars.

Torsion of prismatic bars	References
Simply connected cross-section	[4, 5, 15, 16, 18, 24, 25, 43–46, 50, 61, 79, 82, 85, 90, 93, 95, 112, 113, 122, 123]
Multiply connected cross-section	[5, 43, 51, 57, 76, 77, 82]
Composite	[7, 43, 58, 59, 63, 73, 76, 77, 124]

the whole bar cross-section in the form (7) in Chapter 2. However, sometimes the whole region is divided into several large finite elements with separate solutions in each of them [7, 17, 58, 59, 61, 113]. A different approach consists in repetition of solutions in certain sub-regions due to existence of symmetry axes of the considered region. In case of lack of any symmetry and without divisions into sub-regions, obtaining satisfactory results can sometimes be difficult because of problems with conditioning. For example, Poritsky & Danforth [85] considered torsion of a turbine blade with the help of several numerical methods. The boundary collocation method was an example of a not optimal approach in this case. Because of a large relation of maximal and minimal dimensions of the blade cross-section and the lack of any symmetry, the accurate fulfilment of the boundary conditions between the collocation points was difficult. On the contrary, Descloux & Tolley [17] considered torsion of a bar with a polygonal cross-section. In spite of lack of any symmetry axis, they obtained accurate results, however, after division of the considered region into elements.

Choice of the form of trial functions is connected with a type of system of coordinates. Almost all the authors applied the polar coordinate system (r,θ). The solutions of eqn (1) can then be written as

$$\varphi = 2G\beta\left[\varphi_0 + \sum_{m=1}^{M_f} a_m \varphi_m(r,\theta)\right] \tag{8}$$

where $2G\beta\varphi_0$ is a particular integral of eqn (1), $\varphi_m(r,\theta)$ are trial functions identically fulfilling two-dimensional (2D) Laplace's equation and a_m are constant coefficients calculated in a collocational way to fulfil by φ the boundary conditions. The trial functions applied by different authors to the solution (8) are given in Table 2.

3 Torsion of a bar made of different materials

Let us consider as an example a prismatic bar made of two materials (Fig. 2). In this case, the boundary value problem can be defined as

$$\nabla^T \frac{1}{G_1} \nabla\phi = -2\beta \quad (x,y) \in \Omega_1 \tag{9a}$$

Table 2: Trial functions applied by different authors for solving problems of torsion of prismatic bars.

φ_k	φ_0	References
$r^{4(k-1)}\cos\left[4(k-1)\theta\right]$	$-\dfrac{r^2}{2}$	[4, 15, 60]
$r^{(k-1)}\cos\left[(k-1)\theta\right]$, $r^{(k-1)}\sin\left[(k-1)\theta\right]$	$-\dfrac{r^2}{2}$	[85]
$r^{6(k-1)}\cos\left[6(k-1)\theta\right]$	$-\dfrac{r^2}{2}$	[15]
$r^{3(k-1)}\cos\left[3(k-1)\theta\right]$	$-\dfrac{r^2}{2}$	[15, 60]
$r^{L(k-1)}\cos\left[N(k-1)\theta\right]$ $N = 3,\ 4, 5, 6,...$	$-\dfrac{r^2}{2}$	[79]
$\cos\dfrac{k\pi x}{2a}\cosh\dfrac{k\pi y}{2a}$, where $2a$ is a side of a square	$-\left(x^2 - a^2\right)$	[95]
$r^{2(2k-1)}\sin\left[2(2k-1)\theta\right]$	$-\dfrac{r^2}{4\pi}\left[(\pi-4\theta)\cos 2\theta - \pi\right] + \dfrac{r^2}{\pi}\ln r \sin 2\theta$	[113]
$r^{(2k-1)\pi/2a}\cos\dfrac{(2k-1)\pi\theta}{2a}$ $a \neq \dfrac{\pi}{4}$ $r^{2(2k-1)}\cos\left[2(2k-1)\theta\right]$ for $a = \dfrac{\pi}{4}$, where a is a half of smaller angle of a rhombus	$-\dfrac{r^2}{2}\left(1-\dfrac{\cos 2\theta}{\cos 2a}\right)$ for $a \neq \dfrac{\pi}{4}$ $-\dfrac{r^2}{2}\left[1-\dfrac{4}{\pi}\left(\theta\sin 2\theta - \cos 2\theta \ln\dfrac{r}{b}\right)\right]$ for $a = \dfrac{\pi}{4}$	[58, 90]
$\ln\left[(x-x_k)^2 + (y-y_k)^2\right]$	$\dfrac{1}{2}\left(x^2+y^2\right)$	[82]

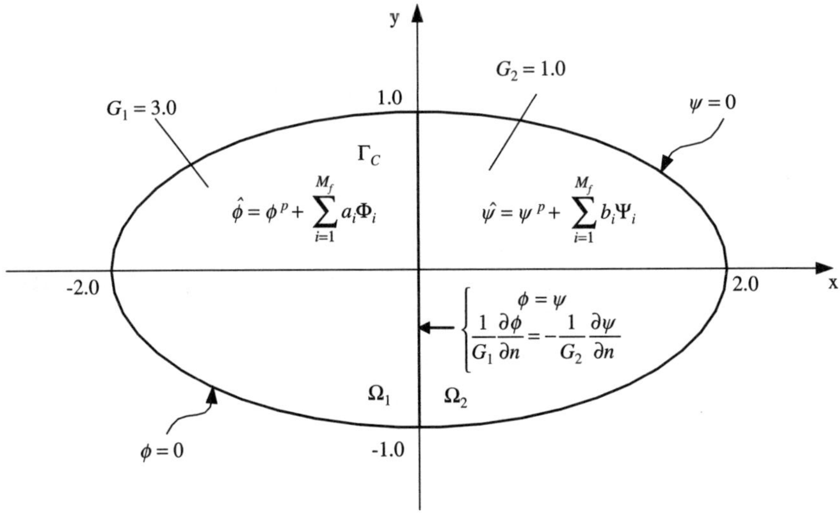

Figure 2: Cross-section of an elliptic bar made of two different materials – example.

$$\nabla^T \frac{1}{G_2} \nabla \psi = -2\beta \quad (x, y) \in \Omega_2 \tag{9b}$$

$$\phi = 0 \quad (x, y) \in \Gamma_1 \tag{9c}$$

$$\psi = 0 \quad (x, y) \in \Gamma_2 \tag{9d}$$

$$\left.\begin{aligned} &\phi = \psi \\ &\frac{1}{G_1}\frac{\partial \phi}{\partial n} = -\frac{1}{G_2}\frac{\partial \psi}{\partial n} \end{aligned}\right\} \quad (x, y) \in \Gamma_c \tag{9e}$$

where on Γ_1 and Γ_2 only the conditions set on the unknown functions are taken into account.

To the solution, we choose boundary integral fit, which is a specific formulation of the boundary collocation method (see Section 2 in Chapter 2). The weighted residual form of (9) can be written similarly to (3) in Chapter 4 [127, 128]:

$$\int_{\Gamma_1} W_v^{(1)} \hat{\phi}\, d\Gamma_1 + \int_{\Gamma_c} \left[W_v^{(2)} \left(\hat{\phi} - \hat{\psi}\right) + W_v^{(3)} \left(\frac{1}{G_1}\hat{\phi}' + \frac{1}{G_2}\hat{\psi}' \right) \right] d\Gamma_c = 0 \tag{10a}$$

$$\int_{\Gamma_c} \left[W_v^{(4)} \left(\frac{1}{G_2}\hat{\psi}' + \frac{1}{G_1}\hat{\phi}' \right) + W_v^{(5)} \left(\hat{\psi} - \hat{\phi}\right) \right] d\Gamma_c + \int_{\Gamma_2} W_v^{(6)} \hat{\psi}\, d\Gamma_2 = 0 \tag{10b}$$

where

$$\hat{\phi} = \sum_{\mu} a_{\mu}^{1} \Phi_{\mu} + \phi^{p} \tag{10c}$$

$$\hat{\psi} = \sum_{\mu} a_{\mu}^{2} \Psi_{\mu} + \psi^{p} \tag{10d}$$

$$\hat{\phi}' \equiv \frac{\partial \hat{\phi}}{\partial n} = \sum_{\mu} a_{\mu}^{1} \Phi'_{\mu} + \phi'^{p} \tag{10e}$$

$$\hat{\psi}' \equiv \frac{\partial \hat{\psi}}{\partial n} = \sum_{\mu} a_{\mu}^{2} \Psi'_{\mu} + \psi'^{p} \tag{10f}$$

ϕ_p, ψ_p – particular solutions.

For the symmetry of the main matrix **K**, the following functions are applied (compare (3a) in Chapter 4):

$$\begin{aligned} W_{\nu}^{(1)} &= W_{\nu}^{(2)} = \Phi_{\nu};\; W_{\nu}^{(3)} = \frac{1}{G_1}\Phi'_{\nu} \\ W_{\nu}^{(5)} &= W_{\nu}^{(6)} = \Psi_{\nu};\; W_{\nu}^{(4)} = \frac{1}{G_2}\Psi'_{\nu}. \end{aligned} \tag{11}$$

In this case, the eqn (10) take the form:

$$\mathbf{K}\,\mathbf{c} = \mathbf{f} \tag{12a}$$

where

$$K_{\nu\mu} = \left\| \begin{array}{c|c} \int\limits_{\Gamma_{1+c}} \Phi_{\nu}\Phi_{\mu}\, d\Gamma_{1+c} + \frac{1}{G_1^2}\int\limits_{\Gamma_c} \Phi'_{\nu}\Phi'_{\mu}\, d\Gamma_c & -\int\limits_{\Gamma_c}\left(\Phi_{\nu}\Psi_{\mu} - \frac{1}{G_1G_2}\Phi'_{\nu}\Psi'_{\mu}\right) d\Gamma_c \\ \hdashline -\int\limits_{\Gamma_c}\left(\Psi_{\nu}\Phi_{\mu} - \frac{1}{G_1G_2}\Psi'_{\nu}\Phi'_{\mu}\right) d\Gamma_c & \int\limits_{\Gamma_{1+c}} \Psi_{\nu}\Psi_{\mu}\, d\Gamma_{1+c} + \frac{1}{G_2^2}\int\limits_{\Gamma_c} \Psi'_{\nu}\Psi'_{\mu}\, d\Gamma_c \end{array} \right\| \tag{12b}$$

$$c_{\mu} = \begin{Bmatrix} a_{\mu}^{1} \\ a_{\mu}^{2} \end{Bmatrix} \tag{12c}$$

$$
f_v = \begin{Bmatrix} -\int\limits_{\Gamma_1} \Phi_v \phi^p d\Gamma_1 + \int\limits_{\Gamma_c} \left[\Phi_v \left(\psi^p - \phi^p\right) - \frac{1}{G_1} \Phi_v' \left(\frac{1}{G_1} \phi'^p + \frac{1}{G_2} \psi'^p \right) \right] d\Gamma_c \\ -\int\limits_{\Gamma_2} \Psi_v \psi^p d\Gamma_2 + \int\limits_{\Gamma_c} \left[\Psi_v \left(\phi^p - \psi^p\right) - \frac{1}{G_2} \Psi_v' \left(\frac{1}{G_2} \psi'^p + \frac{1}{G_1} \phi'^p \right) \right] d\Gamma_c \end{Bmatrix} \tag{12d}
$$

The two half-elliptic areas ($a = 2.0$, $b = 1.0$) were chosen as a numerical example; however, extension to more areas was also considered [128]. For simplicity, we assumed $\beta = 1$ and $G_1 = 3.0$, $G_2 = 1.0$. Figures 3 and 4 show behaviour of the solutions for this case. This is also visible in Table 3, which presents the results of calculations in the connection zone of elements. The integral convergence of solutions is here clearly demonstrated. In Table 4, the convergence of solutions inside the given area can be observed. The poor accuracy of the first results (for $M_f = 3$) is due to the stresses τ_{xz} and τ_{yz} being both approximated only by the linear functions in this case.

4 The Poiseuille flow

4.1 Formulation of the problem

The Poiseuille flow of liquid can be defined as laminar flow in which the current lines are parallel and straight. The liquid is viscous and incompressible. A laminar flow in a straight pipe with circular cross-section is a typical problem of this kind. The equation governing this phenomenon has the form:

$$
\frac{\partial^2 \upsilon(x,y)}{\partial x^2} + \frac{\partial^2 \upsilon(x,y)}{\partial y^2} = \frac{1}{\mu} \frac{dp}{dz} \tag{13}
$$

where $\upsilon(x,y)$ is a velocity along the channel axis, μ is the dynamic viscosity coefficient and dp/dz is a constant pressure gradient along the channel axis.

The boundary conditions, which should be fulfilled by the solution of eqn (13), result from the condition of lack of liquid slip on the solid body wall, i.e.

$$
\upsilon = 0 \quad \text{on } \Gamma_1 \tag{14a}
$$

or from symmetry of the problem

$$
\frac{\partial \upsilon}{\partial n} = 0 \quad \text{on } \Gamma_2 \tag{14b}
$$

where Γ_1 is a channel contour and Γ_2 is a symmetry axis of the flow.

After introduction of non-dimensional variables:

$$
X = \frac{x}{a}, \quad Y = \frac{y}{a}, \quad V = \frac{\upsilon}{-\dfrac{a^2}{\mu}\dfrac{dp}{dz}}, \tag{15}
$$

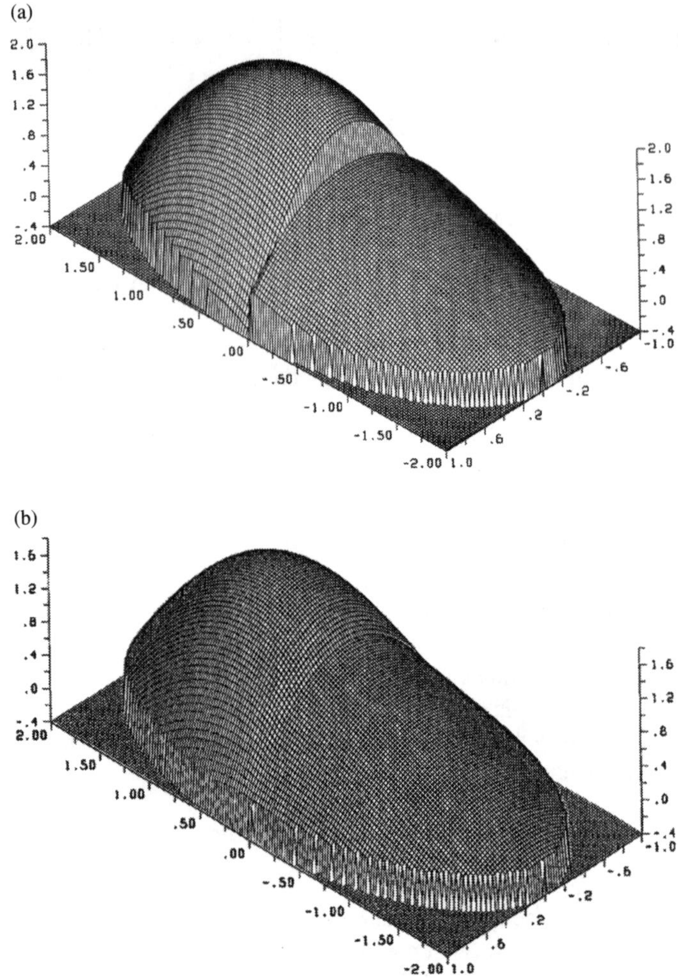

Figure 3: Direct coupling of the two half-elliptic regions – stress function, $G_1 = 3.0$, $G_2 = 1.0$. (a) $M_f = 3$; (b) $M_f = 9$.

eqn (13) can be reduced to the form

$$\frac{\partial^2 V}{\partial X^2} + \frac{\partial^2 V}{\partial Y^2} = -1 \tag{16}$$

and after substituting

$$V = -\frac{X^2 + Y^2}{4} + \varphi(X,Y) \tag{17}$$

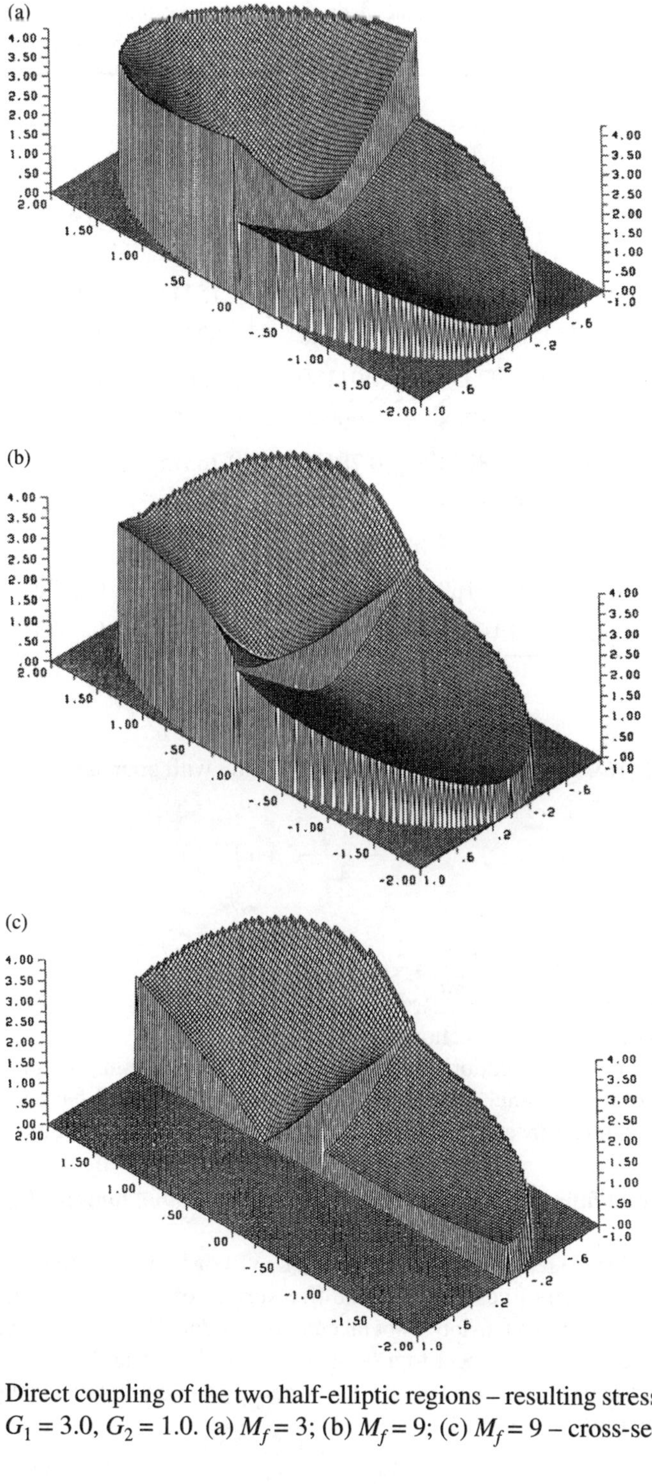

Figure 4: Direct coupling of the two half-elliptic regions – resulting stresses (5.7), $G_1 = 3.0$, $G_2 = 1.0$. (a) $M_f = 3$; (b) $M_f = 9$; (c) $M_f = 9$ – cross-section.

Table 3: Values of functions $\hat{\varphi}$ and $\hat{\psi}$ on the common boundary Γ_c of areas Ω_1 and Ω_2 ($G_1 = 3.0$, $G_2 = 1.0$) – direct coupling of the two half-elliptic elements.

	$M_f = 3$		$M_f = 6$		$M_f = 9$	
y	$\hat{\varphi}$	$\hat{\psi}$	$\hat{\varphi}$	$\hat{\psi}$	$\hat{\varphi}$	$\hat{\psi}$
0.0	1.49541	1.12667	1.18827	1.20396	1.20165	1.19944
0.1	1.47567	1.11711	1.17745	1.19158	1.18928	1.18756
0.2	1.41646	1.08840	1.14481	1.15447	1.15231	1.15188
0.3	1.31777	1.04056	1.08988	1.09281	1.09113	1.09227
0.4	1.17961	0.97359	1.01180	1.00686	1.00626	1.00855
0.5	1.00197	0.88748	0.90943	0.89701	0.89822	0.90056
0.6	0.78486	0.78223	0.78128	0.76375	0.76726	0.76820
0.7	0.52827	0.65785	0.62553	0.60766	0.61308	0.61158
0.8	0.23221	0.51433	0.44001	0.42947	0.43451	0.43108
0.9	–0.10333	0.35168	0.22225	0.22998	0.22909	0.22751
1.0	–0.47835	0.16989	–0.03056	0.01011	–0.00734	0.00224

instead of (16), we obtain the 2D Laplace equation with boundary conditions:

$$\varphi = \frac{X^2 + Y^2}{4} \quad \text{on } \Gamma_1 \tag{18}$$

and

$$\frac{\partial \varphi}{\partial n} = \frac{1}{2}\left(Xn_X + Yn_Y\right) \quad \text{on } \Gamma_2 \tag{19}$$

where on n_X and n_Y are directional cosines.

The boundary collocation method was applied to modelling of the Poiseuille flow in a closed channel, in a channel containing longitudinal bars, in a channel with free liquid surface and in a longitudinal flow along a group of bars (Table 5).

4.2 Optimal filling of a cylindrical pipe in gravitational laminar liquid flow

Let us consider a steady gravitational flow of a viscous liquid in a cylindrical pipe, which is only partly filled (Fig. 5). The free surface of the liquid is plane, which means that the surface tension is not taken into account. The pipe diameter is equal $2a$ and its angle with respect to a horizontal level is γ. The distance of the free surface to the pipe axis is noted as $a \times s$, where s is a non-dimensional parameter of the pipe filling ($-1 \leq s \leq 1$).

Table 4: Convergence of the stress function $\hat{\psi}$ and the resulting stresses τ ((7) in Chapter 5) inside the area Ω_2 – direct coupling (A_A – Aitken extrapolation (10) in Chapter 5).

Coordinates		Stress function $\hat{\psi}$				Resulting stresses τ			
x	y	$M_f = 3$	$M_f = 6$	$M_f = 9$	A_A	$M_f = 3$	$M_f = 6$	$M_f = 9$	A_A
0.00	0.00	1.12667	1.20396	1.19944	1.19969	0.42726	0.62696	0.60539	0.60749
0.00	0.45	0.93292	0.95490	0.95760	0.95798	0.96128	1.21389	1.20291	1.20336
0.30	0.00	0.99461	1.03278	1.03165	1.03168	0.45318	0.52852	0.52364	0.52393
0.60	0.45	0.66102	0.67937	0.67934	0.67934	0.98542	0.99096	0.99081	0.99081
0.90	0.00	0.70714	0.72594	0.72570	0.72571	0.50504	0.52811	0.52773	0.52774
1.20	0.35	0.43454	0.45153	0.45140	0.45141	0.85469	0.84070	0.84008	0.84006
1.50	0.00	0.38856	0.37196	0.37198	0.37198	0.55689	0.66566	0.66562	0.66562

Table 5: Papers in which the boundary collocation method was applied to the Poiseuille flow.

Poiseuille flow	References
Closed channel	[10–13, 56, 78, 92, 94, 98–100, 102, 115, 120, 121, 126]
Closed channel with longitudinal rods	[19, 27, 57, 75, 89]
Flow through bundle of parallel rods	[2, 14, 42, 54, 91, 101]
Channel with free surface	[36, 37, 55, 118]
Flow with heat exchange in closed channel	[10–13, 78, 92, 99, 100]
Flow with heat exchange through bundle of parallel rods	[2, 14, 103]

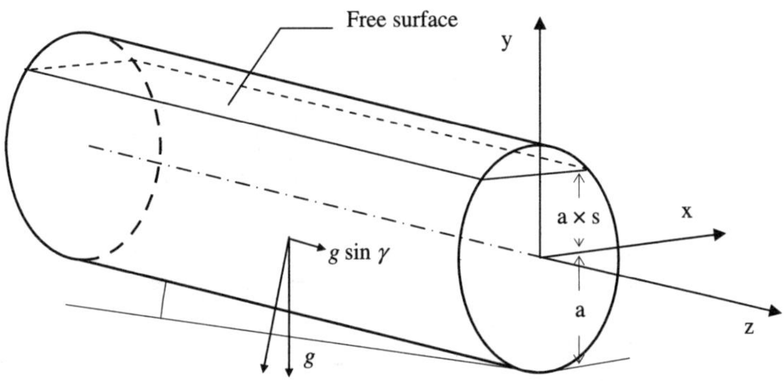

Figure 5: Gravitational flow in partly filled cylindrical pipe.

We assume that the pipe is infinitely long and interaction of gas above the free liquid surface can be neglected. Then the problem is: for what value of the filling parameter s (for $\gamma = const$) maximal flow capacity is reached? Intuitively, it can be supposed that the maximal capacity occurs for $s = 1$, but it is not true. Along the channel walls the velocity of viscous liquid is equal to zero, which slows down the flow. This effect is partly eliminated when the pipe is not full (Fig. 6). Thus, it appears that for a certain values of s, the increase of flow capacity caused by the free liquid surface is larger than its decrease resulted from smaller cross-section of the flowing liquid.

This problem was considered by Greenwell & Wang [26]. They reduced the governing equation to the Laplace form and the investigated region of the channel cross-section they defined in the form of two crossing circles. Then, this region

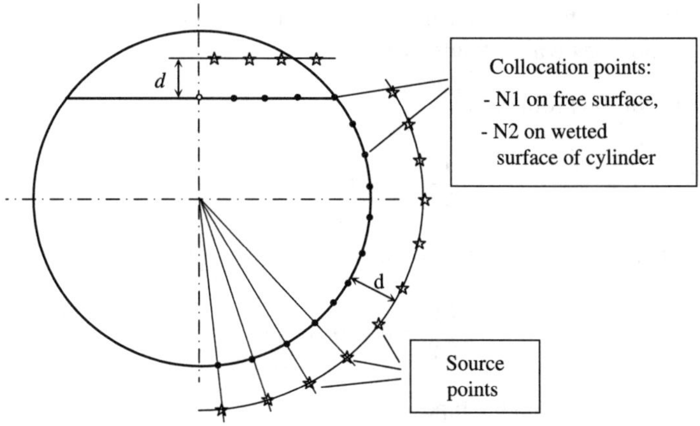

Figure 6: Positions of source and collocation points.

was transformed by conformal mapping to a unit circle and the problem was solved by the integral Poisson formula. The calculations resulted in evaluation of the filling parameter $s = 0.724$ for the maximal flow capacity, which was 23.3% larger than the capacity of the full pipe channel.

The above problem was solved by the first of the monograph authors with the help of the boundary collocation method and fundamental solutions used as trial functions (the Kupradze functions variant). It occurred that this version of the method led to a simple form of the approximate solution, which also appeared to be very accurate.

In case of gravitational flow of incompressible, viscous liquid, the Navier–Stokes set of equations in the assumed system of coordinates (Fig. 5) can be reduced to the form:

$$0 = -\frac{1}{\rho}\frac{dp(y)}{dy} + g\cos\gamma \tag{20a}$$

$$0 = \nu\left(\frac{\partial^2}{\partial x^2} + \frac{\partial^2}{\partial y^2}\right)\upsilon(x,y) + g\sin\gamma \tag{20b}$$

where ν is the kinematic viscosity coefficient, ρ is the liquid density (constant) and g is the acceleration of gravity. In eqn (20a), we can observe that the pressure gradient is different from zero in direction perpendicular to the free surface and does not have any influence on the flow.

After introduction of non-dimensional variables defined by the formulas:

$$X = \frac{x}{a}, \quad Y = \frac{y}{a}, \quad V = \frac{\upsilon}{a^2 g \sin\gamma}, \tag{21}$$

eqn (20b) takes the form (16). This equation should be solved with the following boundary conditions which results from:

- symmetry of the problem:

$$\frac{\partial V}{\partial X} = 0 \quad \text{for } X = 0 \tag{22}$$

- lack of shear stresses on the free surface:

$$\frac{\partial V}{\partial Y} = 0 \quad \text{for } Y = s \tag{23}$$

- lack of the flow slip on the pipe wall:

$$V = 0 \quad \text{for} \begin{cases} X^2 + Y^2 = 1 \\ Y \le s \end{cases} \tag{24}$$

In case of the boundary collocation method, the approximate solution takes the following form:

$$V = -\frac{X^2 + Y^2}{4} + \sum_{j=1}^{M} a_j \ln\left(\left((X - X_j)^2 + (Y - Y_j)^2\right)\right)\left((X + X_j)^2 + (Y - Y_j)^2\right)\Big) \tag{25}$$

where (X_j, Y_j) are coordinates of the source points distributed outside the investigated region. The above form of solution identically fulfils eqn (16) and the symmetry boundary condition (22). The unknown coefficients a_j are determined from the boundary conditions (23, 24). The condition (23) requires calculation of the derivative:

$$\frac{\partial V}{\partial Y} = -\frac{Y}{2} + \sum_{j=1}^{M} a_j \left(\frac{2(Y - Y_j)}{(X - X_j)^2 + (Y - Y_j)^2} + \frac{2(Y - Y_j)}{(X + X_j)^2 + (Y - Y_j)^2} \right). \tag{26}$$

In our solution, the collocation points are distributed along the boundary Γ in an equidistant way (Fig. 6). Their number N can be equal or larger than the number of the Kupradze fundamental functions. The source points are situated outside the investigated region in the equal distance d from its boundary (Fig. 6). We assume that the mutual distances between N_1 collocation points on the free surface are approximately equal to the distances between N_2 points on the pipe wall. The respective system of linear collocation equations on the unknown coefficients $a_1, a_2, \ldots, a_M$ can be written as:

$$\mathbf{Aa} = \mathbf{b} \tag{27a}$$

where

$$A_{ij} = \frac{2(Y_{Ci} - Y_j)}{(X_{Ci} - X_j)^2 + (Y_{Ci} - Y_j)^2} + \frac{2(Y_{Ci} - Y_j)}{(X_{Ci} + X_j)^2 + (Y_{Ci} - Y_j)^2} \quad i = 1,...N_1 \tag{27b}$$

$$A_{ij} = \ln\left[\left((X_{Ci} - X_j)^2 + (Y_{Ci} - Y_j)^2\right)\left((X_{Ci} + X_j)^2 + (Y_{Ci} - Y_j)^2\right)\right] \quad i = N_1 + 1,...N \tag{27c}$$

$$b_i = \frac{1}{2} Y_{Ci}, \quad i = 1,...N_1 \tag{27d}$$

$$b_i = \frac{1}{4}(X_{Ci}^2 + Y_{Ci}^2), \quad i = N_1 + 1,...N. \tag{27e}$$

$N = N_1 + N_2$ is the total number of collocation points and the values X_{ci}, Y_{ci}, $i = 1,..., N$ are coordinates of the collocation points. In our example, $N = M$ collocation points were applied, thus a system of N linear equations (27) was solved.

The flow capacity can be calculated from the formula:

$$Q = \iint_{\Omega} \upsilon \, d\Omega \tag{28}$$

which is convenient to present in the non-dimensional form:

$$\bar{Q} = \frac{Q}{Q_p} \tag{29}$$

where

$$Q_p = \frac{\pi a^4 g \sin\gamma}{8\nu} \tag{30}$$

is the flow capacity in the full pipe. **The boundary collocation method gives the solution in any internal point inside Ω in the simple form of series (25) which is a considerable advantage in comparison to the boundary element approach.** Therefore, integral (28) could have been easy calculated in a very exact way.

In the boundary collocation method, we try to fulfil the boundary conditions as well as possible, therefore the boundary error ε_b (21) (in Chapter 5) can be considered as a measure of the solution error. In the Kupradze formulation, the results depend not only on the number of T-complete functions and collocation points N but also on distance d of the source points from the boundary. However, the solution behaves in a very stable way. Above a certain limit number ($M = N = 16$), the solution error was negligible. Also influence of the distance d was rather small for the large range of its value (Table 6). Dependence of the flow capacity on the non-dimensional distance of the free surface is presented in Table 7. As it was already

Table 6: Absolute value of maximal relative boundary error of V versus filling parameter s and distance d of source points from the contour.

	s			
d	0.0	0.2	0.6	0.9
0.05	0.0058	0.0112	0.0846	0.0628
0.10	0.0183	0.0289	0.0724	0.0613
0.15	0.0200	0.0152	0.0682	0.0612
0.20	0.0173	0.0410	0.0675	0.0611
0.25	0.0277	0.0389	0.0673	0.0611
0.30	0.0250	0.0384	0.0672	0.0613

Table 7: Non-dimensional flow capacity $\tilde{Q}$ versus filling parameter s and distance d of source points.

s	$d = 0.10$	$d = 0.15$
0.0	0.4974	0.5000
0.1	0.6282	0.6321
0.2	0.7650	0.7679
0.3	0.8976	0.9142
0.4	0.9697	1.0432
0.5	1.1671	1.1548
0.6	1.2403	1.2370
0.7	1.2775	1.2765
0.8	1.2597	1.2594
0.9	1.1716	1.1715

mentioned, Greenwell & Wang investigating the same problem [26] obtained $s = 0.724$ as a value for which the relative flow capacity reached its maximal value 1.273. The accurate results for the boundary collocation method were in this case $Q = 1.2779$ for $s = 0.7258$.

5 Singular membrane problem

A singular problem of a membrane extended on a triangular frame (Fig. 7) is the next solution presented in this chapter. The problem is well known and its solution

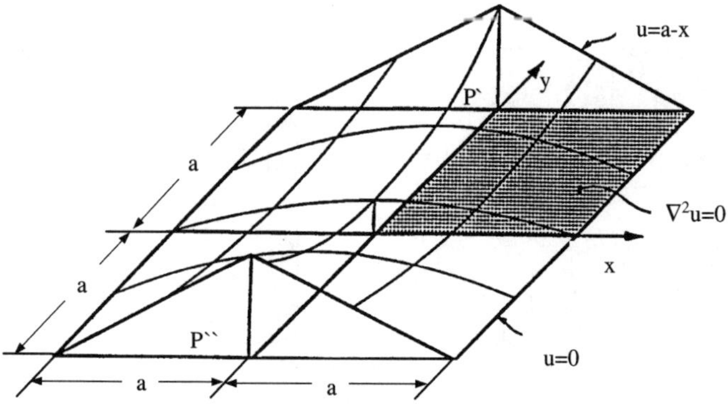

Figure 7: Singular membrane problem.

u_{EX} is given in the form of an infinite series [72]. It was solved by the hybrid-Trefftz approach (see Chapter 4 and [35]), which for a single element covering all the investigated area (dark in Fig. 7) is very similar to the integral fit in the boundary collocation approach. This form of solution is presented in the whole current chapter. For the applied data $a = 1.0$, $M = 32$ of the polynomial Herrera T-function (HT-H), the errors:

$$\varepsilon_u = 10^3 \Delta u \tag{31}$$

$$\varepsilon_e = 10^2 \left[\left(\frac{\partial \Delta u}{\partial x} \right)^2 + \left(\frac{\partial \Delta u}{\partial y} \right)^2 \right]^{1/2} \tag{32}$$

where $\Delta u = u_{COLL} - u_{EX}$ are presented in Fig. 8. **The characteristic feature of the boundary collocation method is here distinctly visible: the largest errors appear near the boundary; in the central part of investigated area a phenomenon of super-convergence of the results can be noticed**. The same example has also been investigated for the fundamental Kupradze logarithmic functions (HT-K). A contour Γ_h homothetic (equidistant) in relation to the contour Γ_e of the investigated area (Fig. 9) was chosen as position of the sources.

The test results presented hereafter show the most characteristic tendencies observed by the authors during the numerical investigations of the HT-H and HT-K approach. The analytical solutions of the example are known [72] and serve as a reference to the numerical example [107]:

$$u(x, y) = \sum_{n=0}^{N_S} \frac{8a \cos\left[(2n+1)\pi x/(2a)\right] \cosh\left[(2n+1)\pi y/(2a)\right]}{(2n+1)^2 \pi^2 \cosh\left[(2n+1)\pi/2\right]} \tag{33}$$

$N_S \to \infty$ (in the example $N_S = 200$).

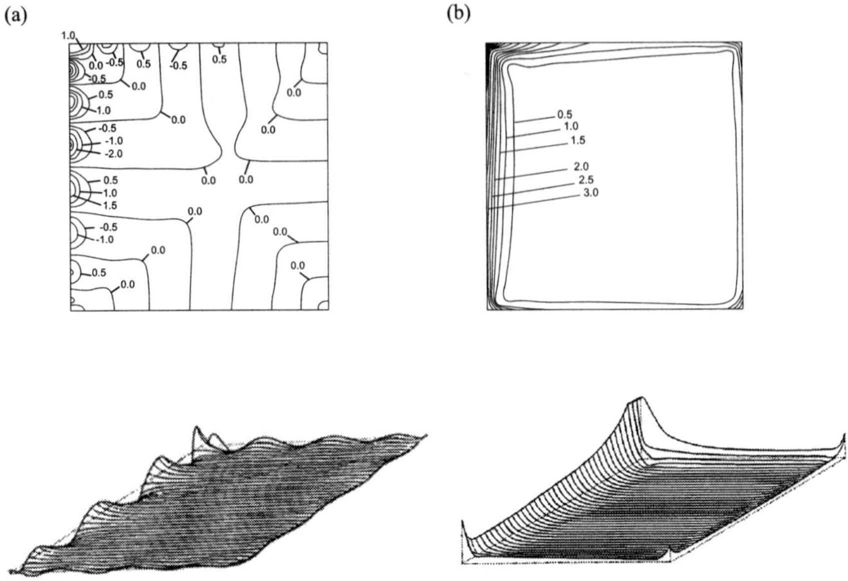

Figure 8: Errors of approximate solution of problem defined in Fig. 7. (a) errors ε_u (31), (b) errors ε_e (32).

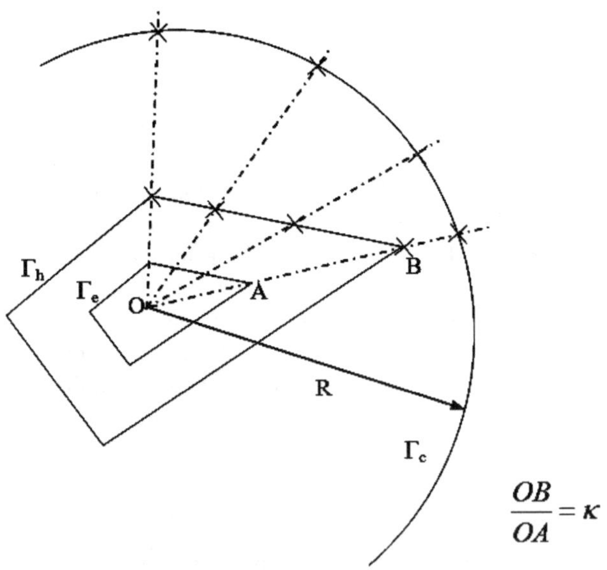

Figure 9: Position of singularities of the Kupradze functions in the HT-K solution.

The norm (see (46) in Chapter 2)

$$\varepsilon_{\text{cond}} = \max_i \left|(\mathbf{H}^{-1}\mathbf{H})_{ii} - 1\right| \tag{34}$$

quantifies the inversion error of the solution matrix **H** to $\mathbf{H}^{-1}$ (compare eqn (15a) in Chapter 4) and is a measure of conditioning of the problem.

The average relative solution error $\overline{\delta}_u$ has here been defined as

$$\overline{\delta}_u = \frac{\sum_{i,j=1}^{K} \left|u_{i,j}^{K,H} - u_{i,j}^{EX}\right|}{K^2 u_c^{EX}} = \frac{1}{K^2}\sum_{i,j=1}^{K} \left|\delta_u\right|_{ij} \tag{35}$$

$u_c^{EX} = a$ exact solution in point P' (Fig. 7), with the control points chosen regularly inside the investigated area (K rows × K points in one row; in all the examples $K = 11$). The generalized internal energy U has approximated the exact value:

$$U^{EX} = \frac{1}{2}\int_\Omega \left[\left(\frac{\partial u^{EX}}{\partial x}\right)^2 + \left(\frac{\partial u^{EX}}{\partial y}\right)^2\right] d\Omega = 0.25a^2 \tag{36}$$

in this case.

Assuming $M = 16$ of the Kupradze logarithmic functions, the calculations have been carried out for various values of the parameter κ of homothety (Fig. 10a). For the HT-K solution between $\kappa \cong 1.4$ and $\kappa \cong 2.5$, the results are stable, slightly better than for HT-H formulation and much better than for a conventional single hierarchic p-element. Increase of the number M of shape functions improves the results, but for the HT-K approach it diminishes the range of the acceptable parameter κ (Fig. 10b).

The inversion error measure ε_k can serve as an indicator of reliability of HT-K solutions. However, it should be noted that, in all the examples carried out by the authors, the range of acceptable final results was broader than the range of stable ε_k.

Figure 11 presents distribution of the relative solution error:

$$\delta_u = \frac{u^{K,H} - u^{EX}}{u_c^{EX}} \tag{37}$$

and the relative measure of the generalized energy density error:

$$\gamma_U = \left\{\frac{\left[\frac{\partial}{\partial x}\left(u^{K,H} - u^{EX}\right)\right]^2 + \left[\frac{\partial}{\partial y}\left(u^{K,H} - u^{EX}\right)\right]^2}{2\rho^{EX}}\right\}^{1/2} \qquad \rho^{EX} = \frac{U^{EX}}{a^2} = 0.25 \tag{38}$$

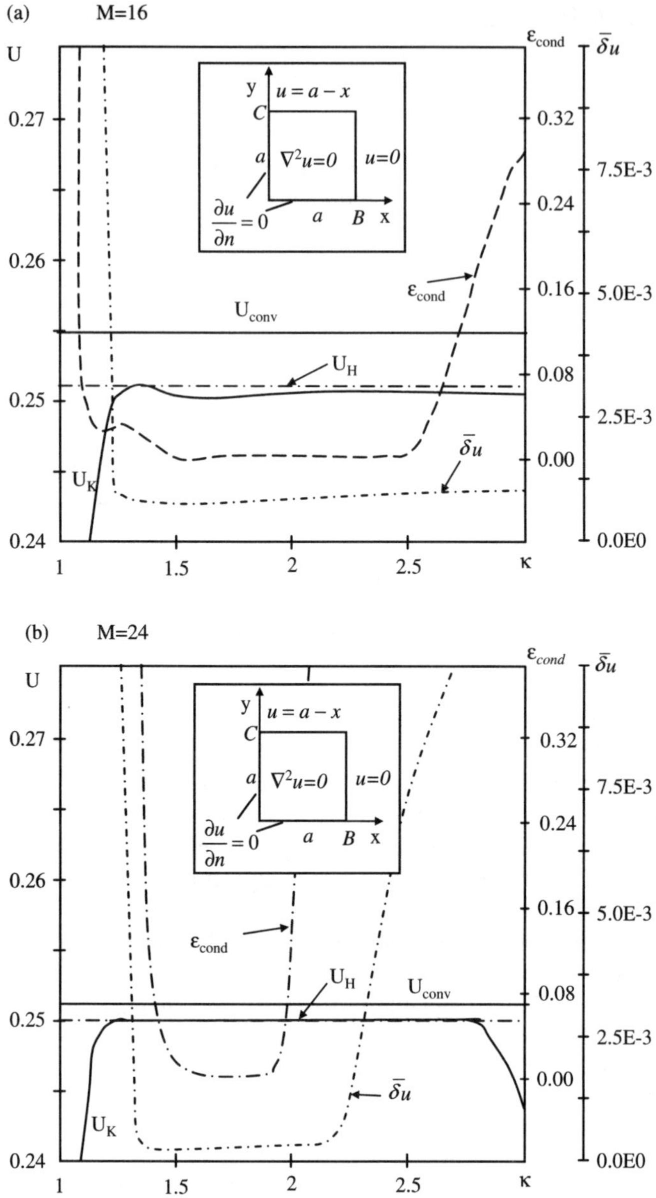

Figure 10: Range of acceptable distance from boundary Γ_e of singularities of the Kupradze functions in HT-K solutions; homothetic contour Γ_h. Generalized energy calculated with help of a single conventional p-element (U_{conv}), HT-H approach (U_H) and HT-K formulation (U_k). $a = 1$, $U^{EX} = 0.25$, $\varepsilon_{\text{cond}}$ (34), $\bar{\delta}_u$ (35). (a) $M = 16$; (b) $M = 24$.

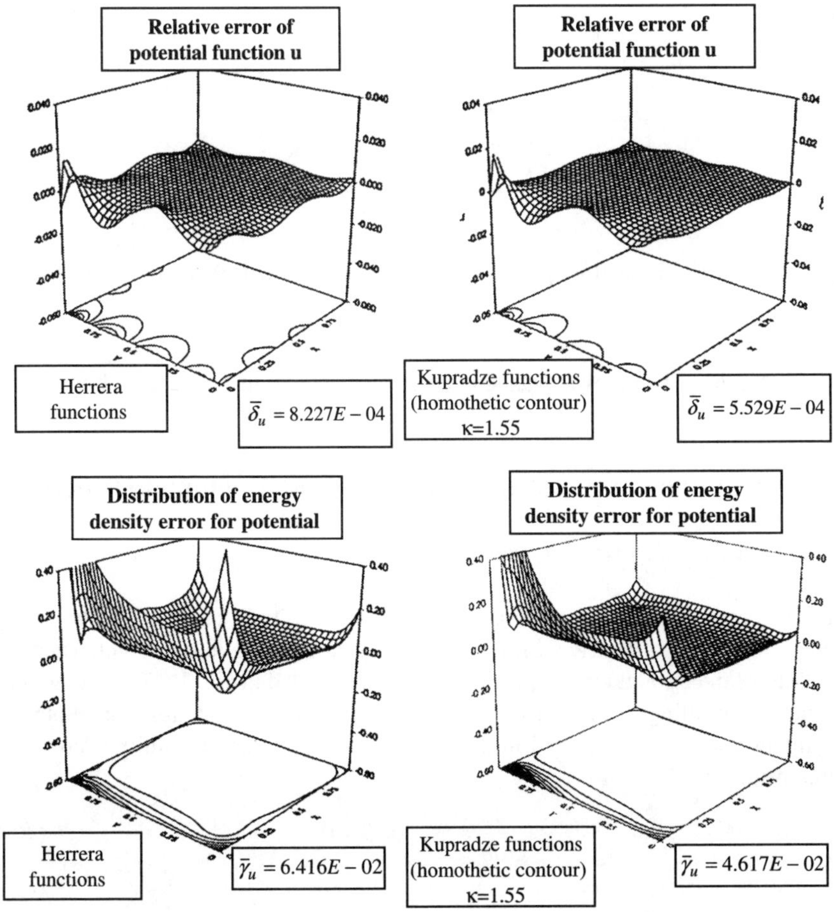

Figure 11: Relative solution error (37) and relative measure of generalized energy density error (38) for potential problem defined in Fig. 7.

with its average value

$$\overline{\gamma}_U = \frac{1}{K^2} \sum_{i,j=1}^{K} \gamma_{Uij}. \tag{39}$$

This figure also demonstrates the characteristic feature of all the Trefftz-type solutions: their super-convergence in the central part of the investigated area. The largest errors are obviously located on its boundary Γ_e.

The HT-K solution was in the above example considerably better than HT-H, but it must be remembered that the calculations were carried out with the optimal parameter κ. Situating the singularities on a circle with the radius R resulted in further improvement of accuracy ($\overline{\delta}_u = 4.65 \cdot 10^{-4}$, $\overline{\gamma}_u = 4.07 \cdot 10^{-2}$ for $(R/a) = 0.85$), but

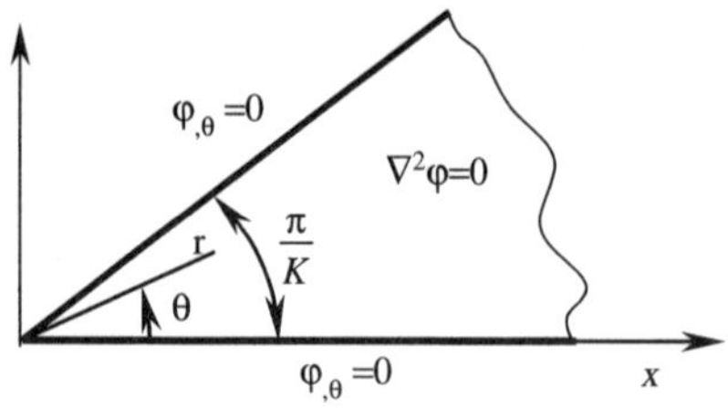

Figure 12: Repeating part of investigated region with K symmetry axes.

the acceptable range of R was smaller than that of κ in the case of homothetic contour.

6 The Laplace-type solutions with special T-complete functions

6.1 Area with K symmetry axes

Let us consider a simply connected region in which the shape and the boundary conditions are symmetric with respect to K axes crossing in one point. Then the solution can be sought in a repeating element presented in Fig. 12. The T-complete function fulfilling the Laplace equation and the boundary conditions along the symmetry axes are presented in this section. The boundary conditions have then the form:

$$\frac{\partial \varphi}{\partial \theta} = 0 \quad \text{for } \theta = 0 \tag{40a}$$

$$\frac{\partial \varphi}{\partial \theta} = 0 \quad \text{for } \theta = \frac{\pi}{K}. \tag{40b}$$

Considering the general form of solution of the 2D Laplace eqn (1) from Chapter 3 and the boundary conditions (40), we obtain:

$$\begin{aligned} \left.\frac{\partial \varphi}{\partial \theta}\right|_{\theta=0} &= B_0 + C_0 \ln r - \sum_{n=1}^{\infty} \lambda_n \left(A_n r^{\lambda_n} + B_n r^{-\lambda_n}\right) \sin\left(\lambda_n 0\right) \\ &+ \sum_{n=1}^{\infty} \lambda_n \left(C_n r^{\lambda_n} + D_n r^{-\lambda_n}\right) \cos\left(\lambda_n 0\right) = 0. \end{aligned} \tag{41}$$

This equation is fulfilled when

$$B_0 = C_0 = C_n = D_n = 0. \tag{42}$$

The solution should be limited for $r = 0$, hence

$$D_0 = B_n = 0. \tag{43}$$

Now the solution takes the form:

$$\varphi = A_0 + \sum_{n=1}^{\infty} A_n r^{\lambda_n} \cos\lambda_n\theta. \tag{44}$$

From condition (40b), we obtain:

$$\left.\frac{\partial\varphi}{\partial\theta}\right|_{\theta=\frac{\pi}{K}} = -\sum_{n=1}^{\infty} \lambda_n A_n r^{\lambda_n} \sin\lambda_n \frac{\pi}{K} = 0, \tag{45}$$

thus n takes only the values

$$n = Kk \quad \text{where } k = 1,2,3,.... \tag{46}$$

Finally, the solution of the presented problem has the form:

$$\varphi = \sum_{k=1}^{\infty} A_{k-1} r^{K(k-1)} \cos K(k-1)\theta. \tag{47}$$

After denoting: $a_1 \equiv A_0$, $a_2 \equiv A_1$,... we can write:

$$\hat{\varphi} = \sum_{m=1}^{M} a_m N_m(r,\theta) \tag{48}$$

where

$$N_1 = 1,\ N_m = r^{K(m-1)} \cos\left[K(m-1)\right],\ m = 2,\ 3,... \tag{49}$$

can be treated as the special purpose T-complete functional system in the above case.

Let us consider a doubly connected region with a circular hole inside (Fig. 13). Let the value of potential ϕ on the circular boundary be equal to one. Thus, the boundary conditions can be written as

$$\varphi = 1 \quad \text{for } r = R \tag{50a}$$

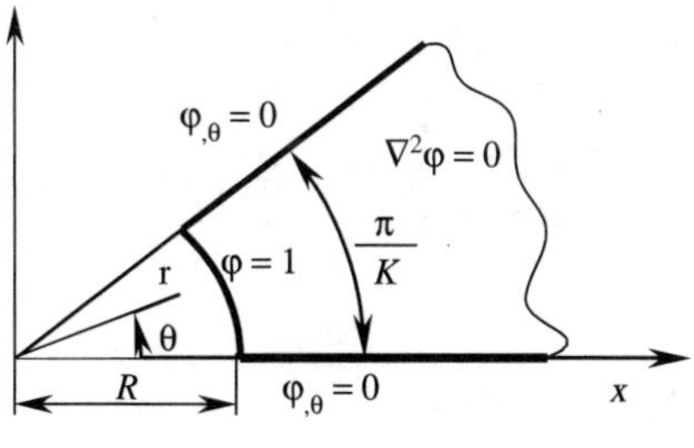

Figure 13: Repeating part of investigated region with K symmetry axes and cylindrical cut-out.

$$\frac{\partial\varphi}{\partial\theta} = 0 \quad \text{for } \theta = 0 \tag{50b}$$

$$\frac{\partial\varphi}{\partial\theta} = 0 \quad \text{for } \theta = \frac{\pi}{K}. \tag{50c}$$

Analogically to the previous case we now obtain:

$$\varphi(r,\theta) = A_0 + D_0 \ln r + \sum_{k=1}^{\infty}\left(A_k r^{Kk} + B_k r^{-Kk}\right)\cos(Kk\theta). \tag{51}$$

From the boundary condition (50a), we have:

$$\varphi(R,\theta) = A_0 + D_0 \ln R + \sum_{k=1}^{\infty}(A_k R^{Kk} + B_k R^{-Kk})\cos(Kk\theta) = 1 \tag{52}$$

which is fulfilled when

$$B_n = -A_n R^{2kK} \text{ and } A_0 = 1 - D_0 \ln R. \tag{53}$$

Finally, the Trefftz functions have in this case the form:

$$\varphi(r,\theta) = 1 + D_0 \ln\frac{r}{R} + \sum_{k=1}^{\infty} A_k\left(r^{Kk} - \frac{R^{2Kk}}{r^{Kk}}\right)\cos(Kk\theta) \tag{54}$$

and after denoting $a_1 \equiv D_0$, $a_2 \equiv A_1$,... the approximate solution can be written in the form:

$$\hat{\varphi} = 1 + \sum_{m=1}^{M} a_m N_m(r,\theta) \tag{55}$$

where

$$N_1 = \ln\frac{r}{R} \tag{56a}$$

$$N_m = \left[r^{K(m-1)} - \frac{R^{2K(m-1)}}{r^{K(m-1)}}\right]\cos\left[K(m-1)\theta\right], \quad m = 2,3,4,.... \tag{56b}$$

6.2 Region with a corner

In an analogical way, we can discuss a problem of a region with an angle $2a$ in a corner and the zero potential φ along both sides of the angle. A symmetric half of

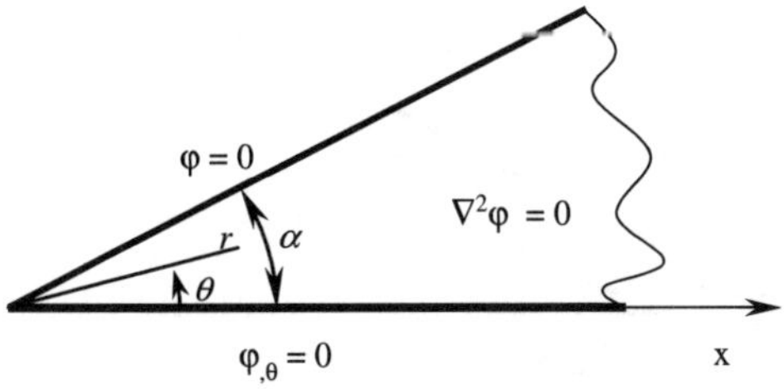

Figure 14: Part of investigated region with corner.

this region is presented in Fig. 14. According to this figure, the boundary conditions on the sides of the angle have the form:

$$\frac{\partial\varphi}{\partial\theta}=0 \quad \text{for } \theta=0 \tag{57a}$$

$$\varphi=0 \quad \text{for } \theta=a. \tag{57b}$$

Hence, in a similar way as before we obtain:

$$\varphi=\sum_{k=1}^{\infty}A_k R^{\frac{\pi}{2a}(2k-1)}\cos\left[\frac{\pi}{2a}(2k-1)\theta\right] \tag{58}$$

or

$$\hat{\varphi}=\sum_{m=1}^{M}a_m N_m(r,\theta), \tag{59}$$

where

$$N_m=r^{\frac{\pi}{2a}(2m-1)}\cos\left[\frac{\pi}{2a}(2m-1)\theta\right], \quad m=1,\ 2,\ 3,.... \tag{60}$$

For a similar problem, but with a constant potential different along the two sides of the angle a, we observe a potential jump in the corner. If we assume the boundary conditions in the form (Fig. 15):

$$\varphi=0 \quad \text{for } \theta=0 \tag{61a}$$

$$\varphi=1 \quad \text{for } \theta=a, \tag{61b}$$

we can write:

$$\hat{\varphi}=\frac{\theta}{a}+\sum_{m=1}^{M}a_m N_m(r,\theta) \tag{62}$$

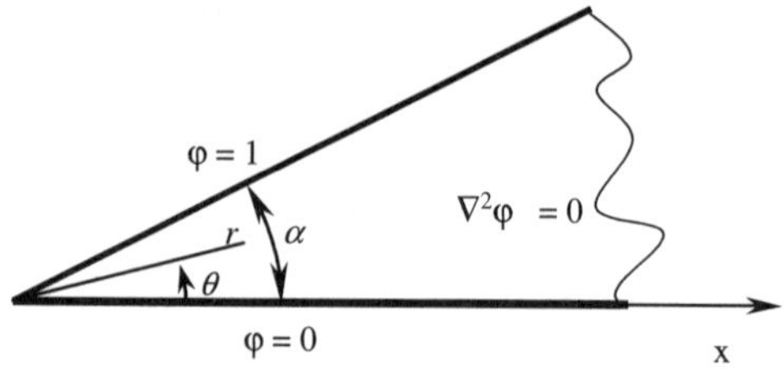

Figure 15: Part of investigated region with corner (jump of potential).

where

$$N_m = r^{\frac{2\pi}{\alpha}m} \sin\left(\frac{\pi}{\alpha} m\theta\right), \quad m = 1,2,3,.... \tag{63}$$

6.3 Composite regions with cylindrical rods

One of the methods for effective determination of thermal conductivity of a composite in form of matrix reinforced by cylindrical rods (fibres) is based on analysis of steady temperature field in repeating fragment of the composite. In case of one-directional reinforcement, this problem is two dimensional (Fig. 16). Publications, in which the boundary collocation method was applied to determination of the steady-state temperature field, are given in Table 8. It can be noticed that among them a considerable group deals with 2D composite regions.

A composite cross-section, which consists of matrix reinforced by cylindrical rods, is visible in Fig. 16. In this figure, a repeating fragment *ABCD* (a cell) has been defined. Depending on the character of thermal contact between the fibres and the matrix, different connectivity conditions can be formed. In case of the boundary collocation approach, these conditions are to be identically fulfilled, i.e. we determine special T-complete functions fulfilling not only the governing equations but also specific connectivity conditions of the problem [57].

Let us consider a case of a composite with ideal contact between fibres and matrix, which means continuity of thermal flux on the interfaces. If the problem

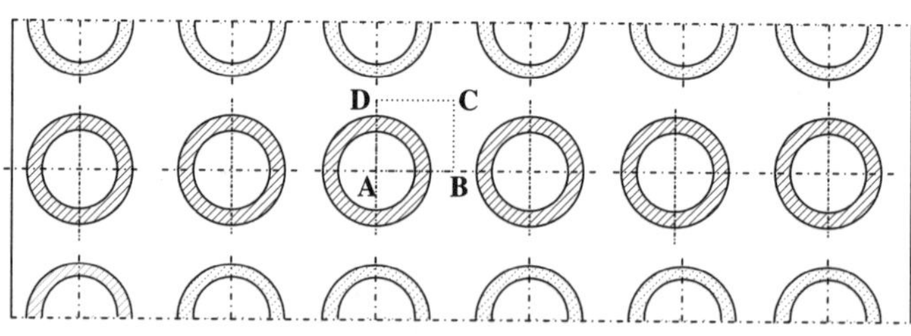

Figure 16: Composite in form of matrix reinforced by parallel bars.

Table 8: Papers in which the boundary collocation method was applied to determination of 2D stationary temperature field.

Stationary temperature field	References
Simply connected region	[22, 34, 38–41, 44, 61, 81, 111, 112]
Multiply connected region	[3, 21, 31, 33, 47–50, 64, 66, 73, 75, 83, 84, 87, 88, 96, 104–106, 108–110, 116]
Composite region	[6, 23, 28, 29, 31, 32, 53, 67–70, 73, 80, 114, 117, 119]

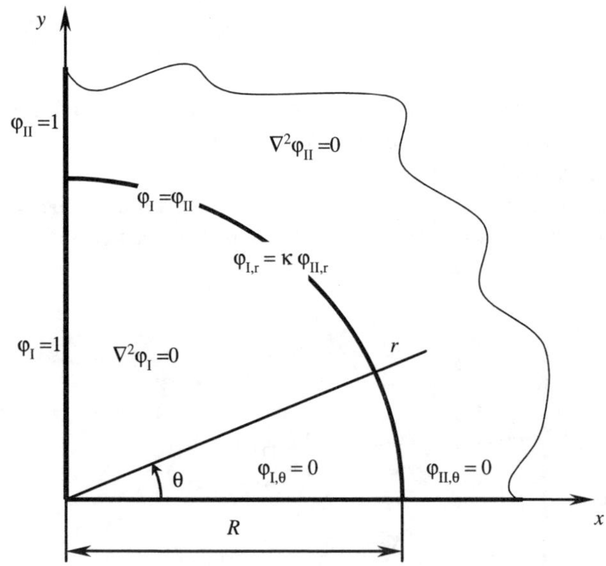

Figure 17: Part of composite region with ideal thermal contact between rods and matrix.

can be considered as symmetric with respect to axis x and, additionally, the temperature is constant along axis y, then the boundary and connectivity conditions take the form (Fig. 17):

$$\frac{\partial \varphi_{\mathrm{I}}}{\partial \theta} = \frac{\partial \varphi_{\mathrm{II}}}{\partial \theta} = 0 \quad \text{for } \theta = 0 \tag{64a}$$

$$\varphi_{\mathrm{I}} = \varphi_{\mathrm{II}} = 1 \quad \text{for } \theta = \frac{\pi}{2} \tag{64b}$$

$$\varphi_{\mathrm{I}} = \varphi_{\mathrm{II}} \quad \text{for } r = R \tag{64c}$$

$$\frac{\partial \varphi_{\mathrm{I}}}{\partial r} = \kappa \frac{\partial \varphi_{\mathrm{II}}}{\partial r} \quad \text{for } r = R. \tag{64d}$$

In the discussed case, in the repeating fragment of the composite we have the fibre region with the temperature field denoted φ_{I} and the matrix region with the field φ_{II}. The approximate solution fulfilling the governing Laplace equations and conditions (64) can be written as ($\hat{\varphi}$ means approximation)

$$\hat{\varphi}_{\mathrm{I}} = 1 + \sum_{k=1}^{M} A_k r^{(2k-1)} \cos\left[(2k-1)\,\theta\right] \tag{65a}$$

$$\hat{\varphi}_{\mathrm{II}} = 1 + \sum_{k=1}^{M} \frac{A_k}{2}\left[(1+\kappa)r^{(2k-1)} + (1-\kappa)\frac{R^{(4k-2)}}{r^{(2k-1)}}\right]\cos\left[(2k-1)\theta\right]. \tag{65b}$$

The unknown constants A_k are determined by collocation conditions along the right and upper boundary of the cell. According to a particular problem, these conditions can have different form.

Let us now consider a case in which a transitional layer exists between the fibre and matrix. In this case, in the composite fragment presented in Fig. 18, we can

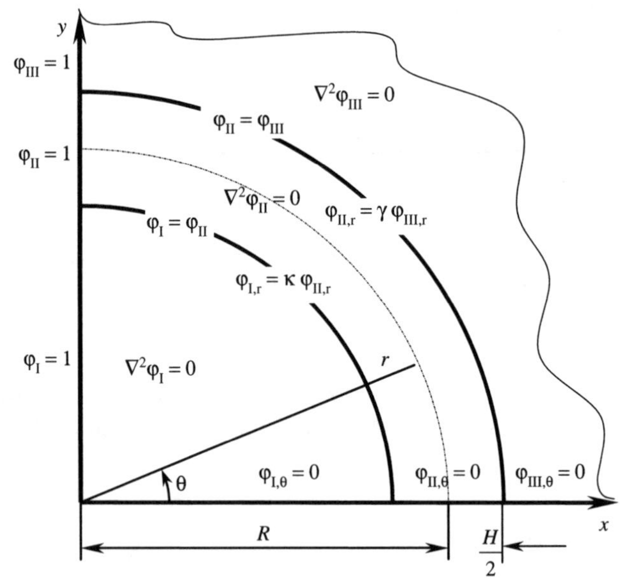

Figure 18: Part of composite region with a transition layer between rods and matrix.

observe three independent temperature fields φ_{I}, φ_{II} and φ_{III} and the boundary and connectivity conditions have the form:

$$\frac{\partial\varphi_{\mathrm{I}}}{\partial\theta}=\frac{\partial\varphi_{\mathrm{II}}}{\partial\theta}=\frac{\partial\varphi_{\mathrm{III}}}{\partial\theta}=0 \quad \text{for } \theta=0 \tag{66a}$$

$$\varphi_{\mathrm{I}}=\varphi_{\mathrm{II}}=\varphi_{\mathrm{III}}=1 \quad \text{for } \theta=\frac{\pi}{2} \tag{66b}$$

$$\varphi_{\mathrm{I}}=\varphi_{\mathrm{II}} \quad \text{for } r=R-\frac{H}{2} \tag{66c}$$

$$\frac{\partial\varphi_{\mathrm{I}}}{\partial r}=\kappa\frac{\partial\varphi_{\mathrm{II}}}{\partial r} \quad \text{for } r=R-\frac{H}{2} \tag{66d}$$

$$\varphi_{\mathrm{II}}=\varphi_{\mathrm{III}} \quad \text{for } r=R+\frac{H}{2} \tag{66e}$$

$$\frac{\partial\varphi_{\mathrm{II}}}{\partial r}=\gamma\frac{\partial\varphi_{\mathrm{III}}}{\partial r} \quad \text{for } r=R+\frac{H}{2}. \tag{66f}$$

The approximate solutions fulfilling the governing equations and conditions (66) can be written as

$$\hat{\varphi}_{\mathrm{I}}=1+\sum_{k=1}^{M}A_k r^{(2k-1)}\cos\left[(2k-1)\theta\right] \tag{67a}$$

$$\hat{\varphi}_{\mathrm{II}}=1+\sum_{k=1}^{M}\frac{A_k}{2}\left[(1+\kappa)r^{(2k-1)}+(1-\kappa)\frac{R^{(4k-2)}}{r^{(2k-1)}}\right]\cos\left[(2k-1)\,\theta\right] \tag{67b}$$

$$\begin{aligned}\hat{\varphi}_{\mathrm{III}}=1+\sum_{k=1}^{M}\frac{A_k}{4}&\left\{\left[(1+\kappa)(1+\gamma)+(1-\kappa)(1-\gamma)\left(\frac{R-H/2}{R+H/2}\right)^{4k-2}\right]r^{2k-1}\right.\\&+\left[(1+\kappa)(1-\gamma)\left(R+\frac{H}{2}\right)^{4k-2}+(1-\kappa)(1+\gamma)\left(R-\frac{H}{2}\right)^{4k-2}\right]\\&\left.\times r^{-(2k-1)}\right\}\cos\left[(2k-1)\theta\right].\end{aligned} \tag{67c}$$

The last example is a composite with non-ideal contact between rods and matrix, which means a certain thermal flux resistance on their interfaces (Fig. 19). Along them except the continuity of the thermal flux, there exists a determined

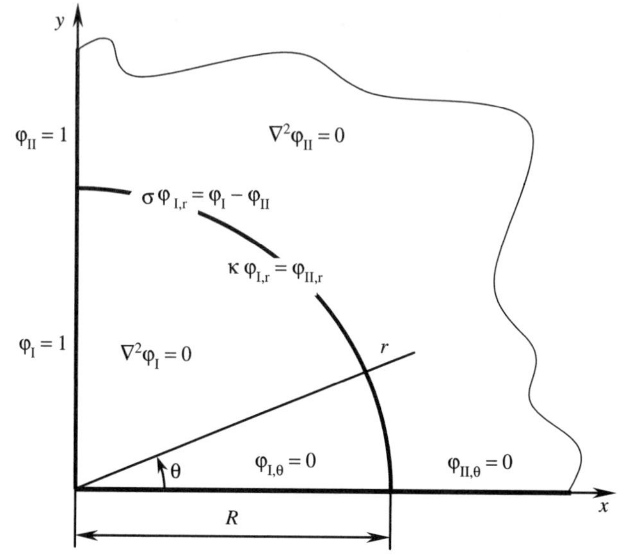

Figure 19: Part of composite region with non-ideal thermal contact between rods and matrix.

temperature jump proportional to the flux. Hence, the necessary conditions take the form:

$$\frac{\partial \varphi_{\mathrm{I}}}{\partial \theta} = \frac{\partial \varphi_{\mathrm{II}}}{\partial \theta} = 0 \quad \text{for } \theta = 0 \tag{68a}$$

$$\varphi_{\mathrm{I}} = \varphi_{\mathrm{II}} = 1 \quad \text{for } \theta = \frac{\pi}{2} \tag{68b}$$

$$\sigma \frac{\partial \varphi_{\mathrm{I}}}{\partial r} = \varphi_{\mathrm{I}} - \varphi_{\mathrm{II}} \quad \text{for } r = R \tag{68c}$$

$$\frac{\partial \varphi_{\mathrm{I}}}{\partial r} = \kappa \frac{\partial \varphi_{\mathrm{II}}}{\partial r} \quad \text{for } r = R \tag{68d}$$

and the sought solutions with the special T-complete functions can be determined as

$$\hat{\varphi}_{\mathrm{I}} = 1 + \sum_{k=1}^{M} A_k r^{(2k-1)} \cos\left[(2k-1)\,\theta\right] \tag{69a}$$

$$\hat{\varphi}_{\mathrm{II}} = 1 + \sum_{k=1}^{M} A_k \frac{1}{2} \Big\{ \left[1 + \kappa - (2k-1)\sigma\right] r^{2k-1} + R^{4k-2} \left[1 - \kappa - (2k-1)\sigma\right] r^{-(2k-1)} \Big\} \cos\left[(2k-1)\theta\right] \tag{69b}$$

At the end it should be noted that in all the cases presented in Section 6, the unknown Trefftz coefficients can be calculated according to the boundary collocation concept with control points along the right and upper sides of the investigated regions.

7 Recent observations

At the end of this chapter, several recent interesting papers concerning the 2D harmonic problems should be commented. In [86] the authors proposed the singular value decomposition (SVD) as a remedy for the ill-conditioning problem of linear sets of equations in the boundary collocation approach. However, several years later, C.S. Chen, H.A. Cho & M.A. Goldberg [8] criticized this proposal showing in numerical experiments that the SVD is for the BCM approach not more reliable than the Gaussian elimination. Since the former process is more expensive, the authors recommend the traditional Gaussian method.

It is well known and also shown in this chapter that the two basic kinds of the Trefftz complete systems applied for the Laplace-type problems are the harmonic Herrera systems and the Kupradze F-functional sets. In paper [9], equivalency of both systems is discussed, as well as differences occurring during their applications.

In paper [65], a modified variant of T-functions for the 2D Laplace equation has been presented. The proposed system takes into account a characteristic dimension of a domain, in which the boundary value problem has been defined. The numerical examples show advantages of the presented T-functions while solving problems of doubly connected objects.

Then, a modified concept of T-functional systems obtained from the double layer potential has been presented in [125]. The characteristic lack of singularities allows in this case to situate the source points on the boundary Γ of the considered region. Eventually, a new interesting variant of fitting boundary conditions – the weighted least square approach – is proposed in [97].

References

[1] Arutjunjan, N.Ch. & Abramjan, B.L., *Twisting of Elastic Solids* (in Russian), GIFML, Moskva, 1963.

[2] Axford, R.A., Two-dimensional, multiregion analysis of temperature fields in reactor tube bundles. *Nuclear Engineering and Design*, **6**, pp. 25–42, 1967.

[3] Balcerzak, M.J. & Raynor, S., Steady state temperature distribution and heat flow in prismatic bars with isothermal boundary conditions. *International Journal of Heat and Mass Transfer*, **3**, pp. 113–125, 1961.

[4] Barta, J., Über die naherungsweise Lösung einiger Zwei-dimensionalier Elastizitatsaufgaben. *Zeitschrift für Angewandte Mathematik und Mechanik (ZAMM)*, **17**, pp. 184–185, 1937.

[5] Beadle, C.W. & Conway, H.D., Stress concentration in twisted structural beams. *Journal of Applied Mechanics*, **30**, pp. 138–141, 1963.

[6] Berger, J.R. & Karageorghis, A., The method of fundamental solutions for heat conduction in layered materials. *International Journal for Numerical Methods in Engineering*, **45**, pp. 1681–1694, 1999.

[7] Bert, C.W. & Chang, S., Analysis of twisting stiffness for a multifiber composite layer. *Journal of Applied Mechanics*, **41**, pp. 658–662, 1974.

[8] Chen, C.S., Cho, H.A. & Golberg, M.A., Some comments on ill-conditioning of the method of fundamental solutions. *Engineering Analysis with Boundary Elements*, **30**, pp. 405–410, 2006.

[9] Chen, J.T., Wu, C.S., Lee, Y.T. & Chen, K.H., On the equivalence of the Trefftz method and method of fundamental solutions for Laplace and biharmonic equations. *Computers and Mathematics with Applications*, **53**, pp. 851–879, 2007.

[10] Cheng, K.C., Analog solution of laminar heat transfer in non-circular ducts by Moire method and point matching. *Journal of Heat Transfer*, **88**, pp. 175–182, 1966.

[11] Cheng, K.C., Dirichlet problems for laminar forced convection with heat sources and viscous dissipation in regular polygonal ducts. *AIChE Journal*, **13**, pp. 1175–1180, 1967.

[12] Cheng, K.C., Laminar flow and heat transfer characteristics in regular polygonal ducts. In: *Proceedings Third International Heat Transfer Conference AIChE*, **1**, pp. 64–76, 1966.

[13] Cheng, K.C., Laminar forced convection in regular polygonal ducts with uniform peripheral heat flux. *Journal of Heat Transfer*, **91**, pp. 159–167, 1969.

[14] Cieszko, M. & Kołodziej, J.A., Determination of heat transfer between bundle of regular rods and fluid in longitudinal laminar flow by means of the boundary collocation method (in Polish). *Journal of Theoretical and Applied Mechanics*, **19**, pp. 605–619, 1981.

[15] Conway, H.D., The approximate analysis of certain boundary value problems. *Journal of Applied Mechanics*, **27**, pp. 275–277, 1960.

[16] Conway, H.D., Torsion of prismatical rods with isosceles triangle and other cross sections. *Journal of Applied Mechanics*, **27**, pp. 209, 1960.

[17] Descloux, J. & Tolley, M.D., Approximation of the Poisson problem and of the eigenvalue problem for the Laplace operator by the method of large singular elements. *Research Report* 81-01, *Seminar für Angewandte Mathematic,* Eidgenossiche Technical Hochschule, 1980.

[18] Deverall, L.I., The approximate analysis of certain boundary values problems. *Journal of Applied Mechanics*, **28**, pp. 156–157, 1961

[19] Dwyer, O.E. & Berry, H.C., Laminar-flow heat transfer in-line flow through unbaffled rod bundles. *Nuclear Science and Engineering*, **42**, pp. 81–88, 1970.

[20] Fan, G.S.A. & Rashed, Y.F., Dipoles formulation for the method of fundamental solutions applied to potential problems. *Advances in Engineering Software*, **38**, pp. 1–8, 2007.

[21] Fend, F.A., Baroody, E.M. & Bell, J.C., An approximate form of the temperature distribution surrounding cooling holes in heat-generating solid. *BMI Report No.* T-42, Battelle Memorial Institute, Columbus, Ohio, October 25, 1950.

[22] France, D.M., Analytical solution to steady-state heat conduction problem with irregularly shaped boundaries. *Journal of Heat Transfer*, **93**, pp. 449–454, 1971.

[23] Goncalves, L.C.C. & Kołodziej, J.A., Determination of effective thermal conductivity in fibrous composites with imperfect thermal contact between constituents. *International Communications in Heat Mass and Transfer*, **20**, pp. 111–121, 1993.

[24] Gorzelańczyk, P. & Kołodziej, J.A., Some remarks concerning the shape of the source contour with application of the method of fundamental solutions to elastic torsion of prismatic rods. *Engineering Analysis with boundary Elements*, **32**, pp. 64–75, 2008.

[25] Gorzelańczyk, P., Tylicki, H. & Kołodziej, J.A., The torsional stiffness of bars with L, [, +, I, and ʃ cross-section, *Steel & Composite Structures*, **7** , pp. 441–456, 2007.

[26] Greenwell, R.N. & Wang, C.Y., Fluid flow through a partially filled cylinder. *Applied Scientific Research*, **36**, pp. 287–307, 1983.

[27] Hagan, S.L. & Ratkowsky, D.A., Laminar flow in cylindrical ducts having regular polygonal shaped cores. *Canadian Journal of Chemical Engineering*, **46**, pp. 387–388, 1968.

[28] Han, L.S. & Cosner, A.A., Effective thermal conductivities of fibrous composites. *Journal of Heat Transfer*, **103**, pp. 387–393, 1981.

[29] Hoffman, R.E. & Ariman, T., Thermal and mechanical stresses in nuclear reactor vessels. *Nuclear Engineering and Design*, **20**, pp. 31–55, 1972.

[30] Holy, Z.J., Three dimensional temperature and thermoelastic stress field in a heat producing sphere due to arbitrary surface heat transfer. *Nuclear Engineering and Design*, **6**, pp. 395–420, 1967.

[31] Hsieh, C.K. & Yang, S.L., A new theory on the critical thickness of insulation. *Journal of Heat Transfer*, **106**, pp. 648–652, 1984.

[32] Hulbert, L.E. & Sampath, S.G., Thermal stress analysis of multiholed composite plates and cylinders by the boundary points-least-squares method. *Journal of Pressure Vessel and Technology*, **96**, pp. 214–219, 1974.

[33] Hulbert, L.E., Solution of thermoelastic problems for plane multiply-connected regions by boundary point least squares. In: *Proceedings of International Conference 'Thermal Stresses and Thermal Fatigue'*, Berkeley Castle, Gloucestershire, England, pp. 268–283, 1969.

[34] Hulbert, L.E., Verification of computer program based on assumed mode methods. In: *Symposium on Engineering Computer Software*, I. Berman, New York, American Society of Mechanical Engineers, pp. 17–25, 1971.

[35] Jirousek, J., Venkatesh, A., Zieliński, A.P. & Rabemanantosa, H., Comparative study of *p*-extensions based on conventional assumed displacement and

hybrid-Trefftz finite element models. *Computers & Structures*, **46**, no. 2, pp. 261–278, 1993.

[36] Jopek, H., Kołodziej, J.A. & Stręk, T., Laminar flow in trapezoidal grooves at finite Bond numbers with shear stress at the liquid-vapor interface by the method of fundamental solutions. *Computer Assisted Mechanics and Engineering Sciences*, **13**, pp. 395–405, 2006.

[37] Kaczmarek, M. & Kołodziej, J.A., Application of boundary collocation method for determination of capillary flow in triangular groove (in Polish). *Zeszyty Naukowe Politechniki Poznańskiej. Maszyny Robocze i Pojazdy*, **22**, pp. 133–138, 1982.

[38] Karageorghis, A. & Fairweather, G., The simple layer potential method of fundamental solutions for certain biharmonic problems. *International Journal for Numerical Methods in Fluids*, **9**, pp. 1221–1234, 1988.

[39] Kita, E. & Kamiya, N., Trefftz method: an overview. *Advances in Engineering Software*, **24**, pp. 3–12, 1995.

[40] Kita, E., Kamiya, N. & Ikeda, Y., An application of Trefftz method to the sensitivity analysis of two-dimensional potential problem. *International Journal for Numerical Methods in Engineering*, **38**, pp. 2209–2224, 1995.

[41] Kita, E., Kamiya, N. & Ikeda, Y., Boundary-type sensitivity analysis scheme based on indirect Trefftz formulation. *Advances in Engineering Software*, **24**, pp. 89–96, 1995.

[42] Kołodziej, J.A. & Dudziak, W., The determination of the stationary incompressible laminar viscous flow past a hexagonal lattice parallel cylindrical bear by means of boundary collocation (in Polish). *Archive of Mechanical Engineering*, **26**, pp. 103–114, 1979.

[43] Kołodziej, J.A. & Fraska, A., Elastic torsion of bars possessing regular polygon in cross-section using BCM. *Computers & Structures*, **84**, pp. 78–91, 2005.

[44] Kołodziej, J.A. & Kleiber, M., Boundary collocation method vs FEM for some harmonic 2-D problems. *Computers & Structures*, **33**, pp. 155–168, 1989.

[45] Kołodziej, J.A. & Musielak, G., Remarks about the boundary collocation used in solving the boundary value problems concerning torsion of prismatic rods (in Polish). *Archive of Mechanical Engineering*, **34**, pp. 111–135, 1987.

[46] Kołodziej, J.A. & Starosta, R., Self-adaptive Trefftz procedure for harmonic problems. *Computer Assisted Mechanics and Engineering Sciences*, **4**, pp. 491–500, 1997.

[47] Kołodziej, J.A. & Stręk, T., Analytical approximations of the shape factors for conductive heat flow in circular and regular polygonal cross-sections. *International Journal of Heat and Mass Transfer*, **44**, pp. 999–1012, 2001.

[48] Kołodziej, J.A. & Stręk, T., Regular polygonal shapes, boundary collocation method, and regression by exponential function. *Zeitschrift für Angewandte Mathematik und Mechanik (ZAMM)*, **80**, pp. S805–S806, 2000.

[49] Kołodziej, J.A. & Uściłowska, A., Regular polygonal shapes, and boundary collocation method. *The International Journal of Mechanical Engineering Education*, **27**, pp. 140–150, 2000.

[50] Kołodziej, J.A. & Uściłowska, A., Trefftz-type procedure for Laplace equation on domains with circular holes, circular inclusions, corners, slits, and symmetry. *Computer Assisted Mechanics and Engineering Sciences*, **4**, pp. 501–519, 1997.

[51] Kołodziej, J.A. & Wachowska, A., Special purpose Trefftz functions for the torsion of bars with regular polygonal cross-section. *Proceedings of International Workshop on Mesh Free Methods,* Institute Seperior Tecnica, Lisboa, pp. 157–162, July 21–23, 2003.

[52] Kołodziej, J.A., *Application of the Boundary Collocation Method to Problems of Mechanics* (in Polish), Poznan University of Technology Press: Poznan, 2001.

[53] Kołodziej, J.A., Determination of effective thermal conductivity of unidirectional fibrous composite by means of boundary collocation method (in Polish). *Theoretical and Applied Mechanics*, **23**, pp. 933–958, 1985.

[54] Kołodziej, J.A., Filtration resistance of a system cylindrical rods at longitudinal laminar flow (in Polish). *Archive of Mechanical Engineering*, **27**, pp. 487–502, 1980.

[55] Kołodziej, J.A., Musielak, G., Kaczmarek, M., Stręk, T. & Starosta, R., Determination of free surface and gravitational flow of liquid in triangular groove. *Computational Mechanics*, **24**, pp. 110–117, 1999.

[56] Kołodziej, J.A., Uściłowska, A. & Ciałkowski, M., Semi-analytical approximations of the laminar friction coefficients for flow in conduits with polygonal cross-section. *Acta Mechanica*, **158**, pp. 127–144, 2002.

[57] Kończak, Z. & Kołodziej, J.A., Application of boundary collocation method in problems for multi-connected regions (in Polish). *Zeszyty Naukowe Politechniki Poznańskiej, Mechanika*, **30**, pp. 171–176, 1984.

[58] Kuo, Y.M. & Conway, H.D., Torsion of reinforced square cylinder. *Journal of Engineering Mechanical Division, Proceedings of ASCM*, **106**, pp. 1341–1347, 1980.

[59] Kuo, Y.M. & Conway, H.D., Torsion of composite rhombus cylinder. *Journal of Applied Mechanics*, **41**, pp. 302–303, 1974.

[60] Leissa, A.W. & Brann, J.H., On the solution of bars having symmetry axes. *Journal of Mechanical Science*, **6**, pp. 45–60, 1964.

[61] Leitao, V.M.A., On the implementation of a multi-region Trefftz-collocation formulation for 2-D potential problems. *Engineering Analysis with Boundary Elements*, **20**, pp. 51–61, 1997.

[62] Li, Z.C., Lu, T.T., Hu, H.Y. & Cheng, A.H.D., Particular solutions of Laplace's equations on polygons and new models involving mild singularities. *Engineering Analysis with Boundary Elements*, **29**, pp. 391–396, 2005.

[63] Liao, W.C., The torsional response and measurement of the *in situ* shear moduli of an edge-cracked thick laminate. *Composite Science and Technology*, **58**, pp. 145–154, 1998.

[64] Lin, C.C. & Ko, C.C., Stress and strength analysis of finite laminate with elliptical hole. *Journal of Composite Materials*, **22**, pp. 373–385, 1988.

[65] Liu, C.S., A highly accurate collocation Trefftz method for solving the Laplace equation in the doubly connected domains. *Numerical Methods for Partial Differential Equations*, **23**, pp. 1–14, 2007.

[66] Lo, C.C., *The solution of plane harmonic and biharmonic boundary value problems in the theory of elasticity*, Ph. D. Thesis, No. 65-2633, Purdue University, 1964.

[67] Lu, S.Y & Kim, S., Effective thermal conductivity of composites containing spheroidal inclusions. *AIChE Journal*, **36**, pp. 927–938, 1990.

[68] Lu, S.Y. & Lin, H.Ch., Effect of interfacial characteristics on effective conductivities of isotropic two-dimensional periodic composites. *Chemical Engineering Science*, **50**, pp. 2611–2631, 1995.

[69] Lu, S.Y. & Lin, H.Ch., Effective conductivity of composites containing aligned spheroidal inclusions of finite conductivity. *Journal of Applied Physics*, **79**, pp. 6761–6769, 1996.

[70] Lu, S.Y., The effective thermal conductivities of composites with 2-D arrays of circular and square cylinders. *Journal of Composite Materials*, **29**, pp. 483–506, 1995.

[71] Lu, T.T., Hu, H.Y. & Li, Z.C., Highly accurate solutions of Motz's and the cracked beam problems. *Engineering Analysis with Boundary Elements*, **28**, pp. 79–96, 2004.

[72] Mukherjee, S. & Morjaria, M., On the efficiency and accuracy of the boundary element method and finite element method. *International Journal for Numerical Methods in Engineering*, **20**, pp. 515–522, 1984.

[73] Naghdi, A.K., Certain heat problems in a rectangular region with multiple cut-outs. *International Journal of Heat and Mass Transfer*, **26**, pp. 1143–1149, 1983.

[74] Naghdi, A.K., Heat conduction in a rectangular parallel piped with multiple cylindrical cavities. *Journal of Heat Transfer*, **107**, pp. 491–494, 1985.

[75] Naghdi, A.K., Solution of Poisson equation in a semicircular region with a hole. *Journal of Applied Mechanics*, **47**, pp. 677–679, 1980.

[76] Naghdi, A.K., Torsion of multihole circular cylinders. *Journal of Applied Mechanics*, **49**, pp. 432–436, 1982.

[77] Naghdi, A.K., Torsion of rectangular bars with multiple cylindrical reinforcement or cavities. *Journal of Applied Mechanics*, **61**, pp. 483–487, 1994.

[78] Neville, C.F., Flow and heat transfer in ducts of arbitrary shape with arbitrary thermal boundary conditions. *Journal of Heat Transfer*, **91**, pp. 588–589, 1969.

[79] Niedenfuhr, F.W. & Leissa, A.W., The torsion of prismatic bars of regular polygonal cross section. *Journal of the Aerospace Science*, **12**, pp. 194–210, 1961.

[80] Nijsing, R., Temperature and heat flux distribution in nuclear fuel element rods. *Nuclear Engineering and Design*, **4**, pp. 1–20, 1966.

[81] Ojalvo, L.U. & Linzer, F.D., Improved point-matching techniques. *Quarterly Journal of Mechanics & Applied Mathematics*, **18**, pp. 41–56, 1965.

[82] Okamoto, M. & Sekiya, T., Some practical solutions for elastic torsion of bars by charge simulation method. *Mechanics Research Communications*, **17**, pp. 47–53, 1990.

[83] Ondris, L., Remarks concerning the solution of stationary plane thermal problems by the boundary point least squares method. *Computer Methods in Applied Mechanics and Engineering*, **31**, pp. 331–336, 1982.

[84] Ondris, L., Solution of some symmetrical plane thermal problems by the boundary point least squares method. *Computer Methods in Applied Mechanics and Engineering*, **28**, pp. 309–325, 1981.

[85] Poritsky, H. & Danforth, C.E., On the torsion problem. In: *Proceedings 3rd U.S. National Congress on Applied Mechanics*, pp. 431–441, 1958.

[86] Ramachandran, P.A., Method of fundamental solutions: singular value decomposition analysis. *Communications in Numerical Methods in Engineering*, **18**, pp. 789–801, 2002.

[87] Rao, K.S., Bapu Rao, M.N. & Ariman, T., Thermal stresses in elastic plates with elliptic holes. *Journal of Engineering Industry*, **96**, pp. 827–832, 1974.

[88] Rao, K.S., Bapu Rao, M.N. & Ariman, T., Thermal stresses in plates with circular holes. *Nuclear Engineering and Design*, **15**, pp. 97–112, 1971.

[89] Ratkowsky, D.A. & Epstein, N., Laminar flow in regular polygonal shaped ducts with circular centred cores. *Canadian Journal of Chemical Engineering*, **46**, pp. 387–388, 1968.

[90] Satttinger, S.S. & Conway, H.D., The solution of certain isosceles triangle and rhombus torsion and plate problem. *International Journal of Engineering Science*, **7**, pp. 221–228, 1965.

[91] Schmidt, J., Longitudinal laminar flow in an array of circular cylinders. *International Journal of Heat and Mass Transfer*, **9**, pp. 925–937, 1966.

[92] Shah, R.K., Laminar flow friction and forced convection heat transfer in ducts of arbitrary geometry. *International Journal of Heat and Mass Transfer*, **18**, pp. 849–862, 1975.

[93] Sherwood, A.A., The application of Fouries series to the solution of the St. Venant torsion problem. *Australian Journal of Applied Science*, **13**, pp. 285–299, 1962.

[94] Shih, F.S., Laminar flow in axisymmetric conduits by a rational approach. *Canadian Journal of Chemical Engineering*, **45**, pp. 287–284, 1967.

[95] Shuleshko, P., Comparative analysis of different collocation methods on the basis of the torsion problem. *Australian Journal of Applied Science*, **12**, pp. 194–210, 1961.

[96] Skiepko, T., Heat transmission across fouled tubes analysed by Trefftz approach. *Computer Assisted Mechanics and Engineering Sciences*, **9**, pp. 239–269, 2002.

[97] Smyrlis, Y.S., The method of fundamental solutions: a weighted least-squares approach. *BIT Numerical Mathematics*, **46**, pp. 163–194, 2006.

[98] Soliman, H.M. & Feingold, A., Analysis of fully developed laminar flow in longitudinal, internally finned tubes. *Chemical Engineering Journal*, **14**, pp. 119–129, 1977.

[99] Soliman, H.M. & Feingold, A., Analysis of heat transfer in internally finned tubes under laminar flow conditions. In: *Proceedings 6th International Heat Transfer Conference*, **2**, pp. 571–576, 1978.

[100] Sparrow, E.M. & Haji-Sheikh, A., Flow and heat transfer in ducts of arbitrary shape with arbitrary thermal boundary conditions. *Journal of Heat Transfer*, **88**, pp. 351–358, 1966.

[101] Sparrow, E.M. & Loeffeler, A.L., Longitudinal laminar flow between cylinders arranged in regular array. *AIChE Journal*, **5**, pp. 323–330, 1959.

[102] Sparrow, E.M., Laminar flow in isosceles triangular ducts. *AIChE Journal*, **8**, pp. 599–604, 1962.

[103] Sparrow, E.M., Loeffler, A.L. & Hubbard, H.A., Heat transfer to longitudinal laminar flow between cylinders. *Journal of Heat Transfer*, **83**, pp. 415–422, 1961.

[104] Sparrow, E.M., Temperature distribution and heat-transfer results for an internally cooled heat-generating solid. *Journal of Heat Transfer*, **82**, pp. 389–392, 1960.

[105] Sultanian, B.K. & Sastri, V.M.K., Effect of geometry on heat conduction in coolant channels of a liquid rocket engine. *Warme – und Stoffübertragung*, **14**, pp. 254–251, 1980.

[106] Sultanian, B.K. & Sastri, V.M.K., Steady heat conduction in a circular duct with circumferentially varying heat transfer coefficients. *Regional Journal of Energy, Heat and Mass Transfer*, pp. 101–110, 1979.

[107] Szybiński, B. & Zieliński, A.P., Alternative T-complete systems of shape functions applied in analytical Trefftz finite elements. *Numerical Methods for Partial. Differential Equations*, **11**, no. 4, pp. 375–388, 1995.

[108] Takeuti, Y. & Sekiya, T., Thermal stresses in a polygonal cylinder with a circular hole under internal heat generation. *Zeitschrift für Angewandte Mathematik und Mechanik (ZAMM)*, **48**, pp. 237–246, 1968.

[109] Takeuti, Y. & Sumi, N., Steady-state thermoelastic problem for a triangular prism with a circular hole under heat generation. *International Journal of Engineering Science*, **5**, pp. 145, 1968.

[110] Takeuti, Y., Approximate formulae temperature distribution in heat generating polygonal cylinder with central circular hole. *Nuclear Engineering and Design*, **8**, pp. 241–246, 1968.

[111] Tolley, M.D. & Wajc, S.J., Approximate solution method of Laplace's equation for a square domain. In: *Advances in Computer Methods for Partial Differential Equations*, **2**, R. Vichnevetsky, Pub. IMACS AICA, pp. 26–33, 1977.

[112] Tolley, M.D., La methode des grands elements finis singuliers: principles et applications. *Bulletin de la Societé Mathematique de Belgique*, **32**, ser. B, pp. 17–28, 1980.

[113] Tolley, M.D., Torsion des barres polygonales. *Bulletin de la classe des Sciences,* Academie Royal de Belgique, **63**, ser. 5, pp. 1977–11, pp. 902–912, 1977.

[114] Wang, C.Y., Conductivity of material containing fibers of rectangular cross section. *Mechanics of Materials*, **17**, pp. 71–77, 1994.

[115] Wang, C.Y., Flow in channel with longitudinal ribs. *Journal of Fluid Engineering*, **116**, pp. 233–237, 1994.

[116] Wang, C.Y., Heat conduction through a thick-walled cylindrical container. *International Journal of Mechanical Sciences*, **36**, pp. 1001–1010, 1994.

[117] Wang, C.Y., Heat transfer across a bilayer composite cylinder with partial debonding. *Journal of Heat Transfer*, **117**, pp. 222–223, 1995.

[118] Wang, C.Y., Maximum flow down a circular sector tube. *Chemical Engineering Science*, **54**, pp. 961–964, 1999.

[119] Wang, C.Y., On the diffusion past a staggered array of rectangular fibers. *Chemical Enginnering Science*, **55**, pp. 6079–6084, 2000.

[120] Wang, C.Y., Stokes flow through a zig-zag channel. *Acta Mechanica*, **126**, pp. 115–125, 1998.

[121] Wang, C.Y., Stokes flow through an array of staggered strips. *Applied Scientific Research*, **59**, pp. 27–41, 1998.

[122] Wang, C.Y., Torsion of a flattened tube. *Mecanica*, **30**, pp. 221–227, 1995.

[123] Wang, C.Y., Torsion of an angle bar. *Mechanics of Structures & Machines*, **24**, pp. 283–294, 1996.

[124] Wang, C.Y., Torsion of polygonal bar with core of different material. *Journal of Engineering Mechanics. Transactions of ASCE*, **125**, pp. 1218–1221, 1999.

[125] Young, D.L., Chen, K.H. & Lee, C.W., Novel meshless method for solving the potential problems with arbitrary domain. *Journal of Computational Physics*, **209**, pp. 290–321, 2005.

[126] Yu, H.S. & Sparrow, E.M., Stratified laminar flow in ducts of arbitrary shape. *AIChE Journal*, **13**, pp. 10–16, 1967.

[127] Zieliński, A.P. & Zienkiewicz, O.C., Generalized finite element analysis with T-complete boundary solution functions. *International Journal for Numerical Methods in Engineering*, **21**, pp. 509–528, 1985.

[128] Zieliński, A.P., Application of T-complete functions method to engineering problems. *Rep. C/R/40/83*, Univ. College of Swansea, 1983.

CHAPTER 7

Problems described by two-dimensional, biharmonic equations

1 Typical problems

In applied mechanics, there are three basic problems which are governed by two-dimensional (2D) biharmonic equations:

- plane creeping flow
- thin plate bending
- 2D elastostatics

In the first case, the current function fulfils the biharmonic equation and its first derivatives are components of velocity flow vector. In the second case, the biharmonic equation is fulfilled by a deflection function of a plate and its non-homogenous part represents the known function of load. In the third case, the biharmonic equation is fulfilled by the stress function and the second derivatives of this function determine three plane stress components. The above basic problems have specific features and their boundary conditions considerably differ from each other. Hence, application of the boundary collocation method to boundary value problems connected with them is below discussed separately.

2 Plane creeping flow. Boundary collocation as a modification of the Trefftz method

A system of equations governing viscous flow of incompressible liquid (called Navier–Stokes set) is non-linear because of its convective term. Neglecting this term is justified when the Reynolds number is very small. Then a non-linear equation set called the Stokes system is obtained, which governs slow flow of viscous liquid.

Boundary problems of 2D creeping flow can be defined in a few different ways. In the first one, the stream function Ψ is used and the problem is defined by the biharmonic equation:

$$\nabla^2\nabla^2\Psi = 0 \quad \text{in } \Omega, \tag{1}$$

with boundary conditions:

$$\left.\begin{array}{l}\Psi=\bar{\Psi} \\ \dfrac{\partial \Psi}{\partial n}=\bar{q}\end{array}\right\} \text{ on } \Gamma_1 \tag{2a}$$

$$\left.\begin{array}{l}\Psi=\bar{\Psi} \\ \dfrac{\partial^2 \Psi}{\partial n^2}=\bar{p}\end{array}\right\} \text{ on } \Gamma_2 \tag{2b}$$

where ∇^2 is the 2D Laplace operator, Ψ is the unknown stream function (velocity components are presented by relations $V_X=\partial\Psi/\partial Y$, $V_Y=-\partial\Psi/\partial X$), $\partial/\partial n$ is a derivative operator normal to the boundary and $\bar{\Psi}$, $\bar{q}$ and $\bar{p}$ are given functions and $\Gamma=\Gamma_1\cup\Gamma_2$.

In the second case, rotation ω and stream function Ψ are used. In such a case, the boundary problem is stated as follows:
the 2D Laplace equation for rotation:

$$\nabla^2\omega=0 \quad \text{in}\,\Omega \tag{3}$$

the 2D Poisson equation for the stream function:

$$\nabla^2\Psi=\omega \quad \text{in}\,\Omega \tag{4}$$

and boundary conditions:

$$\omega=\bar{\omega} \quad \text{on}\,\Gamma_1 \tag{5a}$$

$$\frac{\partial\omega}{\partial n}=\bar{\kappa} \quad \text{on}\,\Gamma_2 \tag{5b}$$

$$\Psi=\bar{\Psi} \quad \text{on}\,\Gamma_1 \tag{5c}$$

$$\frac{\partial\Psi}{\partial n}=\bar{q} \quad \text{on}\,\Gamma_2 \tag{5d}$$

or their other combinations.

A weak formulation of boundary problem (1)–(2) can, for example, be presented in the form of weighted residuum:

$$\begin{aligned}&\int_\Omega W\nabla^2\nabla^2\Psi d\Omega+\int_\Gamma W_1\left(\Psi-\bar{\Psi}\right)d\Gamma \\ &\quad+\int_{\Gamma_1} W_2\left(\frac{\partial\Psi}{\partial n}-\bar{q}\right)d\Gamma+\int_{\Gamma_2} W_3\left(\frac{\partial^2\Psi}{\partial n^2}-\bar{p}\right)d\Gamma=0,\end{aligned} \tag{6}$$

where W, W_1, W_2 and W_3 are weighting functions.

Similarly, a weak boundary problem formulation (3)–(5) can be written as:

$$\int_{\Omega} \tilde{W}_1 \nabla^2 \omega \, d\Omega + \int_{\Omega} \tilde{W}_2 \left(\nabla^2 \Psi - \omega\right) d\Omega + \int_{\Gamma_1} \tilde{W}_3 \left(\omega - \bar{\omega}\right) d\Gamma$$
$$+ \int_{\Gamma_2} \tilde{W}_4 \left(\frac{\partial \omega}{\partial n} - \bar{\kappa}\right) d\Gamma + \int_{\Gamma_1} \tilde{W}_5 \left(\Psi - \bar{\Psi}\right) d\Gamma + \int_{\Gamma_2} \tilde{W}_6 \left(\frac{\partial \Psi}{\partial n} - \bar{q}\right) d\Gamma = 0 \tag{7}$$

where now $\tilde{W}_1$, $\tilde{W}_2$, $\tilde{W}_3$, $\tilde{W}_4$, $\tilde{W}_5$ and $\tilde{W}_6$ are respective weighting functions.

Applying the Trefftz method, we assume a solution of boundary problem (1)–(2) in a form of combination of T-complete functions (compare Table 1 in Chapter 3):

$$\Psi = \sum_{k=1}^{2M} C_k U_k \tag{8}$$

where C_k are unknown coefficients and U_k are trial functions fulfilling the governing equation:

$$\nabla^2 \nabla^2 U_k = 0. \tag{9}$$

Similarly, a solution of the problem (3)–(5) is presented in the form:

$$\omega = \sum_{k=1}^{2M} C_k T_k \tag{10}$$

$$\Psi = \sum_{k=1}^{2M} C_k U_k \tag{11}$$

where C_k are unknown coefficients and T_k and U_k are trial functions which fulfil equations:

$$\nabla^2 T_k = 0. \tag{12a}$$

$$\nabla^2 U_k = T_k. \tag{12b}$$

Substituting solution (8) into eqn (6) and solutions (10) and (11) into eqn (7) we obtain:

$$\int_{\Gamma} W_1 \left(\sum_{k=1}^{2M} C_k U_k - \bar{\Psi}\right) d\Gamma + \int_{\Gamma_1} W_2 \left(\sum_{k=1}^{2M} C_k \frac{\partial U_k}{\partial n} - \bar{q}\right) d\Gamma$$
$$+ \int_{\Gamma_2} W_3 \left(\sum_{k=1}^{2M} C_k \frac{\partial^2 U_k}{\partial n^2} - \bar{p}\right) d\Gamma = 0 \tag{13}$$

and

$$\int_{\Gamma_1} \tilde{W}_3 \left(\sum_{k=1}^{2M} C_k T_k - \overline{\omega} \right) d\Gamma + \int_{\Gamma_2} \tilde{W}_4 \left(\sum_{k=1}^{2M} C_k \frac{\partial T_k}{\partial n} - \overline{\kappa} \right) d\Gamma$$
$$+ \int_{\Gamma_1} \tilde{W}_5 \left(\sum_{k=1}^{2M} C_k U_k - \overline{\Psi} \right) d\Gamma + \int_{\Gamma_2} \tilde{W}_6 \left(\sum_{k=1}^{2M} C_k \frac{\partial U_k}{\partial n} - \overline{q} \right) d\Gamma = 0. \tag{14}$$

Depending of the choice of weighting functions, W_1, W_2 and W_3 or $\tilde{W}_3$, $\tilde{W}_4$, $\tilde{W}_5$ and $\tilde{W}_6$, we obtain various variants of the Trefftz method. The two of them are presented below: the typical boundary collocation method and the opposite-weighting method.

When we choose weighting functions in the form:

$$W_1 = \delta\ (P_j), \quad P_j \in \Gamma, \ j = 1, ..., N \tag{15a}$$

$$W_2 = \delta\ (Q_j), \quad Q_j \in \Gamma_1, \ j = 1, ..., N_1 \tag{15b}$$

$$W_3 = \delta\ (S_j), \quad S_j \in \Gamma_2, \ j = 1, ..., N_2 \tag{15c}$$

where $\delta(P_j)$, $\delta(Q_j)$ and $\delta(S_j)$ are Dirac's δ-functions, we obtain the boundary collocation formulation. After substitution of those functions to (12b), the equation set for unknown coefficients C_k can be written as follows:

$$\sum_{k=1}^{2M} C_k U_k (P_j) = \overline{\Psi}(P_j), \quad j = 1, 2, ..., N \tag{16a}$$

$$\sum_{k=1}^{2M} C_k \frac{\partial U_k (Q_j)}{\partial n} = \overline{q}(Q_j), \quad j = 1, 2, ..., N_1 \tag{16b}$$

$$\sum_{k=1}^{2M} C_k \frac{\partial^2 U_k (Q_j)}{\partial n^2} = \overline{p}(S_j), \quad j = 1, 2, ..., N_2. \tag{16c}$$

If $N + N_1 + N_2 = 2M$, the number of equations is equal to the number of unknowns and we deal with a simple boundary collocation method. If, however, $N + N_1 + N_2 > 2M$, the set (16) can be solved in the least square sense.

A survey of publications in which the boundary collocation method is used to solve 2D creeping flow problems is presented in Table 1. In this table, also papers for axisymmetric creeping flow are gathered. Three-dimensional Stokes flow

Table 1: Papers in which the boundary collocation method was applied to 2D and axisymmetric creeping flow.

Creeping flow	References
Plane(2D)	[3, 11, 14, 31, 41, 45–48, 63, 86, 101, 106, 95, 110, 112, 123, 124, 127, 128, 130, 131, 135]
Axisymmetric	[9, 15, 17–20, 27, 36, 49, 50, 52–57, 60, 61, 65, 67–71, 74–76, 82, 91–94, 97, 125, 126, 133, 136],

problems were presented in the following papers: [28, 32–35, 40, 51, 58, 59, 66, 72, 73, 97, 113, 114, 120, 129, 134]. They will be discussed in detail in Chapter 10.

If we consider weighting functions as

$$W_1 = \frac{\partial U_j}{\partial n}, \quad W_2 = -U_j, \quad W_3 = \frac{\partial^2 U_j}{\partial n^2} \qquad (17)$$
$$j = 1, 2, \ldots, 2M$$

and substitute those expressions in eqn (13), we obtain the opposite-weighting approach:

$$\int_\Gamma \frac{\partial U_j}{\partial n}\left(\sum_{k=1}^{2M} C_k U_k - \bar{\Psi}\right) d\Gamma - \int_{\Gamma_2} U_j \left(\sum_{k=1}^{2M} C_k \frac{\partial U_k}{\partial n} - \bar{q}\right) d\Gamma$$
$$+ \int_{\Gamma_2} \frac{\partial^2 U_j}{\partial n^2}\left(\sum_{k=1}^{2M} C_k \frac{\partial^2 U_k}{\partial n^2} - \bar{p}\right) d\Gamma = 0 \quad j = 1, 2, \ldots, 2M. \qquad (18)$$

The matrix form of eqn (18) is following:

$$\mathbf{K} \cdot \mathbf{C} = \mathbf{f} \qquad (19)$$

where $\mathbf{C} = \left[C_1, C_2, \ldots C_{2M}\right]^T$ and matrix $\mathbf{K}$ of the system and vector $\mathbf{f}$ of the right-hand side are defined by

$$K_{jk} = \int_\Gamma \frac{\partial U_j}{\partial n} U_k d\Gamma - \int_{\Gamma_1} U_j \frac{\partial U_k}{\partial n} d\Gamma + \int_{\Gamma_2} \frac{\partial^2 U_j}{\partial n^2}\frac{\partial U_k}{\partial n^2} d\Gamma \qquad (20a)$$

$$f_j = \int_\Gamma \frac{\partial U_j}{\partial n} \bar{\Psi} d\Gamma - \int_{\Gamma_1} U_j \bar{q} d\Gamma + \int_{\Gamma_2} \frac{\partial^2 U_j}{\partial n^2} \bar{p} d\Gamma. \qquad (20b)$$

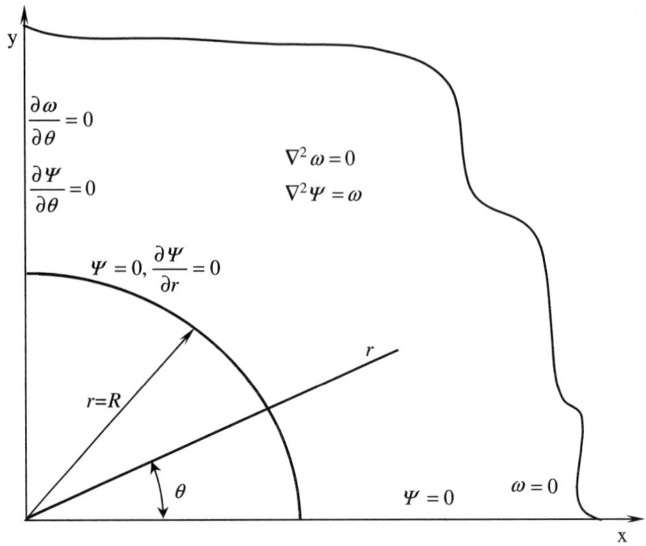

Figure 1: Area near cylinder with two symmetrical axes.

3 Special purpose trial functions for 2D creeping flow

3.1 Bi-symmetrical axis flow in a cylinder

Let us consider flow in vicinity of a cylinder of radius R as shown in Fig. 1 and assume that the flow has two symmetrical axes, one parallel to the flow direction and the other one perpendicular. The governing equations and boundary conditions in the given case are as follows:

$$\frac{\partial^2 \omega}{\partial r^2} + \frac{1}{r}\frac{\partial \omega}{\partial r} + \frac{1}{r^2}\frac{\partial^2 \omega}{\partial \theta^2} = 0 \tag{21}$$

$$\frac{\partial^2 \Psi}{\partial r^2} + \frac{1}{r}\frac{\partial \Psi}{\partial r} + \frac{1}{r^2}\frac{\partial^2 \Psi}{\partial \theta^2} = \omega \tag{22}$$

$$\Psi = 0 \quad \text{for } r = R \tag{23a}$$

$$\frac{\partial \Psi}{\partial r} = 0 \quad \text{for } r = R \tag{23b}$$

$$\Psi = 0 \quad \text{for } \theta = 0 \tag{23c}$$

$$\omega = 0 \quad \text{for } \theta = 0 \tag{23d}$$

$$\frac{\partial \Psi}{\partial \theta} = 0 \quad \text{for } \theta = \frac{\pi}{2} \tag{23e}$$

$$\frac{\partial \omega}{\partial \theta} = 0 \quad \text{for } \theta = \frac{\pi}{2}. \tag{23f}$$

The general solution of eqn (21) in a polar system of coordinates is

$$\omega = A_1 + A_2 \ln r + A_3\theta + A_4\theta \ln r + \sum_{k=1}^{\infty}\left[\left(B_k r^k + C_k r^{-k}\right) \cos(k\theta) + \left(D_k r^k + E_k r^{-k}\right)\sin(k\theta)\right] \tag{24}$$

where A_1, A_2, A_3, A_4, B_k, C_k, D_k, E_k are constants of integration.

Boundary condition (23d) will be fulfilled if

$$A_1 = A_2 = A_4 = B_k = C_k = 0, \tag{25}$$

thus

$$\omega = A_3\theta + \sum_{k=1}^{\infty}\left(D_k r^k + E_k r^{-k}\right)\sin(k\theta). \tag{26}$$

The derivative of (26) with respect to θ takes the form:

$$\frac{\partial \omega}{\partial \theta} = A_3 + \sum_{k=1}^{\infty} k\left(D_k r^k + E_k r^{-k}\right)\cos(k\theta). \tag{27}$$

Taking $A_3 = 0$ and substituting $\theta = \pi/2$ from boundary condition (23f), we obtain:

$$\sum_{k=1}^{\infty} k\left(D_k r^k + E_k r^{-k}\right)\cos k\frac{\pi}{2} = 0 \tag{28a}$$

which can be fulfilled when

$$k = 1,3,5,7,\ldots \tag{28b}$$

Hence, not loosing generalization we can write:

$$\omega = \sum_{k=1}^{\infty}\left(D_k r^{2k-1} + E_k r^{-(2k-1)}\right)\sin\left[(2k-1)\theta\right]. \tag{29}$$

The solution of the eqn (22) can be written as:

$$\Psi = \Psi_h + \Psi_{ps} \tag{30a}$$

$$\Psi_h = F_1 + F_2\theta + F_3 \ln r + F_4\theta \ln r + \sum_{k=1}^{\infty}\left[(G_k r^k + H_k r^{-k})\sin(k\theta) + (P_k r^k + Q_k r^{-k})\cos(k\theta)\right] \tag{30b}$$

where F_1, F_2, F_3, F_4, P_k, H_k, P_k, Q_k are constants and Ψ_{ps} is a particular solution of non-homogenous equation (22). Thus,

$$\frac{\partial^2\Psi_{ps}}{\partial r^2} + \frac{1}{r}\frac{\partial\Psi_{ps}}{\partial r} + \frac{1}{r^2}\frac{\partial^2\Psi_{ps}}{\partial\theta^2} = \sum_{k=1}^{\infty}\left(D_k r^{2k-1} + E_k r^{-(2k-1)}\right)\sin\left[(2k-1)\theta\right]. \tag{31}$$

By the procedure of substituting, it can easily be checked that

$$\Psi_{ps} = \left(\frac{1}{8}r^3 D_1 + \frac{1}{2} r \ln r\, E_1\right)\sin\theta + \sum_{k=2}^{\infty}\left[\frac{1}{8k} D_k r^{2k+1} - \frac{1}{8(k-1)} E_k r^{-2k+3}\right]\sin\left[(2k-1)\theta\right]. \tag{32}$$

In this sense, the general solution of eqn (22) is

$$\begin{aligned}\Psi &= F_1 + F_2\theta + F_3 \ln r + F_4\theta \ln r \\ &+ \sum_{k=1}^{\infty}\left[\left(G_k r^k + H_k r^{-k}\right)\sin(k\theta) + \left(P_k r^k + Q_k r^{-k}\right)\cos(k\theta)\right] \\ &+ \left(\frac{1}{8}r^3 D_1 + \frac{1}{2} r \ln r E_1\right)\sin\theta \\ &+ \sum_{k=2}^{\infty}\left[\frac{1}{8k} D_k r^{2k+1} - \frac{1}{8(k-1)} E_k r^{-2k+3}\right]\sin\left[(2k-1)\theta\right].\end{aligned} \tag{33}$$

Taking boundary conditions (23c) and (23e) into consideration, we have

$$F_1 = F_2 = F_3 = F_4 = P_k = Q_k = 0 \tag{34}$$

and general solution (33) is reduced to

$$\begin{aligned}\Psi &= \sum_{k=1}^{\infty}\left[\left(G_k r^k + H_k r^{-k}\right)\sin(k\theta)\right] + \left(\frac{1}{8}r^3 D_1 + \frac{1}{2} r \ln r\, E_1\right)\sin\theta \\ &+ \sum_{k=2}^{\infty}\left[\frac{1}{8k} D_k r^{2k+1} - \frac{1}{8(k-1)} E_k r^{-2k+3}\right]\sin\left[(2k-1)\theta\right].\end{aligned} \tag{35}$$

Considering conditions (23a) and (23b), we obtain the following relations between the constants:

$$G_1 = -\frac{1}{4}R^2 D_1 - \frac{1}{2}E_1\left(\ln R + \frac{1}{2}\right) \tag{36a}$$

$$H_1 = \frac{1}{8}R^4 D_1 + \frac{1}{4}R^2 E_1 \tag{36b}$$

$$G_k = \frac{R^{-4(k-1)}}{8(k-1)(2k-1)}E_k - \frac{R^2}{4(2k-1)}D_k \tag{36c}$$

$$H_k = \frac{R^{4k}}{8k(2k-1)}D_k + \frac{R^2}{4(2k-1)}E_k. \tag{36d}$$

After substituting formulas (36) into eqn (35), we can present the stream function Ψ as a function of constants D_1, E_1, D_k, E_k, namely:

$$\begin{aligned}\Psi = &\left[\frac{1}{8}\left(r^3 - 2R^2 r + R^4 r^{-1}\right)\sin\theta\right]D_1 + \left[\frac{1}{4}\left(R^2 r^{-1} + 2r\ln\frac{r}{R} - r\right)\sin\theta\right]E_1 \\ &+ \sum_{k=2}^{\infty}\frac{(2k-1)r^{2k+1} - 2kR^2 r^{2k-1} + R^{4k}r^{-(2k-1)}}{8k(2k-1)}\sin\left[(2k-1)\theta\right]D_k \\ &+ \sum_{k=2}^{\infty}\frac{R^{-4(k-1)}r^{2k+1} + 2(k-1)R^2 r^{-2k+1} - (2k-1)r^{-2k+3}}{8(k-1)(2k-1)}\sin\left[(2k-1)\theta\right]E_k.\end{aligned} \tag{37}$$

Now we introduce symbols:

$$\begin{aligned} &C_1 = D_1 \\ &C_2 = D_2 \\ &\quad\vdots \\ &C_k = D_k \quad (k = 2,3,\ldots,M) \\ &C_{M+1} = E_1 \\ &C_{M+2} = E_2 \\ &\quad\vdots \\ &C_{M+k} = E_k \quad (k = 2,3,\ldots,M) \end{aligned} \tag{38}$$

and after cutting infinite series to M first expressions, we can write:

$$\omega = \sum_{k=1}^{2M} C_k T_k(r,\theta) \tag{39a}$$

$$\Psi = \sum_{k=1}^{2M} C_k U_k(r,\theta,R) \tag{39b}$$

where $T_k(r,\theta)$ and $U_k(r,\theta,R)$ are functions given by:

$$T_k(r,\theta) = r^{2k-1}\sin\left[(2k-1)\theta\right], \quad k=1,2,...,M \tag{39c}$$

$$T_{M+k}(r,\theta) = r^{(2k-1)}\sin\left[(2k-1)\theta\right], \quad k=1,2,...,M \tag{39d}$$

$$U_1(r,\theta,R) = \frac{1}{8}\left(r^3 - 2R^2 r + R^4 r^{-1}\right)\sin\theta \tag{39e}$$

$$U_k(r,\theta,R) = \frac{(2k-1)r^{2k+1} - 2kR^2 r^{2k-1} + R^{4k} r^{-(2k-1)}}{8k(2k-1)}\sin\left[(2k-1)\theta\right] \tag{39f}$$

$$U_{M+1}(r,\theta,R) = \frac{1}{4}\left(R^2 r^{-1} + 2r\ln\frac{r}{R} - r\right)\sin\theta \tag{39g}$$

$$U_{M+k}(r,\theta,R) = \frac{R^{-4(k-1)} r^{2k+1} - (2k-1) r^{-2k+3} + 2(k-1)R^2 r^{-(2k-1)}}{8(2k-1)(k-1)} \times \sin\left[(2k-1)\theta\right] \quad k=2,3,...,M \tag{39h}$$

Constants C_k must be defined from the remaining boundary conditions.

An example of a boundary problem, in which trial functions (39) can be used, is creeping flow through a cylinder set arranged according to the square grid (Fig. 2). After selecting the *OABC* fragment from the flow area, the boundary conditions should approximately be fulfilled along the *AB* and *BC* lines, while on the remaining boundaries they are fulfilled exactly.

3.2 Symmetrical flow in vicinity of cylinder

Let us consider an area in vicinity of a cylinder with radius R, and velocity field symmetric with respect to main flow direction (Fig. 3). Now, the governing equations

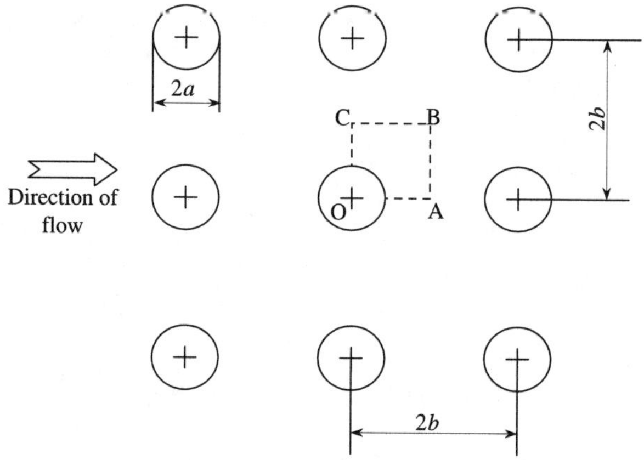

Figure 2: Bi-periodical flow across cylinder set.

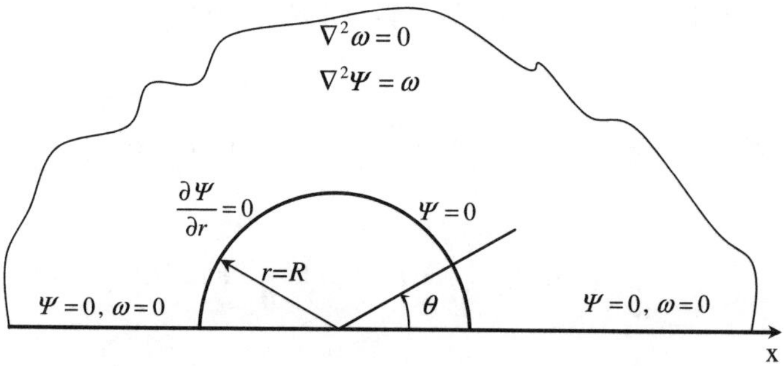

Figure 3: Symmetrical flow in vicinity of a cylinder.

are identical with equations of former cases (21)–(22) and the boundary conditions are as follows:

$$\Psi = 0 \quad \text{for } r = R \tag{40a}$$

$$\frac{\partial \Psi}{\partial r} = 0 \quad \text{for } r = R \tag{40b}$$

$$\Psi = 0 \quad \text{for } \theta = 0 \tag{40c}$$

$$\omega = 0 \quad \text{for } \theta = 0 \tag{40d}$$

$$\frac{\partial \Psi}{\partial \theta} = 0 \quad \text{for } \theta = \pi \tag{40e}$$

$$\frac{\partial \omega}{\partial \theta} = 0 \quad \text{for } \theta = \pi. \tag{40f}$$

Applying a procedure similar to the one presented in the former section, one can derive the following solution which fulfils governing equations (21)–(22) and boundary conditions (40):

$$\omega = \sum_{k=1}^{2M} C_k T_k(r,\theta) \tag{41a}$$

$$\Psi = \sum_{k=1}^{2M} C_k U_k(r,\theta,R) \tag{41b}$$

$$T_k = r^k \sin(k\theta), \quad k = 1,2,\ldots,M \tag{41c}$$

$$T_{M+k} = r^{-k} \sin(k\theta), \quad k = 1,2,\ldots,M \tag{41d}$$

$$U_1(r,\theta,R) = \frac{1}{8}\left(r^3 - 2R^2 r + R^4 r^{-1}\right)\sin\theta \tag{41e}$$

$$U_k(r,\theta,R) = \frac{kr^{k+2} + R^{2(k+1)}r^{-k} - (k+1)R^2 r^k}{4k(k+1)}\sin(k\theta), \quad k = 2,3,\ldots,M \tag{41f}$$

$$U_{M+1}(r,\theta,R) = \frac{1}{4}\left(R^2 r^{-1} + 2r\ln\frac{r}{R} - r\right)\sin\theta \tag{41g}$$

$$U_{M+k}(r,\theta,R) = \frac{R^{-2(k-1)}r^k - kr^{-k+2} + (k-1)R^2 r^{-k}}{4k(k-1)}\sin(k\theta), \quad k = 2,3,\ldots,M. \tag{41h}$$

Such functions can be used, for example, for solving a boundary problem concerning creeping flow of a cylinder situated between two parallel walls as shown in Fig. 4. In this case, constants C_k can be defined from approximately fulfilled boundary conditions on *AGFE* boundary.

3.3 Flow in vicinity of rotating cylinder

Consider a cylinder with radius *R*, which rotates with the constant angle velocity $\tilde{\Omega} = 1$. We assume that the flow is symmetrical with respect to axis *X*, as shown in Fig. 5. In the discussed case, together with eqns (21) and (22), the following boundary conditions should be fulfilled:

$$\Psi = \Psi_1 \quad \text{for } r = R \tag{42a}$$

$$\frac{\partial \Psi}{\partial r} = 1 \quad \text{for } r = R \tag{42b}$$

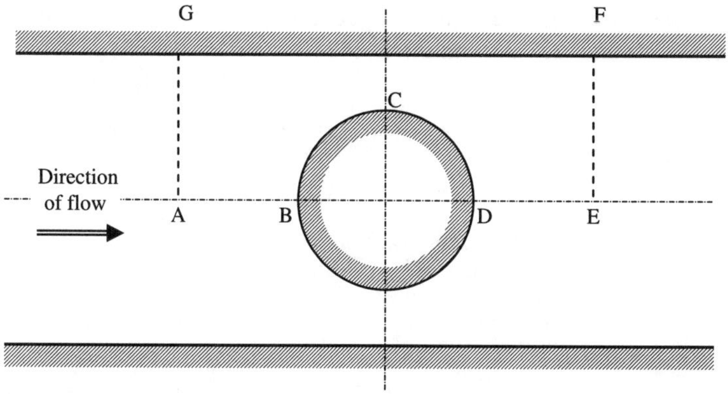

Figure 4: Flow near a cylinder situated between two parallel walls.

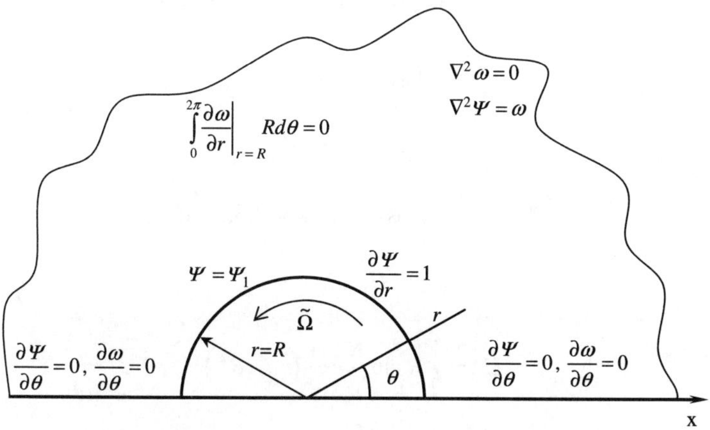

Figure 5: Flow around a rotating cylinder.

$$\int_0^{2\pi} \frac{\partial \omega}{\partial r}\Big|_{r=R} r d\theta = 0 \tag{42c}$$

$$\frac{\partial \omega}{\partial \theta} = 0 \quad \text{for } \theta = 0 \tag{42d}$$

$$\frac{\partial \Psi}{\partial \theta} = 0 \quad \text{for } \theta = 0 \tag{42e}$$

$$\frac{\partial \Psi}{\partial \theta} = 0 \quad \text{for } \theta = \pi \tag{42f}$$

$$\frac{\partial \omega}{\partial \theta} = 0 \quad \text{for } \theta = \pi. \tag{42g}$$

Applying the above method explained above we can obtain the solution, which fulfils eqns (21) and (22) and conditions (42a–g):

$$\omega = \sum_{k=2}^{2M} C_k T_k(r,\theta) \tag{43a}$$

$$\Psi = R \ln \frac{r}{R} + \sum_{k=1}^{2M} C_k U_k(r,\theta,R) \tag{43b}$$

where

$$T_2 = 1 \tag{43c}$$

$$T_k = r^{-(k-1)} \cos\left[(k-1)\theta\right], \quad k = 3,4,\ldots,M \tag{43d}$$

$$T_{M+1} = r\cos\theta \tag{43e}$$

$$T_{M+2} = r^{-1}\cos\theta \tag{43f}$$

$$T_{M+1+k} = r^k \cos(k\theta), \quad k = 2,3,\ldots,M \tag{43g}$$

$$U_1 = 1 \tag{43h}$$

$$U_2 = \frac{1}{2} R^2 \ln \frac{R}{r} + \frac{1}{4}\left(r^2 - R^2\right) \tag{43i}$$

$$U_k = \frac{r^{k-1} R^{-2(k-2)} + (k-2) R^2 r^{-k+1} - (k-1) r^{-k+3}}{4(k-1)(k-2)} \cos\left[(k-1)\theta\right], \quad k = 3,4,\ldots,M \tag{43j}$$

$$U_{M+1} = \frac{1}{8}\left(r^3 - 2R^2 r + R^4 r^{-1}\right)\cos\theta \tag{43k}$$

$$U_{M+2} = \frac{1}{4}\left(2r \ln \frac{r}{R} - r + R^2 r^{-1}\right)\cos\theta \tag{43l}$$

$$U_{M+k} = \frac{r^{-k+1} R^{2k} - kR^2 r^{k-1} - (k-1) r^{k+1}}{4(k-1)k} \cos\left[(k-1)\theta\right], \quad k = 3,4,\ldots,M. \tag{43m}$$

The first constant, i.e. C_1, is an unknown value of the stream function on the cylinder, hence $C_1 = \Psi_1$. Functions presented by formulas (43c–m) can also be

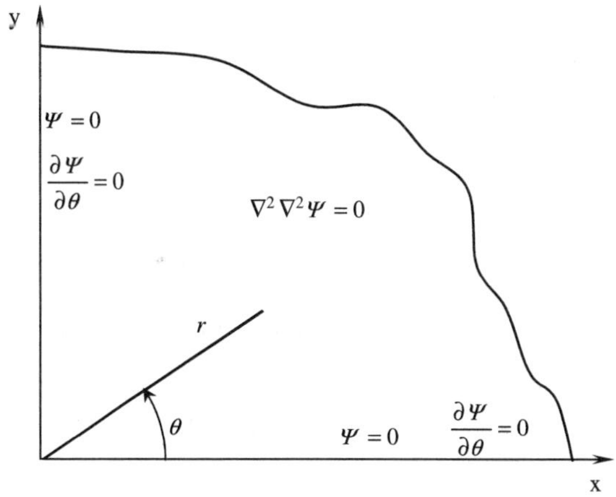

Figure 6: Flow within a stiff corner.

used, for example, in case of modelling of creeping flow between eccentric cylinders, from which outer one is motionless and the inner one rotates with constant velocity (flow in a bearing).

3.4 Flow around a stiff corner

Let us consider flow around a stiff corner with an angle $\pi/2$ as shown in Fig. 6. The governing equation and boundary conditions are as follows:

$$\left(\frac{\partial^2}{\partial r^2}+\frac{1}{r}\frac{\partial}{\partial r}+\frac{1}{r^2}\frac{\partial^2}{\partial \theta^2}\right)^2 \Psi(r,\theta)=0 \tag{44}$$

$$\Psi=0 \quad \text{for } \theta=0 \tag{45a}$$

$$\frac{\partial \Psi}{\partial \theta}=0 \quad \text{for } \theta=0 \tag{45b}$$

$$\Psi=0 \quad \text{for } \theta=\frac{\pi}{2} \tag{45c}$$

$$\frac{\partial \Psi}{\partial \theta}=0 \quad \text{for } \theta=\frac{\pi}{2}. \tag{45d}$$

Moffatt [87], considering flow in vicinity of stiff corners, proposed a solution of eqn (44) in the formula:

$$\Psi=r^{\lambda_k+1} f_k(\theta) \tag{46}$$

where λ_k are real or complex constants called solution exponents. Following Moffatt's approach, we assume the solution of eqn (44) in the form:

$$\Psi(r,\theta) = \beta_0 r f_0(\theta) + \beta_1 r^2 f_1(\theta) + \sum_{k=2}^{\infty} \beta_k r^{\lambda_k+1} f_k(\theta) \tag{47}$$

where β_0, β_1, β_k are constants and

$$f_0(\theta) = A_0 \cos\theta + B_0 \sin\theta + C_0\theta\cos\theta + D_0\theta\sin\theta \tag{48a}$$

$$f_1(\theta) = A_1 \cos 2\theta + B_1 \sin 2\theta + C_1\theta + D_1 \tag{48b}$$

$$\begin{aligned} f_k(\theta) = {} & A_k\cos\left[(\lambda_k+1)\theta\right] + B_k\sin\left[(\lambda_k+1)\theta\right] \\ & + C_k\cos\left[(\lambda_k-1)\theta\right] + D_k\sin\left[(\lambda_k-1)\theta\right] \\ k = {} & 2,3,.... \end{aligned} \tag{48c}$$

Functions $f_k(\theta)$ introduce arbitrary constants A_0, B_0, C_0, D_0, A_1, B_1, C_1, D_1, A_k, B_k, C_k and D_k which must be defined from respective boundary conditions.

Conditions (45a) will be fulfilled if:

$$\begin{gathered} A_0 = 0,\ A_1 + D_1 = 0,\ A_k + C_k = 0 \\ k = 2,3,... \end{gathered} \tag{49}$$

Calculating derivatives of functions $f_k(\theta)$ with respect to θ, we obtain

$$f_0'(\theta) = -A_0\sin\theta + B_0\cos\theta + C_0(\cos\theta - \theta\sin\theta) + D_0(\sin\theta + \theta\cos\theta) \tag{50a}$$

$$f_1'(\theta) = -2A_1\sin 2\theta + 2B_1\cos 2\theta + C_1 \tag{50b}$$

$$\begin{aligned} f_k'(\theta) = {} & -A_k(\lambda_k+1)\sin\left[(\lambda_k+1)\theta\right] + B_k(\lambda_k+1)\cos\left[(\lambda_k+1)\theta\right] \\ & - C_k(\lambda_k-1)\sin\left[(\lambda_k-1)\theta\right] + D_k(\lambda_k-1)\cos\left[(\lambda_k-1)\theta\right]. \end{aligned} \tag{50c}$$

Introduction of formulas (50) into boundary condition (45b) leads to the following relations:

$$B_0 + C_0 = 0,\ 2B_1 + C_1 = 0 \tag{51a}$$

$$\begin{gathered} B_k(\lambda_k+1) + D_k(\lambda_k-1) = 0 \\ k = 2,3,.... \end{gathered} \tag{51b}$$

Condition (45c) will be fulfilled if:

$$B_0 + D_0 \frac{\pi}{2} = 0 \tag{52a}$$

$$-A_1 + C_1 \frac{\pi}{2} + D_1 = 0 \tag{52b}$$

$$\begin{aligned} &A_k \cos\left[(\lambda_k + 1)\frac{\pi}{2}\right] + B_k \sin\left[(\lambda_k + 1)\frac{\pi}{2}\right] + C_k \cos\left[(\lambda_k - 1)\frac{\pi}{2}\right] \\ &+ D_k \sin\left[(\lambda_k - 1)\frac{\pi}{2}\right] = 0. \end{aligned} \tag{52c}$$

Condition (45d) leads to:

$$-A_0 - C_0 \frac{\pi}{2} + D_0 = 0 \tag{53a}$$

$$-2B_1 + C_1 = 0 \tag{53b}$$

$$\begin{aligned} &-A_k (\lambda_k + 1)\sin\left[(\lambda_k + 1)\frac{\pi}{2}\right] + B_k (\lambda_k + 1)\cos\left[(\lambda_k + 1)\frac{\pi}{2}\right] \\ &- C_k (\lambda_k - 1)\sin\left[(\lambda_k - 1)\frac{\pi}{2}\right] + D_k (\lambda_k - 1)\cos\left[(\lambda_k - 1)\frac{\pi}{2}\right] = 0 \\ &k = 2, 3, \ldots \end{aligned} \tag{53c}$$

From (49), (51a), (52a) and (53a) we get:

$$A_0 = B_0 = C_0 = D_0. \tag{54a}$$

Similarly, from formulas (49), (51a), (52b), (53b), it results that

$$A_1 = B_1 = C_1 = D_1. \tag{54b}$$

Equations (52c) and (53c) express C_k and D_k through A_k and B_k, namely:

$$C_k = -A_k \tag{54c}$$

$$D_k = -\frac{(\lambda_k + 1)}{(\lambda_k - 1)} B_k. \tag{54d}$$

Now eqns (52c) and (53c) can be written as

$$A_k\left\{\cos\left[(\lambda_k+1)\frac{\pi}{2}\right]-\cos\left[(\lambda_k-1)\frac{\pi}{2}\right]\right\}$$
$$+B_k\left\{\sin\left[(\lambda_k+1)\frac{\pi}{2}\right]-\frac{\lambda_k+1}{\lambda_k-1}\sin\left[(\lambda_k-1)\frac{\pi}{2}\right]\right\}=0 \quad (55a)$$

$$-A_k\left\{(\lambda_k+1)\sin\left[(\lambda_k+1)\frac{\pi}{2}\right]-(\lambda_k-1)\sin\left[(\lambda_k-1)\frac{\pi}{2}\right]\right\}$$
$$+B_k\left\{(\lambda_k+1)\cos\left[(\lambda_k+1)\frac{\pi}{2}\right]-(\lambda_k+1)\cos\left[(\lambda_k-1)\frac{\pi}{2}\right]\right\}=0 \quad (55b)$$

which can be expressed as

$$A_k(\lambda_k-1)\sin\frac{\lambda_k\pi}{2}=B_k\lambda_k\cos\frac{\lambda_k\pi}{2} \quad (56a)$$

$$A_k\lambda_k\cos\frac{\lambda_k\pi}{2}=-B_k(\lambda_k+1)\sin\frac{\lambda_k\pi}{2}. \quad (56b)$$

After dividing both sides of the above equations, we obtain

$$\lambda_k^2\cos^2\frac{\lambda_k\pi}{2}=-(\lambda_k^2-1)\sin^2\frac{\lambda_k\pi}{2}. \quad (57)$$

This formula leads to two transcendental equations in the form:

$$\sin\frac{\lambda_k\pi}{2}=\lambda_k \quad (58a)$$

$$\sin\frac{\lambda_k\pi}{2}=-\lambda_k. \quad (58b)$$

The solution of these equations is given in Table 2.

From eqn (56b), we get

$$B_k=-A_k\frac{\lambda_k\operatorname{ctg}\frac{\lambda_k\pi}{2}}{(\lambda_k-1)}. \quad (59)$$

Table 2: Roots of eqn (58) determined numerically.

k	a_k	b_k
1	0.0	0.0
2	1.0	0.0
3	2.7396	1.1190
4	4.8083	1.4639
5	6.8451	1.6816
6	8.8688	1.8424
7	10.8856	1.9702
8	12.8981	2.0764
9	14.9079	2.1673
10	16.9158	2.2468
20	36.9528	2.7409

After accepting the above results in solution (47), we obtain:

$$\Psi = \sum_{k=1}^{2M} F_k U_k(r,\theta) \tag{60}$$

where $U_k(r,\theta)$ are functions given by:

$$U_k(r,\theta) = \mathrm{Re}\left\{ \begin{array}{l} r^{\lambda_k+1}\left[\cos(\lambda_k+1)\theta - \cos(\lambda_k-1)\theta\right] \\ +\dfrac{\lambda_k \mathrm{ctg}\dfrac{\lambda_k \pi}{2}}{(\lambda_k+1)}\left[\sin(\lambda_k+1)\theta - \dfrac{\lambda_k+1}{\lambda_k-1}\sin(\lambda_k-1)\theta\right] \end{array} \right\}, \quad k = 1,2,\ldots,2M \tag{61}$$

and $\lambda_k = a_k + ib_k$ are complex numbers, which are solutions of the transcendental equations (58).

The functions presented in formulas (60) and (61) can also be used to define creeping flow in a cavern, as shown in Fig. 7. In this case, the whole area can be divided into four large parts. For two of these subregions (element III and element IV in Fig. 7), solutions (60)–(61) can be applied.

3.5 Stick-slip problem

Let us consider an area in vicinity of an edge of a half-infinite plate (Fig. 8). The main flow direction is parallel to the edge of the plate. The governing

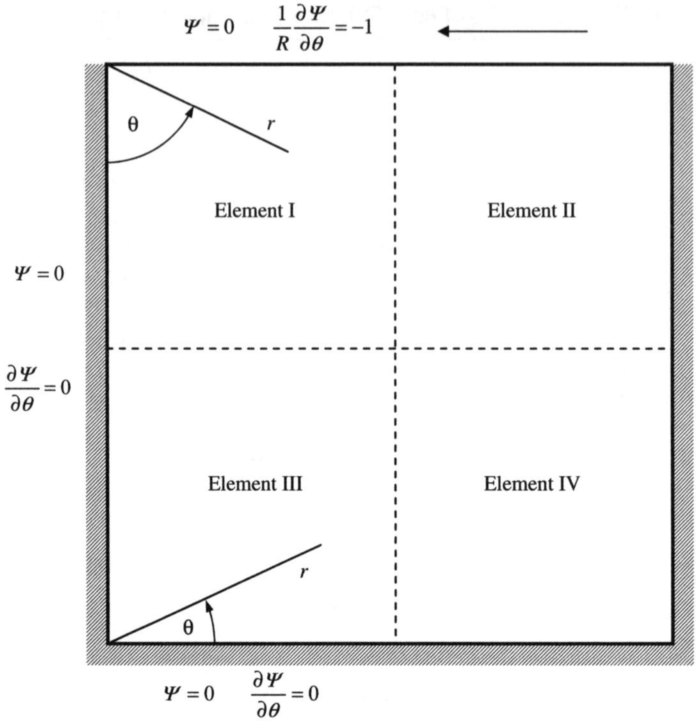

Figure 7: Flow in a cavern.

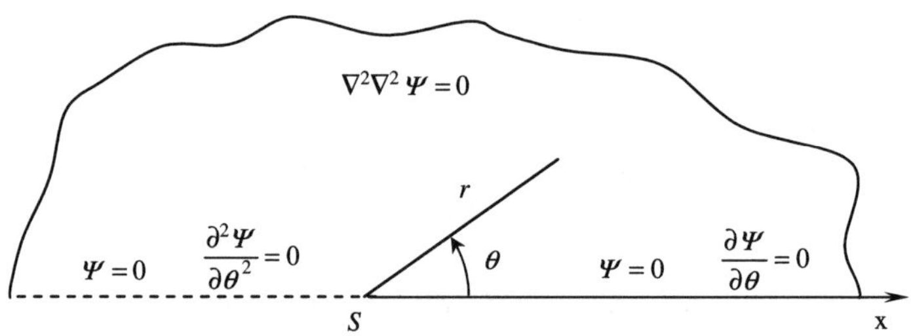

Figure 8: Flow around a half-infinite plate.

equation is identical with (44), but the conditions on the boundary sides are as follows:

$$\Psi = 0 \quad \text{for } \theta = 0 \tag{62a}$$

$$\frac{\partial \Psi}{\partial \theta} = 0 \quad \text{for } \theta = 0 \tag{62b}$$

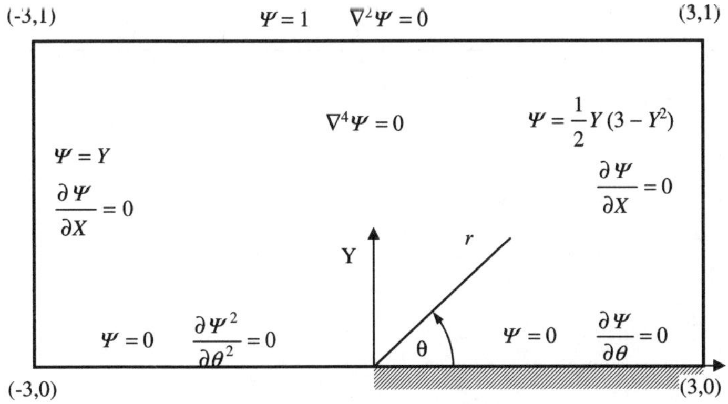

Figure 9: Stick-slip problem solved by Richardson [108].

$$\Psi = 0 \quad \text{for } \theta = \pi \tag{62c}$$

$$\frac{\partial^2 \Psi}{\partial \theta^2} = 0 \quad \text{for } \theta = \pi. \tag{62d}$$

After applying a procedure similar to one presented earlier, we obtain the following solution, which fulfils eqn (44) and boundary conditions (62):

$$\Psi = \sum_{k=1}^{2M} C_k U_k(r,\theta) \tag{63a}$$

where

$$U_k(r,\theta) = r^{\lambda_k+1}\left\{\cos\left[(\lambda_k+1)\theta\right] - \cos\left[(\lambda_k-1)\theta\right]\right\} \tag{63b}$$

$$U_{M+k}(r,\theta) = r^{\lambda_k+1}\left\{\sin\left[(\lambda_k+1)\theta\right] - \frac{\lambda_k+1}{\lambda_k-1}\sin\left[(\lambda_k-1)\theta\right]\right\}, \quad k = 1,2,\ldots,M \tag{63c}$$

where

$$\lambda_1 = \frac{1}{2},\quad \lambda_2 = \frac{3}{2},\quad \lambda_3 = 2,\quad \lambda_4 = \frac{5}{2},\quad \lambda_5 = 3,\quad \lambda_6 = \frac{7}{2},\quad \lambda_7 = 4,\ldots. \tag{63d}$$

Functions (63) can be applied, for example, to solve a boundary value problem presented in Fig. 9 and called 'stick-slip', which has been solved by Richardson [108]. It should, however, be noted that the notion of a stick-slip problem exists also in dry friction, but it has there a different meaning [8].

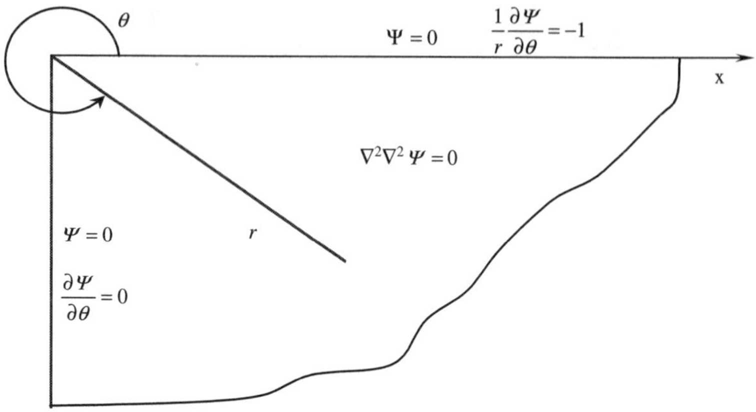

Figure 10: Flow within a rectangular corner with one moving wall.

3.6 Flow in vicinity of a rectangular corner, one wall of which moves at constant velocity

Let us consider area of a corner at an angle $\pi/2$, one wall of which moves at constant velocity $\upsilon = 1$, as shown in Fig. 10. The governing equation is identical with the stiff corner equation (44) and the boundary conditions are as follows:

$$\Psi = 0 \quad \text{for } \theta = 0 \tag{64a}$$

$$\frac{1}{r}\frac{\partial \Psi}{\partial \theta} = -1 \quad \text{for } \theta = 0 \tag{64b}$$

$$\Psi = 0 \quad \text{for } \theta = -\frac{\pi}{2} \tag{64c}$$

$$\frac{\partial \Psi}{\partial \theta} = 0 \quad \text{for } \theta = -\frac{\pi}{2}. \tag{64d}$$

After applying derivation analogical to the one presented earlier, we obtain the following trial functions fulfilling eqn (44) and boundary conditions (64):

$$\Psi = r\frac{1}{1-\left(\frac{\pi}{2}\right)^2}\left[\left(\frac{\pi}{2}\right)^2 \sin\theta - \theta\cos\theta + \frac{\pi}{2}\theta \ \sin\theta\right] + \sum_{k=1}^{2M} F_k U_k\left(r,\theta\right) \tag{65}$$

where $U_k\ (r,\theta)$ are defined by (61) and $\lambda_k = a_k + ib_k$ are complex numbers, which fulfil transcendental equations (58). Functions (61) can also be applied to model creeping flow in the cavern shown in Fig. 7 but now in elements I and II.

4 Thin plate bending

4.1 Boundary problem formulation

The governing equation of a thin plate bending has a form:

$$\nabla^4 w = \frac{q}{D} \tag{66}$$

where w is plate deflection, D is a plate stiffness and q is plate loading.

The plate stiffness D is defined as

$$D = \frac{Eh^3}{12\left(1-v^2\right)} \tag{67}$$

where h is the plate thickness and v is Poisson's ratio.

Equation (66) should be solved with the proper boundary conditions which may be expressed by: deflection w, boundary slope $\partial w/\partial n$, bending moment M_n and effective shear force V_n. The latter two quantities are expressed by means of deflection w as follows:

$$M_n(w) = D\left[\frac{\partial^2 w}{\partial n^2} + v\frac{\partial^2 w}{\partial s^2}\right] \tag{68a}$$

$$V_n(w) = D\left[\frac{\partial\left(\nabla^2 w\right)}{\partial n} - (1-v)\frac{\partial}{\partial s}\left(\frac{1}{\rho}\frac{\partial w}{\partial s} - \frac{\partial^2 w}{\partial s \partial n}\right)\right] \tag{68b}$$

where ρ is a curvature radius of the boundary contour and s, n are tangential and normal directions to the plate boundary, respectively.

Typical boundary conditions, depending on the type of plate fastening, give on the edge of a structure values equal to zero for two out of four functions w, $\partial w/\partial n$, M_n, V_n and can be written as follows:

- for clamped edge of the plate:

$$w = \frac{\partial w}{\partial n} = 0 \tag{69a}$$

- for simple support of the plate:

$$w = M_n(w) = 0 \tag{69b}$$

Table 3: Papers in which the boundary collocation method was applied to bending of thin plates.

Bending of thin plates	References
Simply connected	[1, 6, 10, 21–25, 29, 30, 42, 43, 46, 77–81, 84, 85, 89, 90, 102, 104, 105, 109, 115–118, 121, 122, 132]
Multiply connected	[4, 5, 13, 37–39, 44, 83, 96, 98–100, 111]
Infinite	[62, 64, 83, 119]

- on free edge of the plate:

$$V_n(w) = M_n(w) = 0 \tag{69c}$$

- on a symmetry line of the plate:

$$V_n(w) = \frac{\partial w}{\partial n} = 0$$

$$V_n(w) = D\left[\frac{\partial^3 w}{\partial n^3} + (2-\nu)\frac{\partial^3 w}{\partial s^2 \partial n}\right]. \tag{69d}$$

The boundary collocation method has been applied by many authors to determine thin plate deflection (Table 3). One of the first known applications of that method concerns deflection of a regularly supported roof modelled by a thin plate [119]. In the construction of trial functions, applied by various authors, at least three methods can be noticed. Some authors used general solutions of biharmonic equations in a polar coordinate system [6, 23–26, 30, 79, 84]. Others used the method of fundamental solutions. In this case, the trial functions are fundamental solutions of the biharmonic problems, singularities of which are placed outside the considered area [10, 43, 46]. Some others, in case of analysing polygonal plate deflection, used the so-called large finite elements [42, 115, 116, 121, 122]. Also the influence T-functions expressed by trigonometric series can in this case be applied (see Section 3 in Chapter 3).

4.2 Construction of special purpose trial functions

In all the cases considered below, the plate is regularly loaded by constant, continuous load q. After introducing dimensionless deflection W:

$$W = \frac{wD}{qa^4}, \tag{70a}$$

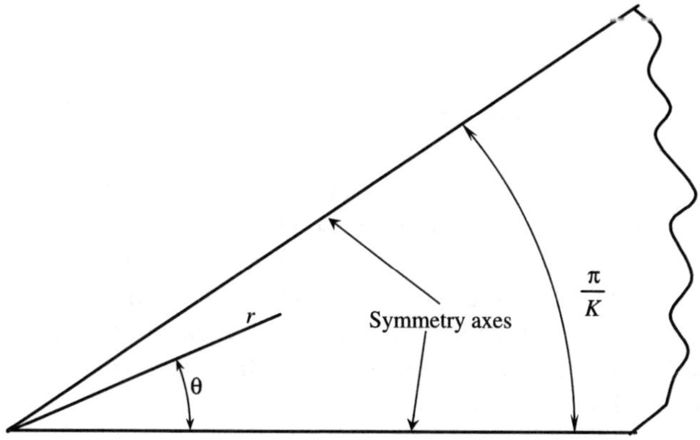

Figure 11: Repeated plate fragment with symmetry axes.

a particular solution of non-homogenous equation (67) is equal:

$$W_{ps} = \frac{r^4}{64} \tag{70b}$$

Here, $r = \bar{r}/a$ is a non-dimensional radius and a is a characteristic dimension of the plate (compare explanations after eqn (8) in Chapter 2).

Let us consider a simply connected area of a homogenously loaded plate, which has K symmetry axes. Then, determination of the plate deflection can be limited to the repeating area bounded by symmetry lines (Fig. 11).

The boundary conditions on these lines are

$$\frac{\partial W}{\partial \theta} = \frac{\partial^3 W}{\partial \theta^3} = 0 \quad \text{for } \theta = 0 \tag{71a}$$

$$\frac{\partial W}{\partial \theta} = \frac{\partial^3 W}{\partial \theta^3} = 0 \quad \text{for } \theta = \frac{\pi}{K} \tag{71b}$$

and the solution is finite for $r = 0$.

Let us now consider the general form solution in polar coordinates (4) (in Chapter 3) with boundary conditions (71). The fact that the solution has to be finite for $r = 0$ and the cut of the infinite series to its $2M$ first expressions gives:

$$W = \frac{r^4}{64} + \sum_{k=1}^{2M} A_k T_k \left(r, \theta, K\right) \tag{72}$$

where A_k are unknown constants and

$$T_k = 1.0 \quad \text{for } k = 1 \tag{73a}$$

$$T_k = r^{K(k-1)} \cos\left[K\left(k-1\right)\theta\right] \quad \text{for } k = 2, 3, \ldots, M \tag{73b}$$

$$T_k = r^2 \quad \text{for } k = M+1 \tag{73c}$$

$$T_k = r^{K(k-M-1)+2} \cos\left[K(k-M-1)\theta\right] \quad \text{for } k = M+2,\ M+3,\dots,2M. \tag{73d}$$

The unknown coefficients A_k can be determined by fulfilling boundary conditions on the right edge of the repeated area (see Fig. 11).

Consider now a doubly connected area of a homogenously loaded plate, which has K symmetry axes. The hole in the plate centre is clamped. Then, determination of the plate deflection can be limited to the area repeated and bounded by symmetry lines (Fig. 12). In the considered case, the boundary conditions on the hole edge are in the following form:

$$W = 0 \quad \text{for } r = R \tag{74a}$$

$$\frac{\partial W}{\partial r} = 0 \quad \text{for } r = R \tag{74b}$$

and the other conditions result from the symmetry:

$$\frac{\partial W}{\partial \theta} = \frac{\partial^3 W}{\partial \theta^3} = 0 \quad \text{for } \theta = 0 \tag{75a}$$

$$\frac{\partial W}{\partial \theta} = \frac{\partial^3 W}{\partial \theta^3} = 0 \quad \text{for } \theta = \frac{\pi}{K}. \tag{75b}$$

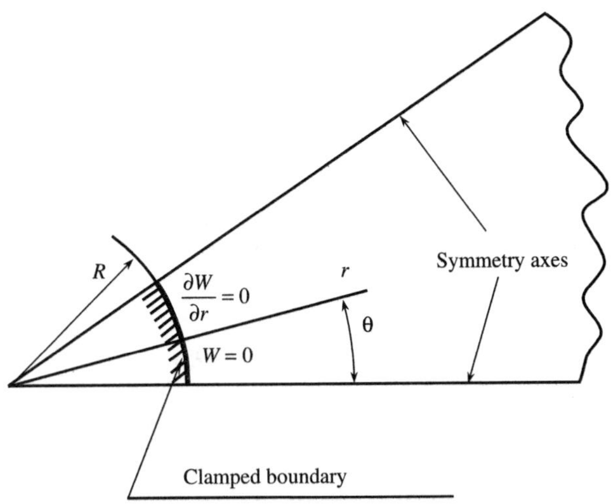

Figure 12: Repeated plate fragment with symmetry axes and clamped hole in the centre.

The solution, which exactly fulfils the differential equation and boundary conditions (74) and (75), is now:

$$W = \frac{1}{64}\left(r^2 - R^2\right)^2 + \sum_{k=1}^{2M} A_k T_k\left(r,\theta,R,K\right) \tag{76a}$$

where

$$T_1 = \left(\frac{1}{2} + \ln\frac{r}{R} - \frac{r^2}{2R^2}\right) \tag{76b}$$

$$T_k = \frac{Kkr^{Kk+2} + R^{2(Kk+1)}R^{-Kk} - \left(Kk+1\right)R^2 r^{Kk}}{Kk}\cos\left(Kk\theta\right) \tag{76c}$$

for $k = 2,3,\ldots,M$

$$T_k = \left(\frac{R^2}{2} - \frac{r^2}{2} + r^2\ln\frac{r}{R}\right) \quad \text{for } k = M+1 \tag{76d}$$

$$T_{M+1+j} = \frac{Kjr^{-Kj+2} - R^{-2(Kj-1)}r^{Kj} - \left(Kj+1\right)R^2 r^{-Kj}}{Kj}\cos\left(Kj\theta\right), \tag{76e}$$

$j = 2,3,\ldots,M.$

Let us consider again a doubly connected area of a homogenously loaded plate with K symmetry axes, but now with a simply supported hole in its centre. Then again the determination of the plate deflection can be limited to the separated area bounded by symmetry lines (Fig. 13). The boundary conditions along the symmetry lines are now as follows:

$$\frac{\partial W}{\partial \theta} = \frac{\partial^3 W}{\partial \theta^3} = 0 \quad \text{for } \theta = 0 \tag{77a}$$

$$\frac{\partial W}{\partial \theta} = \frac{\partial^3 W}{\partial \theta^3} = 0 \quad \text{for } \theta = \frac{\pi}{K} \tag{77b}$$

and the boundary conditions on the hole edge:

$$W = 0 \quad \text{for } r = R \tag{78a}$$

$$\frac{\partial^2 W}{\partial r^2} + \nu\left(\frac{1}{r}\frac{\partial W}{\partial r} + \frac{1}{r^2}\frac{\partial^2 W}{\partial \theta^2}\right) = 0 \quad \text{for } r = R. \tag{78b}$$

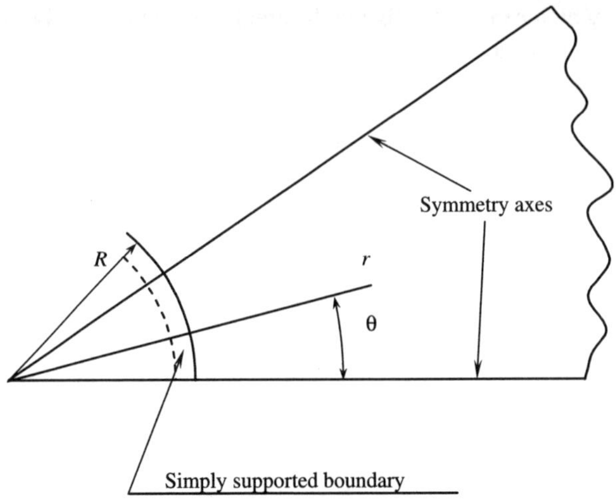

Figure 13: Repeated plate fragment with symmetry axes and simply supported hole in the centre.

The solution that exactly fulfils the differential equation (66) and boundary conditions (77) and (78) takes the form:

$$W = \frac{r^4(1+v)+R^4(5+v)-2R^2r^2(3+v)}{64(1+v)} + \sum_{k=1}^{2M} A_k T_k(r,\theta,R,v,K) \tag{79a}$$

where

$$T_1 = \frac{r^2(1-v)-\left[2(1+v)\ln R+3+v\right]r^2+2(1+v)r^2\ln R}{2(1+v)R^2} \tag{79b}$$

$$T_k = \frac{(1+v)(Kk+1)R^2r^{Kk}-(2Kk+1+v)R^{2Kk}r^{-Kk}+Kk(1-v)r^{Kk+2}}{Kk(1+v)}, \tag{79c}$$

$k = 2,\ldots,M$

$$T_{M+1} = \frac{2(3+v)R^2-\left[2(1+v)\ln R+3+v\right]r^2+2(1+v)r^2\ln R}{2(1+v)} \tag{79d}$$

$$T_{M+1+j} = \frac{(1+v)(Kj-1)R^2r^{-Kj}-(2Kj-vKj-1-v)R^{-2Kj+2}r^{Kj}+Kj(1-v)r^{-Kj+2}}{Kj(1-v)}$$

for $j = 2,\ldots,M$. (79e)

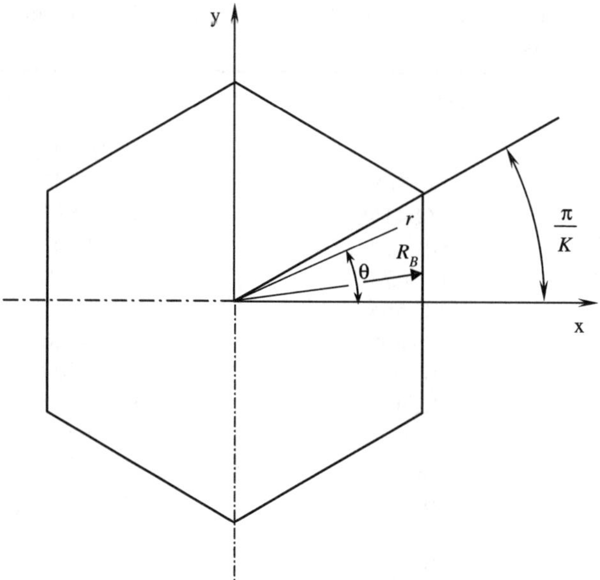

Figure 14: Regular, polygonal plate, $K = 6$.

4.3 Application of the special purpose trial functions. Example

Let us consider a regular, polygonal plate, homogenously loaded on the surface. Let K be the number of the polygon (Fig. 14). We shall consider both the case of a clamped plate and a one simply supported on its edge.

In these cases, the boundary conditions resulting from the symmetry are

$$\frac{\partial W}{\partial \theta} = \frac{\partial^3 W}{\partial \theta^3} = 0 \quad \text{for } \theta = 0 \tag{80a}$$

$$\frac{\partial W}{\partial \theta} = \frac{\partial^3 W}{\partial \theta^3} = 0 \quad \text{for } \theta = \frac{\pi}{K}. \tag{80b}$$

And of course the solution is bounded for r=0.

Moreover, for the simply supported plate, there are conditions on the right edge of the repeated fragment:

$$W = 0 \text{ and } \nabla^2 W = 0 \quad \text{for } r = R_b = \frac{\cos\dfrac{\pi}{K}}{\cos\theta} \tag{81a}$$

Table 4: Non-dimensional deflection (70a) in centre of plate – regular polygon with K sides

K	Clamped plate	Simply supported plate
3		0.0052083
4	0.0055888	0.016251
5	0.0084156	0.024881
6	0.010314	0.030707
8	0.012474	0.037329
10	0.013565	0.040661
16	0.014805	0.044426
20	0.015099	0.045309
32	0.015419	0.046267
40	0.015493	0.046487
50	0.015541	0.046627
64	0.015574	0.046724
80	0.015592	0.046779
128	0.015612	0.046838
Circular plate	$1/64 = 0.015625$	$1/64\left((5+\nu)/(1+\nu)\right)\big|_{\nu=0,3} = 0,063702$

and for the clamped plate:

$$W = 0 \text{ and } \frac{\partial W}{\partial x} = 0 \quad \text{for } r = R_b = \frac{\cos\frac{\pi}{K}}{\cos\theta}. \tag{81b}$$

In both cases, we accepted form (79) of the solution. It exactly fulfils the differential equation (66) and the respective symmetry conditions. Unknown constants A_k come from boundary conditions (81a) or (81b) depending on way of the plate fastening. In the considered cases, we obtained sufficiently accurate solutions for only a few collocation points. The solutions were accurate in the sense that the maximal error in fulfilment of the boundary conditions was very small.

In Table 4, the dimensionless deflection in the centre of investigated plate, obtained by means of the method described above, is presented. Similarly, like for other numerical methods (FEM and BEM), we face here the regular polygon paradox, which means that deflection of a simply supported regular polygon does not lead to a circular plate deflection with increase in the number of the polygon sides. It should be noticed that deflection of a simply supported circular plate depends on Poisson's ratio, whereas the polygonal plate deflection does not depend on that coefficient. This paradox is well known in literature and was noticed by such authors as Rao & Rajaiah [105], Rajaiah & Rao [103], Murray [88], Rhee & Atluri [107], Ahmed & Dey [2] and Batista de Paiva [7]. Generally speaking, it results from different character of boundary conditions in both cases of the Kirchhoff plate model.

In this section we have discussed two first problems defined by the 2D biharmonic equation. The third problem, plane elastostatics, is very different from the physical and numerical point of view and will be discussed in the next chapter.

References

[1] Achonina, S.I., Computing of thin plates by collocation method (in Russian). *Sbornik Trudov Leningradskogo Instituta*, **326**, pp. 13–17, 1971.

[2] Ahmed, S. & Dey, S.S., The circle-polygon paradox in the light of the boundary element method. *Computers & Structures*, **29**, pp. 157–163, 1988.

[3] Alves, C.J.S. & Silvestre, A.L., Density results using Stokeslets and method of fundamental solutions for the Stokes equations. *Engineering Analysis with Boundary Elements*, 2**8**, pp. 1245–1252, 2004.

[4] Artemov, V.V. & Belkin, N.I., On bending of square plate with stiffening ring (in Russian). *Sbornik Trudov Vsesojuznogo Zaočnogo Politechničeskogo Instituta*, **81**, pp. 165–174, 1973.

[5] Artemov, V.V., Bending of square plates with central cylindrical holes supported by elastic rings (in Russian). *Sbornik Trudov Vsesojuznogo Zaočnogo Politechničeskogo Instituta*, **73**, pp. 87–91, 1972.

[6] Barta, J., Über die naherungsweise Lösung einiger Zwei-dimensionalier Elastizitatsaufgaben. *Zeitschrift für Angewandte Mathematik und Mechanik (ZAMM)*, **17**, pp. 184–185, 1937.

[7] Batista de Paiva J., Boundary element analysis of the paradox circle-polygon in plate bending. *Engineering Analysis with Boundary Elements*, **11**, pp. 157–163, 1993.

[8] Bogacz, R. & Ryczek, B., Dry friction self-excited vibrations analysis and experiment. *Engineering Transactions*, **45**, pp. 487–504, 1997.

[9] Bowen, B.D. & Masliyah, J.H., Drag force on isolated axisymmetric particles in Stokes flow. *Canadian Journal of Chemical Engineering*, **51**, pp. 8–15, 1973.

[10] Burges, G. & Mahajerin, E., A numerical method for laterally loaded thin plates. *Computer Methods in Applied Mechanics and Engineering*, **49**, pp. 1–15, 1985.

[11] Burgess, G. & Mahajerin, E., A comparison of the boundary element and superposition methods. *Computers & Structures*, **19**, pp. 516, 697–705, 1984.

[12] Chapman, A.M. & Higdom, J.J.L., Oscillatory Stokes flow in periodic porous media. *Physics of Fluids A*, **4**, pp. 2099–2116, 1992.

[13] Checumov, R.A. & Simonov, V.S., Bending of plates with circular holes of large diameters (in Russian). *Sbornik Trudov Belgorodskogo Techničeskogo Instituta. Stroitelstvo Materialov*, **10**, pp. 10–16, 1975.

[14] Chen, C.W., Young, D.L., Tsai, C.C. & Murugesan, K., The method of fundamental solutions for inverse 2D Stokes problems. *Computational Mechanics*, **37**, pp. 2–14, 2005.

[15] Chen, F. & Wang, C.Y., Convective instability in saturated porous enclosure with a vertical insulating baffle. *International Journal of Heat Mass Transfer*, **7**, pp. 1897–1904, 1993.

[16] Chen, J.T., Wu, C.S., Lee, Y.T. & Chen, K.H., On the equivalence of the Trefftz method and method of fundamental solutions for Laplace and biharmonic equations. *Computers and Mathematics with Applications*, **53**, pp. 851–879, 2007.

[17] Chen, L.S. & Keh, H.J., Hydrodynamic interaction of two freely suspensed droplets in linear flow fields. *Journal of Colloid and Interface Science*, **204**, pp. 66–76, 1998.

[18] Chen, S.B. & Keh, H.J., Axisymmetric electrophoresis of multiple colloidal spheres. *Journal of Fluid Mechanics*, **238**, pp. 251–276, 1992.

[19] Chen, S.B., Axisymmetric motion of multiple composite spheres: Solid core with permeable shell, under creeping flow conditions. *Physics of Fluids*, **10**, pp. 1550–1563, 1998.

[20] Chen, S.H. & Keh, H.J., Axisymmetric thermophoresis of multiple aerosol spheres. *Aerosol Science & Technology*, **24**, pp. 21–35, 1996.

[21] Chen, S.S.H. & Pickett, G., Bending of uniform plates arbitrary shapes with mixed boundary conditions. In: *Proceedings of 10th Midwestern Mechanics Conference,* Colorado State University, Fort Collons, Colorado, August, 1967.

[22] Conway, H.D. & Farnham, K.A., Deflection of uniformly loaded circular plates with combination of clamped, simply supported and free boundary conditions. *Journal of Mechanical Science*, **9**, pp. 661–671, 1967.

[23] Conway, H.D., On some system of equations encountered in thin plate and elasticity theory. *Journal of Applied Mechanics*, **28**, pp. 143–144, 1961.

[24] Conway, H.D., The approximate analysis of certain boundary value problems. *Journal of Applied Mechanics*, **27**, pp. 275–277, 1960.

[25] Conway, H.D., The bending, buckling, and flexural vibration of simply supported polygonal plates by point matching. *Journal of Applied Mechanics*, **28**, pp. 288–291, 1961.

[26] Conway, H.D., Triangular plates analysed by point matching. *Journal of Applied Mechanics*, **29**, pp. 755–756, 1962.

[27] Dagan, Z., Pfeffer, P. & Weinbaum, S., Axisymmetric stagnation flow of a spherical particle near a finite planar surface at zero Reynolds number. *Journal of Fluid Mechanics*, **122**, pp. 273–294, 1982.

[28] Dagan, Z., Weinbaum, S. & Pfeffer, P., General theory for the creeping motion of a finite sphere along the axis of circular orifice. *Journal of Fluid Mechanics*, **127**, pp. 143–170, 1982.

[29] Denisov, N.G., On a certain variant of collocation method (in Russian). *Trudy Industrialnogo Instituta*, **40**, pp. 90–94, 1974.

[30] Deverall, L.I., The approximate analysis of certain boundary values problems. *Journal of Applied Mechanics*, **28**, pp. 156–157, 1961.

[31] Diez, J., Betalu, S. & Gratton, R., The crumbing of a viscous prism with an inclined free surface. *Archive of Applied Mechanics*, **68**, pp. 407–421, 1998.

[32] Ganatos, P., Pfeffer, R. & Weinbaum, S., A numerical solution technique for three-dimensional Stokes flows with application to the motion of strongly interacting sphere in a plane. *Journal of Fluid Mechanics*, **84**, pp. 79–111, 1978.

[33] Ganatos, P., Pfeffer, R. & Weinbaum, S., A strong interaction theory for the creeping motion of a sphere between plane parallel boundaries. Part 1. Perpendicular motion. *Journal of Fluid Mechanics*, **99**, pp. 739–753, 1980.

[34] Ganatos, P., Pfeffer, R. & Weinbaum, S., A strong interaction theory for the creeping motion of a sphere between plane parallel boundaries. Part 2. Parallel motion. *Journal of Fluid Mechanics*, **99**, pp. 755–785, 1980.

[35] Ganatos, P., Weinbaum, S. & Pfeffer, R., Gravitational and zero-drag motion of a sphere of arbitrary size in an inclined channel at low Reynolds number. *Journal of Fluid Mechanics*, **124**, pp. 27–43, 1982.

[36] Gluckman, M.J., Pfeffer, R. & Weinbaum, S., A new technique for treating multiparticle slow viscous flow: axisymmetric flow past sphere and spheroids. *Journal of Fluid Mechanics*, **50**, pp. 705–740, 1971.

[37] Guseva, E.B. & Belkin, V., Stresses and displacements in bending of rectangular plates with central hole (in Russian). *Izvesti Vyshebnych Uchebnych Zavedenij Stroitelstvo i Arkhitektura,* no. **4**, pp. 44–47, 1973.

[38] Guseva, E.B., Bending of rectangular plates with circular hole by point force approach (in Russian). *Sbornik Trudov Vsesojuznogo Zaočnogo Politechničeskogo Instituta*, **81**, pp. 175–181, 1973.

[39] Guseva, E.B., Bending of rectangular plates with circular hole by point force approach (in Russian). *Sbornik Trudov Vsesojuznogo Zaočnogo Politechničeskogo Instituta*, **81**, pp. 182–188, 1973.

[40] Hassonjee, Q., Ganatos, P. & Pfeffer, R., A strong-interaction theory for the motion of arbitrary three-dimensional clusters of spherical particles at low Reynolds number. *Journal of Fluid Mechanics*, **197**, pp. 1–37, 1988.

[41] Hills, C.P., Eddy structures induced within a wedge by a honing circular arc. *Theoretical and Computational Fluid Dynamics*, **15**, pp. 1–10, 2001.

[42] Janssens, P. & Tolley, M.D., On the deformation of elastic plates. *Zeitschrift für Angewandte Mathematic und Physik (ZAMP)*, **30**, pp. 234–242, 1979.

[43] Johnson, D., Plate bending by a boundary point method. *Computers & Structures*, **26**, pp. 673–680, 1987.

[44] Kalinin, H.G. & Pankov, A.I., Stress distribution in rectangular plate with central cylindrical hole (in Russian). *Sbornik Trudov RKIIGA*, **241**, Ryga, 1972.

[45] Karageorghis, A. & Fairweather, G., The Almansi method of fundamental solutions for solving biharmonic problems. *International Journal for Numerical Methods in Engineering*, **26**, pp. 1665–1682, 1988.

[46] Karageorghis, A. & Fairweather, G., The method of fundamental solutions for the numerical solution of the biharmonic equation. *Journal of Computational Physics*, vol. **69**, pp. 434–459, 1987.

[47] Karageorghis, A. & Fairweather, G., The method of fundamental solutions for the solutions of nonlinear plane potential problems. *IMA Journal of Numerical Analysis*, **9**, pp. 231–242, 1989.

[48] Karageorghis, A., The method of fundamental solution for steady-state free boundary problems. *Journal of Computational Physics*, **98**, pp. 119–128, 1992.

[49] Keh, H.J. & Chen, J.B., Particle interactions in electrophoresis. V. Motion of multiple spheres with thin but finite electrical double layers. *Journal of Colloid and Interface Science*, **158**, pp. 199–222, 1993.

[50] Keh, H.J. & Chen, L.S., Droplet interactions in axisymmetric thermocapillary motion. *Journal of Colloidal and Interface Science*, **151**, pp. 1–16, 1992.

[51] Keh, H.J. & Chen, S.H., Thermophoresis of an arbitrary three-dimensional array of interacting arbitrary spheres. *Journal of Aerosol Science*, **27**, pp. 1035–1061, 1996.

[52] Keh, H.J. & Chiou, J.Y., Electrophoresis of a colloidal sphere in a circular cylindrical pore. *AIChE Journal*, **42**, pp. 1397–1406, 1996.

[53] Keh, H.J. & Jan, J.S., Boundary effects on diffusiophoresis and electrophoresis: motion of a colloidal sphere normal to plane wall. *Journal of Colloidal and Interface Science*, **183**, pp. 458–475, 1996.

[54] Keh, H.J. & Lien, L.C., Electrophoresis of a colloidal sphere along the axis of a circular orifice or circular disk. *Journal of Fluid Mechanics*, **224**, pp. 305–333, 1991.

[55] Keh, H.J. & Lin, Y.M., Particle interactions in diffusiophoresis: Axisymmetric motion of multiple spheres in nonelectrolyte gradients. *Langmuir*, **10**, pp. 3010–3017, 1994.

[56] Keh, H.J. & Luo, S.C., Particle interactions in diffusiophoresis: Axisymmetric motion of multiple spheres in electrolyte gradients. *Langmuir*, **12**, pp. 657–667, 1996.

[57] Keh, H.J. & Tseng, C.H., Slow motion of an arbitrary axisymmetric body along its axis of revolution and normal to plane surface. *International Journal of Multiphase Flow*, **20**, pp. 185–210, 1994.

[58] Keh, H.J. & Tseng, Y.K., Slow motion of multiple droplets in arbitrary three-dimensional configurations. *AIChE Journal*, **38**, pp. 1881–1904, 1992.

[59] Keh, H.J. & Yang, F.R., Particle interaction in electrophoresis. IV. Motion of arbitrary three-dimensional clusters of spheres. *Journal of Colloidal and Interface Science*, **145**, pp. 362–389, 1991.

[60] Keh, H.J. & Yang, F.R., Particle interactions in electrophoresis. III. Axisymmetric motion of multipole spheres. *Journal of Colloidal and Interface Science*, **139**, pp. 105-, 1990.

[61] Keh, H.J., Chen, J.B. & Kim, S., Axisymmetric electrophoresis of coaxial spheroids. *Colloids and Surfaces A: Physicochemical and Engineering Aspects*, **75**, pp. 147–162, 1993.

[62] Kobayashi, Y. & Fumio, F., Analysis of point-supported rectangular plates under uniform load by boundary collocation method. *Transactions of Japan Society of Civil Engineering*, **12**, pp. 111–112, 1980.

[63] Kołodziej, J.A. & Mendes, C.A., T-functions for 2-D creeping flow on domains with circular cylinders, corners, and possessing symmetry. *Computer Assisted Mechanics and Engineering Sciences,* vol. **8**, pp. 359–384, 2001.

[64] Kreller, H. & Rubert, A., Eine analytische Lösung mit Randkollokation der Kirchhoffschen Plattendifferentialgleichung für die unendich ausgedehnte Vollplatte mit gelenkiger Lochrandlagerung. *Ingenieur-Archiv*, **54**, pp. 62–72, 1984.

[65] Kucaba-Piętal, A., Flow past a sphere moving towards a wall in micropolar fluid. *Journal of Theoretical and Applied Mechanics*, **37**, pp. 301–318, 1999.

[66] Kucaba-Piętal, A., Nonaximmetric Stokes flow past a torus in the presence of a wall. *Archives of Mechanics*, **38**, pp. 647–663, 1986.

[67] Kucaba-Piętal, A., Axisymmetric quasi-stationary motion of a torus in presence of a wall; Stokes approximation (in Polish). *Engineering Transactions*, **31**, pp. 151–161, 1983.

[68] Kucaba-Piętal, A., Resistance coefficient of a particle moving in the presence of a deformed wall. *Journal of Theoretical and Applied Mechanics*, **35**, pp. 813–828, 1997.

[69] Kucaba-Piętal, A., Stokes flow past a torus in the presence of a wall. *Zeitschrift für Angewandte Mathematik und Mechanik (ZAMM)*, **68**, pp. T309–T311, 1988.

[70] Kucaba-Piętal, A., The torque and drag acting on torus moving slowly in a direction perpendicular to a plane wall in viscous fluid. *Bulletin of the Polish Academy of Science*, **36**, pp. 501–511, 1988.

[71] Kuo, J. & Keh, H.J., Effects of adsorbed polymers on the axisymmetric motion of two colloidal spheres. *Journal of Colloid and Interface Science*, **195**, pp. 353–367, 1997.

[72] Kuo, J. & Keh, H.J., Motion of a colloidal sphere covered by a layer of adsorbed polymers normal to a plane surface. *Journal of Colloid and Interface Science*, **210**, pp. 296–308, 1999.

[73] Larson, R.E. & Higdon, J.J.L., A periodic grain consolidation model of porous media. *Physics of Fluids A*, **1**, pp. 38–46, 1989.

[74] Leichtberg, S., Pfeffer, R. & Weinbaum, S., Stokes flow past finite coaxial clusters of spheres in a circular cylinder. *International Journal of Multiphase Flow*, **3**, pp. 147–169, 1976.

[75] Leichtberg, S., Weinbaum, S. & Pfeffer, R., A new hydrodynamic interaction mechanism for the formation of rouleaux. *Biorheology*, **13**, pp. 165–179, 1976.

[76] Leichtberg, S., Weinbaum, S., Pfeffer, R. & Gluckman, M.J., A study of unsteady force at low Reynolds number: A strong interaction theory for the coaxial setting. *Philosophical Transactions of Royal Society*, **A282**, pp. 585–610, 1976.

[77] Leissa, A.W. & Clausen, W.E., Deflection of a circular plate having mixed boundary conditions. *AIAA Journal*, **5**, pp. 2287–2288, 1967.

[78] Leissa, A.W. & Niedenfuhr, F.W., Bending of square plate with two adjacent edges free and others clamped or simply supported. *AIAA Journal*, **1**, pp. 116–120, 1963.

[79] Leissa, A.W. & Niedenfuhr, F.W., A study of the cantilevered square plate subjected to a uniform loading. *Journal of Aerospace Science*, **29**, pp. 162–169, 1962.

[80] Leissa, A.W., Claussen, W.E., Hulbert, L.E. & Hopper, A.T., Comparison of approximate method for solution of plate bending problems. *AIAA Journal*, **7**, pp. 920–928, 1969.

[81] Leissa, A.W., Lo, C.C. & Niedenfuhr, F.W., Uniformly loaded plates of regular polygonal shape. *AIAA Journal*, **3**, pp. 566–567, 1965.

[82] Liao, W.H. & Krueger, D.A., Multiple expansion calculation of slow viscous flow about spheroids of different sizes. *Journal of Fluid Mechanics*, **96**, pp. 223–241, 1980.

[83] Lo, C.C. & Leissa, A.W., Bending of plates with circular holes. *Acta Mechanica*, **4**, pp. 64–78, 1967.

[84] Lo, C.C., *Bending of Rectangular Plates with All Edges Clamped*, M.Sc. Thesis, Ohio State University, 1960.

[85] Lo, C.C., Niedenfuhr, F.W. & Leissa, A.W., Further studies in the application of the point matching technique to plate bending and other harmonic and biharmonic boundary value problem. *AFFDT-TR-64-113*, pp. 175, June, 1965.

[86] Mendes, A.C. & Kołodziej, J.A., An adaptive boundary collocation method for creeping flow between eccentric cylinders. *Advances in Fluid Mechanics Series*, **9**, pp. 85–91, *First International Conference on Advances in Fluid Mechanics*, AFM 96, 1996.

[87] Moffatt, H.K., Viscous and resistance eddies near a sharp corner. *Journal of Fluid Mechanics*, **18**, pp. 1–18, 1964.

[88] Murray, N.W., The polygon-circle paradox and convergence in thin plate theory. In: *Proceedings Cambridge Philosophical Society*, **73**, pp. 279–282, 1973.

[89] Nash, W.A., Several approximate analysis of the bending of a rectangular cantilever plate by uniform pressure. *Journal of Applied Mechanics*, **19**, pp. 33–36, 1952.

[90] Neidenfuhr, F.W., Leissa, A.W. & Lo, C.C., A study of the point matching method as applied to thermally and transversally loaded plate and other boundary value problems. *Technical Report* No. AFFDL-TR-159, AD-610146, Oct. U.S. Air Force Flight Dynamics Lab, 1964.

[91] O'Brien, V., Form factors for deformed spheroids in Stokes flow. *AIChE Journal*, **14**, pp. 870–875, 1968.

[92] O'Brien, V., Stokes flow about deformed spheroids. *International Journal of Engineering Science*, **4**, pp. 925–937, 1966.

[93] O'Brien, V., Deformed spheroids in Stokes flow. *Appl. Phys. Lab. Rep.*, TG-716, Johns Hopkins University, Silver Spring, 1965.

[94] O'Brien, V., Eggs and other deformed spheroids in Stokes flow. *APL Technical Digest*, **4**, pp. 11–16, 1965.

[95] Ogata, H., Aname, K., Sugita, M. & Okano, D., A fundamental solution method for viscous flow problems with obstacles in a periodic array. *Journal of Computational and Applied Mathematics*, **152**, pp. 411–425, 2003.

[96] Ozaki, E., On an appropriate analysis method of rectangular plates with a circular hole. *Transactions of the Japanese Society of Irrigation, Drainage and Reclamation Engineering,* no. 31, pp. 29–34, 1970.

[97] Pakdel, P. & Kim, S.T., On the capabilities of the double-layer representation for Stokes Flows 1. Analytical solutions. *Engineering Analysis with Boundary Elements*, **13**, pp. 339–347, 1994.

[98] Pankov, A.I., Bending of rectangular plate with central cylindrical hole (in Russian). *Vsesojuznyj Sbornik Voprosov Dinamiki i Prochnosti*, **26**, Riga, Zinatne, pp.1–13, 1973.

[99] Pankov, A.I., On influence of distribution of collocation points on satisfying boundary conditions (in Russian). *Vsesojuznyj Sbornik Voprosov Dinamiki i Prochnosti*, **31**, Riga, Zinatne, pp. 156–163, 1975.

[100] Pirogow, I.M. & Belkin, N.I., Distribution of stresses in triangular bending plate with circular hole (in Russian). *Sbornik Trudov Vsesojuznogo Zaočnogo Politechničeskogo Instituta*, **59**, pp. 70–76, 1970.

[101] Poullikkas, A., Karageorghis, A., Georgiu, G. & Ascough, J., The method of fundamental solutions for Stokes flows with free surface. *Numerical Methods for Partial Differential Equations*, **14**, pp. 667–678, 1998.

[102] Rajaiah, K. & Rao, A.K., Collocation solution for point-supported square plates. *Journal of Applied Mechanics*, **45**, pp. 424–425, 1978.

[103] Rajaiah, K. & Rao, A.K., On limiting cases in the flexure of simply supported regular polygonal plates. In: *Proceedings of Cambridge Philosophical Society*, **65**, pp. 831–834, 1969.

[104] Rajaiah, K., Flexure of elastically restrained rhombic plates. *Journal of Aeronautical Society India*, **35**,pp. 119–124, 1983.

[105] Rao, A.K. & Rajaiah, K., Polygon-circle paradox of simply supported thin plates under uniform pressure. *AIAA Journal*, **6**, pp. 155–156, 1968.

[106] Ratkowsky, D.A. & Roten, Z., Viscous flow in a rectangular cut out. *Physics of Fluids*, **11**, pp. 2761–2763, 1968.

[107] Rhee, H.C. & Atluri, S.P., Polygon-circle paradox in the finite element analysis of bending of a simply supported plate. *Computers & Structures*, **22**, pp. 553–558, 1986.

[108] Richardson, S., A 'stick-slip' problem related to the motion of free jet at low Reynolds numbers. In: *Proceedings of Cambridge Philosophical Society*, **67**, pp. 477–489, 1970.

[109] Robinson, N.I., Some boundary methods of solution for bending of plates. *Rev. Roum. Sci. Tech. Ser. Mec. Appl.*, **18**, pp. 901–924, 1973.

[110] Sangani, A.S. & Acrivos, A., Slow flow past periodic arrays of cylinders with application to heat transfer. *International Journal of Multiphase Flow*, **9**, pp. 193–206, 1982.

[111] Sekiya, T., An approximate solution in the problem of elastic plates with an arbitrary external form and a circular hole. In: *Proceedings of 5th Japan National Congress on Applied Mechanics*, pp. 95-98, 1955.

[112] Shankar, P.N., The eddy structure in Stokes flow in a cavity. *Journal of Fluid Mechanics*, **250**, pp. 371–383, 1993.

[113] Shankar, P.N., Three-dimensional eddy structure in a cylindrical contained. *Journal of Fluid Mechanics*, **342**, pp. 97–118, 1997.

[114] Shankar, P.N., Three-dimensional Stokes flow in a cylindrical container. *Physics of Fluids*, **10**, pp. 540–549, 1998.

[115] Skala, D. & Tolley, M.D., Flexion des plaques minces polygonales. *Bulletin de la Classe des Sciences,* ser. 5, **67**, pp. 372–384, 1981.

[116] Skala, D., Application de la methode des grands elements finis singuliers à des problems de plaques. *Memoire de fin d'etudes,* Université Libre de Bruxelles, 1979.

[117] Stoljarov, N.A., Error estimation in solution of problem of plate bending by means of collocation method (in Russian). *Rasčët Prostranstvennych Konstrukcji,* no. 7, pp. 72–75, 1977.

[118] Thorne, J., Square plates fixed at points. *Journal of Applied Mechanics*, **15**, pp. 73–79, 1948.

[119] Tolke, F., Über Spannungszustande in dunnen Rechteckplatten. *Ingenieur-Archiv*, **5**, pp. 187–237, 1934.

[120] Tozeren, H., Torque on eccentric spheres flowing in tubes. *Journal of Applied Mechanics*, **49**, pp. 279–283, 1982.

[121] Wah, T., Elastic quadrilateral plates. *Computers & Structures*, **10**, pp. 457–466, 1979.

[122] Wah, T., Partially loaded elastic plates. *International Journal of Mechanical Science*, **24**, pp. 89–93, 1982.

[123] Wang, C.Y., Low Reynolds number flow due to a rotating finned shaft in cylinder. *Zeitschrift für Angewandte Mathematic und Physik (ZAMP)*, **48**, pp. 439–450, 1997.

[124] Wang, C.Y., Rotation of a cylinder in a casing at zero Reynolds number. *Journal of Applied Mechanics*, **62**, pp. 662–666, 1995.

[125] Wang, C.Y., Settling of discs inside a vertical fluid-filled tube. *Applied Scientific Research*, **56**, pp. 43–51, 1996.

[126] Wang, C.Y., Slow rotation of a disc in a fluid-filled casing. *Acta Mechanica*, **94**, pp. 97–103, 1992.

[127] Wang, C.Y., Stokes flow through a zig-zag channel. *Acta Mechanica*, **126**, pp. 115–125, 1998.

[128] Wang, C.Y., Stokes flow through an array of staggered strips. *Applied Scientific Research*, **59**, pp. 27–41, 1998.

[129] Wang, C.Y., Stokes flows through a thin screen with patterned holes. *AIChE Journal*, **40**, pp. 419–423, 1994.

[130] Wang, C.Y., Stokes flows through a transversally finned channel. *Journal of Fluids Engineering*, **119**, pp. 110–114, 1997.

[131] Wang, C.Y., The Stokes drag due to the sliding of a smooth plate over a finned plate. *Physics of Fluids*, **6**, pp. 2248–2252, 1994.

[132] Warren, W.E., Bending of rhombic plates. *AIAA Journal*, **2**, pp. 16–168, 1964.

[133] Wu, W.Y. & He, X.Y., The Stokes flow of the rotating double spheres and multiple Spheres. *Applied Mathematics and Mechanics*, **8**, pp. 123–136, 1987.

[134] Yan, Z., Weinbaum, S., Ganatos, P. & Pfeffer, R., The three-dimensional hydrodynamic interaction of a finite sphere with a circular orifice at low Reynolds number. *Journal of Fluid Mechanics*, **174**, pp. 39–68, 1987.

[135] Young, D.L., Chen, C.W., Fan, C.M., Murugesan, K. & Tsai, C.C., The method of fundamental solutions for Stokes flow in rectangular cavity with cylinders. *European Journal of Mechanics B/Fluids*, **24**, pp. 703–716, 2005.

[136] Yuan, F. & Wu, W., The Stokes flow of an arbitrary prolate axisymmetric body towards an infinite plane wall. *Applied Mathematics and Mechanics*, **8**, pp. 17–30, 1987.

CHAPTER 8

Two-dimensional problem of elastostatics

1 Formulation of the 2D elastic problem

Introduction of a stress function is a basic solution method of a two-dimensional (2D) problem of elastostatics. The governing equation system leads then to one biharmonic equation for the stress function Φ, which in polar coordinates takes the form:

$$\left(\frac{\partial^2}{\partial r^2}+\frac{1}{r}\frac{\partial}{\partial r}+\frac{1}{r^2}\frac{\partial^2}{\partial \theta^2}\right)^2 \Phi = 0 \tag{1}$$

where non-zero stress components and a displacement vector can be calculated as follows:

$$\sigma_{rr} = \frac{1}{r}\frac{\partial \Phi}{\partial r}+\frac{1}{r^2}\frac{\partial^2 \Phi}{\partial \theta^2} \tag{2a}$$

$$\sigma_{\theta\theta} = \frac{\partial^2 \Phi}{\partial r^2} \tag{2b}$$

$$\sigma_{r\theta} = \frac{1}{r^2}\frac{\partial \Phi}{\partial \theta}-\frac{1}{r}\frac{\partial^2 \Phi}{\partial \theta \partial r} \tag{2c}$$

$$u_r = \frac{1}{E}\int\left(\sigma_{rr}-\nu\sigma_{\theta\theta}\right)dr \tag{2d}$$

$$u_\theta = \frac{r}{E}\int\left(\sigma_{\theta\theta}-\nu\sigma_{rr}\right)d\theta-\frac{1}{E}\iint\left(\sigma_{rr}-\nu\sigma_{\theta\theta}\right)drd\theta \tag{2e}$$

where σ_{rr}, $\sigma_{\theta\theta}$, $\sigma_{r\theta}$ are stress components, u_r, u_θ are displacement components, E is the Young modulus and ν is the Poisson ratio.

The 2D elastic problem can also be formulated by a complex variable function [79]. Then the boundary problem solution depends on two holomorphic functions $\phi(z)$ and $\psi(z)$, which are related to the stress components and displacement vector in the following way:

$$\sigma_{rr} + \sigma_{\theta\theta} = 2\left[\phi(z) + \overline{\phi(z)}\right] \tag{3a}$$

$$\sigma_{rr} + i\sigma_{r\theta} = \phi(z) + \overline{\phi(z)} - \overline{z\phi'(z)} - \frac{\overline{z}}{z}\psi(z) \tag{3b}$$

$$2G(u_r + iu_\theta) = \int i\left[\kappa\phi(z) - \overline{\phi(z)} + \overline{z\phi'(z)} + \frac{\overline{z}}{z}\overline{\psi(z)}\right]dz \tag{3c}$$

$$z = x + iy$$

where $\kappa = (3-v)/(1+v)$ for plane stress and $\kappa = 3-4v$ for plane strain.

To solve 2D elastostatic problems, the boundary collocation method was used by many authors. In Table 1, their papers have been given and divided according to the particular topics. The authors used either the stress function or the complex variable formulations. In the first case, the approximate solution was obtained with help of the T-complete systems, which exactly fulfilled the governing equations and respective boundary conditions on a part of the boundary (as shown in the former section for creeping flow). In the second case, the holomorphic functions were adequately chosen for the discussed problem. Cheung *et al.* [15], for example, have applied two holomorphic functions and have determined the stress intensity coefficient in a rectangular plate with two side cracks using:

$$\phi(z) = i\sqrt{z^2 - c^2}\sum_{k=1}^{m} E_k z^k + \sum_{k=1}^{m} F_k z^k \tag{4}$$

Table 1: Survey of papers in which boundary collocation method has been applied to 2D elastostatic problem.

Problem	References
Determination of stress intensity factors	[2–10, 13–17, 22, 26, 29–34, 41, 43–47, 54, 55, 60, 61, 63–65, 69, 71, 73, 75, 77, 78, 81, 82, 84, 87, 89, 98–101, 103, 104, 106–108, 110–116, 119, 121–123]
Contact problems	[11, 18–21, 37, 72, 74–76, 94, 95]
Determination of composite properties	[1, 12, 27, 28, 35, 36, 38–40, 42, 53, 62, 64, 66–68, 83, 85, 86, 102, 105, 109, 116–118, 120]
Others	[23–25, 48, 70, 90–93, 127]

$$\psi(z)=i\sqrt{z^2-c^2}\sum_{k=1}^{m}E_k z^k-\sum_{k=1}^{m}F_k z^k \tag{5}$$

where E_k, F_k are constants defined from adequate boundary conditions and $c = b-a$, where $2b$ is width of the plate and a is length of the crack.

2 Example: tension of square plate with a central hole

A solution of a boundary value problem of a square plate with a central hole subjected to homogenous tension (Fig. 1) was a first example of application of the stress function approach.

The following boundary conditions should then supplement eqn (1):

$$\sigma_{rr}=0 \quad \text{for } r=R \tag{6a}$$

$$\sigma_{r\theta}=0 \quad \text{for } r=R \tag{6b}$$

$$\frac{\partial\Phi}{\partial\theta}=\frac{\partial^3\Phi}{\partial\theta^3}=0 \quad \text{for } \theta=0 \tag{6c}$$

$$\frac{\partial\Phi}{\partial\theta}=\frac{\partial^3\Phi}{\partial\theta^3}=0 \quad \text{for } \theta=\frac{\pi}{2} \tag{6d}$$

$$\sigma_{XX}=\sigma_{rr}\cos^2\theta+\sigma_{\theta\theta}\sin^2\theta-2\sigma_{r\theta}\sin\theta\cos\theta=\sigma_0 \quad \text{for } X=1 \tag{6e}$$

$$\sigma_{XY}=(\sigma_{rr}-\sigma_{\theta\theta})\sin\theta\cos\theta+\sigma_{R\theta}\cos 2\theta=0 \quad \text{for } X=1 \tag{6f}$$

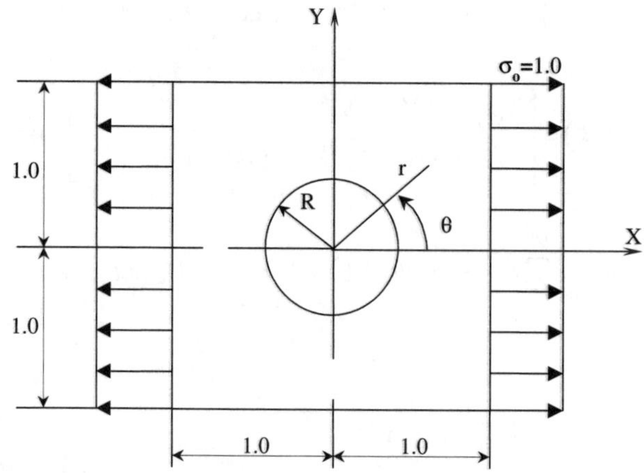

Figure 1: Homogenously loaded square plate with a hole in its centre.

$$\sigma_{YY} = \sigma_{rr} \sin^2 \theta + \sigma_{\theta\theta} \cos^2 \theta + 2\sigma_{r\theta} \sin\theta \cos\theta = 0 \quad \text{for } Y = 1 \tag{6g}$$

$$\sigma_{XY} = (\sigma_{rr} - \sigma_{\theta\theta}) \sin\theta \cos\theta + \sigma_{r\theta} \cos 2\theta = 0 \quad \text{for } Y = 1. \tag{6h}$$

The boundary conditions on the hole edge (6a and b) and symmetry conditions (6c and d) can exactly be fulfilled by adequately chosen special trial T-functions. After algebraic transformations similar to those presented in Section 3 (in Chapter 7), we have obtained stress formulas that exactly fulfil conditions (6a–d):

$$\sigma_{rr} = C_1 2\left(1 - \frac{R^2}{r^2}\right) + \sum_{k=1}^{M_f} C_{k+1} H_1(r,\theta,R,k) + \sum_{k=1}^{M_f} C_{M_f+k+1} H_2(r,\theta,R,k) \tag{7a}$$

where

$$H_1(r,\theta,R,k) = \begin{bmatrix} (2k+1)(2k-1)R^2 r^{2k-2} \\ -(2k+1)R^{4k+2} r^{-2k-2} - 2\left(2k^2 - k - 1\right) r^{2k} \end{bmatrix} \cos(2k\theta) \tag{7b}$$

$$H_2(r,\theta,R,k) = \begin{bmatrix} (2k-1)R^{-4k+2} r^{2k-2} \\ +\left(4k^2 - 1\right)R^2 r^{-2k-2} - 2\left(2k^2 + k - 1\right) r^{-2k} \end{bmatrix} \cos(2k\theta) \tag{7c}$$

$$\sigma_{\theta\theta} = C_1 2\left(1 + \frac{R^2}{r^2}\right) + \sum_{k=1}^{M_f} C_{k+1} H_3(r,\theta,R,k) + \sum_{k=1}^{M_f} C_{M_f+k+1} H_4(r,\theta,R,k) \tag{7d}$$

where

$$H_3(r,\theta,R,k) = \begin{bmatrix} -\left(4k^2 - 1\right)R^2 r^{2k-2} \\ +(2k+1)R^{4k+2} r^{-2k-2} + 2(k+1)(2k+1) r^{2k} \end{bmatrix} \cos(2k\theta) \tag{7e}$$

$$H_4(r,\theta,R,k) = \begin{bmatrix} 2(k-1)(2k-1) r^{-2k} \\ -(2k-1)R^{-4k+2} r^{2k-2} - \left(4k^2 - 1\right)R^2 r^{-2k-2} \end{bmatrix} \cos(2k\theta) \tag{7f}$$

$$\sigma_{r\theta} = \sum_{k=1}^{M_f} C_{k+1} H_5(r,\theta,R,k) + \sum_{k=1}^{M_f} C_{M_f+k+1} H_6(r,\theta,R,k) \tag{7g}$$

where

$$H_5(r,\theta,R,k) = \begin{bmatrix} 2k(2k+1) r^{2k} \\ -\left(4k^2 1\right)R^2 r^{2k-2} - (2k+1)R^{4k+2} r^{-2k-2} \end{bmatrix} \sin(2k\theta) \tag{7h}$$

$$H_6(r,\theta,R,k)=\begin{bmatrix}(4k^2-1)R^2r^{-2k-2}\\ -2k(2k-1)r^{-2k}-(2k-1)R^{-4k+2}r^{2k-2}\end{bmatrix}\sin(2k\theta). \tag{7i}$$

Unknown constants C_k, $k = 1, 2,\ldots, 2M_f + 1$ have been determined fulfilling boundary conditions (6e–6h) in a finite number of points on the boundaries $X = 1$ and $Y = 1$. The exemplary values of the constants are given in Table 2. It is worth noticing that those constants quickly decrease and for a sufficiently accurate solution you can use only a few first terms. A reasonable number of the constants depends on the size of hole radius R. With increase of relation of the hole to the plate width, more constants should be taken into account (Table 2). In the choice of necessary number of constants C_k easy observation of the error on boundary Γ between the collocation points is very helpful. Examples of maximal values of the relative percentage error in fulfilling boundary conditions (6e–6h) have been shown in Table 3.

For an infinite plate with a circular hole, subjected to homogenous tension in infinity, the exact solution has the form:

$$\frac{\sigma_{rr}}{\sigma_0}=\frac{1}{2}\left(1-\frac{R^2}{r^2}\right)+\frac{1}{2}\left(1+3\frac{R^4}{r^4}-4\frac{R^2}{r^2}\right)\cos(2\theta) \tag{8a}$$

$$\frac{\sigma_{\theta\theta}}{\sigma_0}=\left(1+\frac{R^2}{r^2}\right)-\frac{1}{2}\left(1+3\frac{R^4}{r^4}\right)\cos(2\theta) \tag{8b}$$

$$\frac{\sigma_{r\theta}}{\sigma_0}=-\frac{1}{2}\left(1-3\frac{R^4}{r^4}+2\frac{R^2}{r^2}\right)\sin(2\theta) \tag{8c}$$

$$\frac{\sigma_{XX}}{\sigma_0}=1-\frac{R^2}{r^2}\left(\frac{3}{2}\cos(2\theta)+\cos(4\theta)\right)+\frac{3}{2}\frac{R^4}{r^4}\cos(4\theta) \tag{8d}$$

$$\frac{\sigma_{XY}}{\sigma_0}=-\frac{R^2}{r^2}\left(\frac{1}{2}\cos(2\theta)-\cos(4\theta)\right)-\frac{3}{2}\frac{R^4}{r^4}\cos(4\theta) \tag{8e}$$

$$\frac{\sigma_{XY}}{\sigma_0}=-\frac{R^2}{r^2}\left(\frac{1}{2}\sin(2\theta)+\sin(4\theta)\right)+\frac{3}{2}\frac{R^4}{r^4}\sin(4\theta). \tag{8f}$$

However, it cannot be a reference solution to the finite plate. Obviously, comparison of both results could lead to serious errors. To illustrate this opinion, the stress values in two characteristic plate points, obtained by means of the boundary collocation method, are presented in Table 4. For the infinite plate, the maximal, non-dimensional stress is equal $\sigma_{XX}/\sigma_0 = 3$ for $X = 0$, $Y = R$. In Table 4, we can see that only for small relations of the diameter to width of the plate, the solution for an infinite plate is near to the results for the finite plate.

Table 2: Constants C_k in expressions (7a, d, g) determined by the boundary collocation method in least square variant; number of equations $N_e = 78$.

k	$R = 0.4$ $M_f = 7$	$R = 0.6$ $M_f = 10$	$R = 0.8$ $M_f = 10$
1	0.2920839417	0.3712442671	0.6164400542
2	0.1181535149	0.5658356420	4.5431316533
3	0.0051040328	0.0167913964	0.0753034711
4	0.0108850138	0.0479635077	0.3756363576
5	0.0004716157	0.0020457911	0.0137986560
6	0.0009904749	0.0047577730	0.0444419398
7	0.0000093929	0.0001453347	0.0016279159
8	0.0000273609	0.0002874045	0.0038212070
9	0.1538658496	0.0000121628	0.0001365175
10	0.0000793602	0.0000057427	0.0001219788
11	0.0000062330	0.0000006068	0.0000052485
12	0.0000000058	0.9146940541	9.8891857992
13	0.0000000004	0.0023624731	0.0440457915
14	−3.7627E−13	0.0011997729	0.1071167441
15	−1.6263E−14	0.0000063676	0.0017813179
16		0.0000018779	0.0018234812
17		6.461E−10	0.0000154397
18		−7.189E−10	−0.0000044962
19		1.023E−11	−0.0000002289
20		−5.611E−12	−0.0000010800
21		−1.497E−15	−0.0000000129

Also a characteristic stress value in the unbounded disc is the stress $\sigma_{XX} = \sigma_0$ in section $X = 0$, $Y \rightarrow \infty$. In Table 4, this value in section $X = 0$, $Y = 1$ of the finite plate has been presented. It is considerably different from $\sigma_{XX} / \sigma_0 = 1$ and for larger R takes a negative sign.

3 Application of the T-complete Herrera and Kupradze functions

In the next example, a variant of the single hybrid Trefftz element is applied (see Chapter 4 and Section 5 in Chapter 6). Application of this version of the boundary

Table 3: Maximum values of relative percentage error in fulfilling boundary conditions between collocations points.

R	A	B	C	D
0.5	0.2611	0.2359	0.0838	0.0021
0.6	0.5608	0.5714	0.2325	0.0046
0.7	1.9038	2.5507	1.1958	0.0196
0.8	13.896	15.751	12.211	0.1298

$A-$ $\max\left(\left|\sigma_{XX}-\sigma_0\right|/\sigma_0\right)\times 100$ on the boundary $X=1$,

$B-$ $\max\left(\left|\sigma_{XY}\right|/\sigma_0\right)\times 100$ on the boundary $X=1$,

$C-$ $\max\left(\left|\sigma_{YY}\right|/\sigma_0\right)\times 100$ on the boundary $Y=1$,

$D-$ $\max\left(\left|\sigma_{XY}\right|/\sigma_0\right)\times 100$ on the boundary $Y=1$;

Table 4: Stresses σ_{XX} in two characteristic points of the plate (Fig. 1).

R	σ_{XX} for $X=0$, $Y=R$	σ_{XX} for $X=0$, $Y=1$
0.05	3.0214	0.9924
0.1	3.0865	0.9692
0.2	3.3579	0.8650
0.3	3.8801	0.6049
0.4	4.7826	0.1323
0.5	6.3888	–0.8222
0.6	9.4807	–2.8050
0.7	16.320	–7.6361
0.8	34.288	–21.602

collocation method to a 2D elasticity problem defined in Fig. 2 leads to similar conclusions as in the examples presented in Section 5 in Chapter 6. The displacement error function is here defined as

$$\delta_u=\left\{\frac{\left(u^{K,H}-u^{EX}\right)^2+\left(\upsilon^{K,H}-\upsilon^{EX}\right)^2}{\left(u_C^{EX}\right)^2+\left(\upsilon_C^{EX}\right)^2}\right\}^{1/2} \tag{9}$$

where u^K, u^H and υ^K, υ^H are displacement determined with the help of the Kupradze and Herrera functions, respectively, and $\left\{u_C^{EX},\upsilon_C^{EX}\right\}$ is the exact solution in point C (Fig. 2).

(a)

(b)

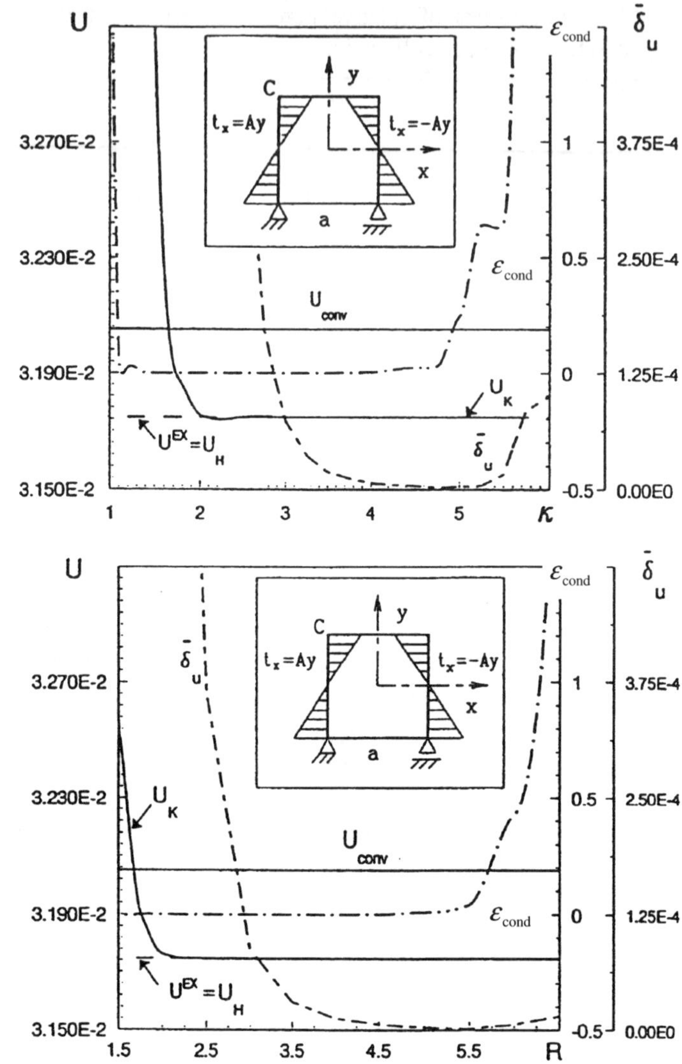

Figure 2: Range of acceptable distance from the element boundary of singularities of Kupradze functions in HT-K solutions. Kupradze functions with singularities situated (a) on a homothetic contour and (b) on a circle (Fig. 9 in Chapter 6): $M_f = 12$, $a = 2$, $A = 100$. Generalized energy calculated with help of a single conventional p-element (U_{conv}), HT-H approach (U_H) and HT-K formulation (U_K), $U^{EX} = 0.031746$, $u_C = 9.5238 \cdot 10^{-4}$, $\upsilon_C = 0$, ε_{cond} – defined in ((34) in Chapter 6), $\bar{\delta}_u$ (10).

The average value of this function takes the form:

$$\bar{\delta}_u = \frac{1}{K^2}\sum_{i,j=1}^{K} \delta_{uij}. \tag{10}$$

The relative measure of the energy density error is given by the following relation:

$$\gamma_U = \left\{ \frac{1}{E\rho^{EX}} \left[\frac{1}{2}\left(\sigma_x^{K,H} - \sigma_x^{EX}\right)^2 + \frac{1}{2}\left(\sigma_y^{K,H} - \sigma_y^{EX}\right)^2 \right.\right.$$
$$\left.\left. -\nu\left(\sigma_x^{K,H} - \sigma_x^{EX}\right)\left(\sigma_y^{K,H} - \sigma_y^{EX}\right) + (1+\nu)\left(\tau_{xy}^{K,H} - \tau_{xy}^{EX}\right)^2 \right] \right\}^{1/2} \tag{11}$$

$$\rho^{EX} = \frac{U^{EX}}{a^2}, \quad U^{\mathrm{EX}} = 0.031746 \text{ – exact elastic energy of the object,}$$

and

$$\bar{\gamma}_U = \frac{1}{K^2}\sum_{i,j=1}^{K} \gamma_{Uij}. \tag{12}$$

The exact solution has in this case the simple form:

$$u^{EX}(x,y) = \frac{A}{E}(1 - xy), \quad \upsilon^{EX}(x,y) = \frac{A}{2E}\left(x^2 + \nu y^2 - \nu - 1\right). \tag{13}$$

Thus, the HT-H method with polynomial T-functions is here capable to present practically exact results with the help of the first several terms. On the other hand, the Kupradze functions are in the logarithmic form (discussed in detail in Chapter 12) and, thus, are unable to model solution (13) exactly. In spite of this, the results obtained with the F-functions are very accurate (Fig. 3). However, it is interesting to note that in this very regular case of an elastic 2D problem, high accuracy has been obtained for larger values of parameters κ and R than in the Laplace problems in Section 5 in Chapter 6.

The last example studies sensitivity of the hybrid-Trefftz results to object distortion (Fig. 4). It must be emphasized that generally this type of element is much less sensitive to such distortion than are classical finite elements [49]. Table 5 shows a comparison of two different types of the Trefftz elements. The global relative energy error has here been calculated as

$$\bar{\delta}_U = \frac{\left|U^{FE} - U^{EX}\right|}{U^{EX}} \cdot 100\% \tag{14}$$

and the displacement error is defined by (10) but in [%]. The HT-K solution again appears here to be slightly better than HT-H, but this advantage is limited to a certain range of parameters κ and R. On the other hand, inversion of the matrix

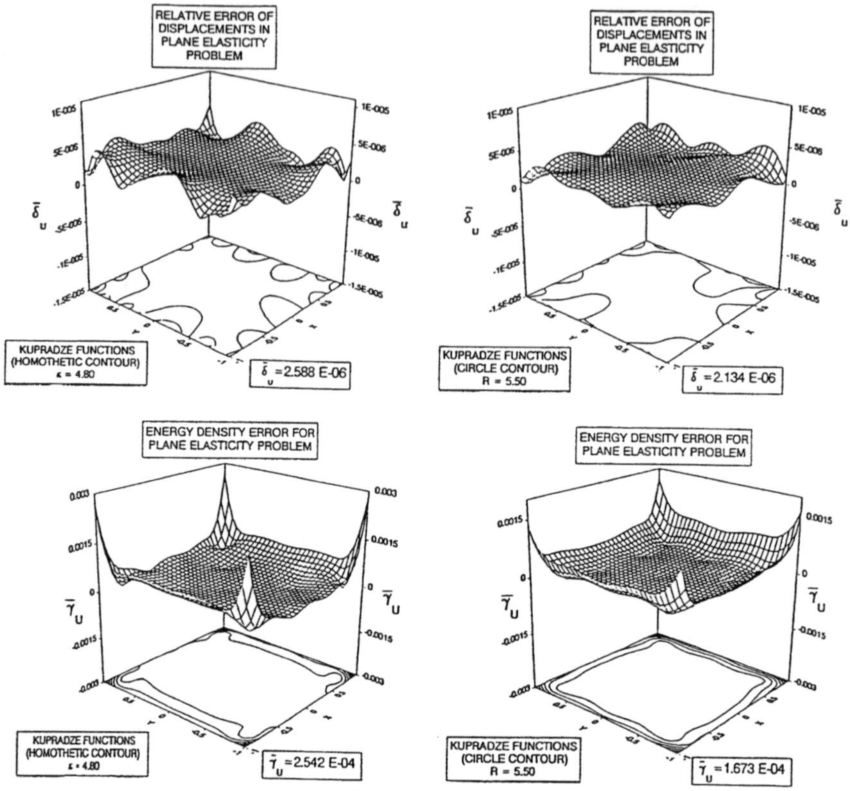

Figure 3: Relative solution error (9) and relative measure of energy density error (11) for elastic problem defined in Fig. 2.

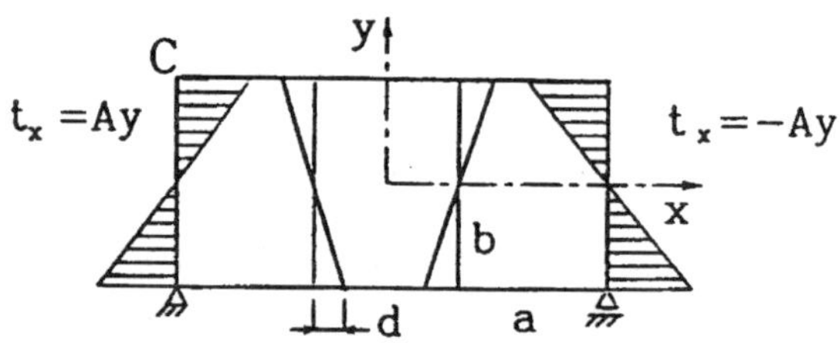

Figure 4: Investigation of area with distortion d (three hybrid-Trefftz elements).

H is much more stable in HT-H formulation (compare $\varepsilon_{\text{cond}}$ definition (34) (in Chapter 6).

The authors investigated properties of two different types of shape functions: general solutions of homogenous equations (in HT-H elements) and fundamental solutions with singularities outside the element (HT-K). As it can be observed,

Table 5: Sensitivity of HT K and HT H solutions (Fig. 4 in this chapter and Fig. 9 in Chapter 6) to mesh distortion: $M_f = 12$, $a = b = 2$, $A = 200$.

Distortion d	Solution type		$\varepsilon_{\text{cond}}$ ((34) in Chapter 6)	$\bar{\delta}_U[\%]$	$\bar{\delta}_u[\%]$
	HT-H		0.2E–12	0.000	0.025
		$\kappa = 2.50$	1.0E–5	0.001	0.011
$d = 0.0$	HT-K	$\kappa = 4.00$	0.7E–1	0.000	0.019
		$R = 2.50$	0.3E–8	0.001	0.009
		$R = 4.00$	0.5E–4	0.000	0.034
	HT-H		0.2E–13	1.266	0.171
		$\kappa = 2.50$	0.2E–4	1.246	0.188
$d = 0.1$	HT-K	$\kappa = 4.00$	0.6E–2	1.226	0.186
		$R = 2.50$	0.2E–8	1.126	0.186
		$R = 4.00$	0.2E–4	1.326	0.187
	HT-H		0.1E–14	5.018	0.397
		$\kappa = 2.50$	0.2E–7	4.605	0.367
$d = 0.2$	HT-K	$\kappa = 4.00$	0.7E–3	4.551	0.362
		$R = 2.50$	0.2E–8	4.677	0.369
		$R = 4.00$	0.2E–6	4.579	0.364
	HT-H		0.1E–11	10.409	0.606
		$\kappa = 2.50$	0.2E–7	9.515	0.507
$d = 0.3$	HT-K	$\kappa = 4.00$	0.2E–2	9.447	0.508
		$R = 2.50$	0.2E–7	9.665	0.528
		$R = 4.00$	0.3E–4	9.443	0.512

the solution with Kupradze functions are, as a rule, more accurate than those presented by HT-H approach. Also, it should be noted that in some cases application of the Herrera system is difficult or impossible (e.g. elements far from the symmetry axis in 3D axisymmetric problems [49]). However, the HT-K solutions require choice of position of the shape function singularities. Although in most cases the range of stable results is relatively large (for varying parameters κ and R), this range is problem dependent and depends also on the number of shape functions used in the solutions. For application of the above functions, the authors would suggest the choice $\kappa = (2.0\text{–}2.5)$ and observation of the control parameter $\varepsilon_{\text{cond}}$. This procedure can be carried out automatically by a suitable subroutine. However, to suggest a possibly optimal algorithm of choice of the parameters κ or R, careful investigations are necessary.

4 Optimization of helical spring cross-section

4.1 T-complete system

Main limitation of the boundary collocation approach is existence of the Trefftz-type functions fulfilling particular differential equations. There are different

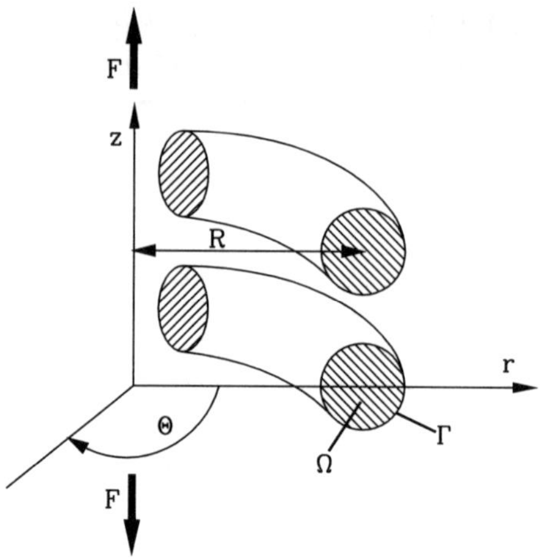

Figure 5: Scheme of helical spring (lift of the spring has been neglected).

possibilities of derivation of such functions [125]. A good example of such a procedure is given in [56–59] for the equation on stress function **Φ** in a helical spring (see Fig. 5; lift of the spring has been neglected) [97]:

$$\frac{\partial^2 \Phi}{\partial r^2} + \frac{\partial^2 \Phi}{\partial z^2} - \frac{3}{r}\frac{\partial \Phi}{\partial r} + 2C = 0 \tag{15}$$

where C is a constant dependent on the load F, while

$$\tau_{r\theta} = \frac{GR^2}{r^2}\frac{\partial \Phi}{\partial z}, \quad \tau_{\theta z} = -\frac{GR^2}{r^2}\frac{\partial \Phi}{\partial r} \tag{16}$$

where G is the Kirchhoff modulus and $\tau_{r\theta}$, $\tau_{\theta z}$ are tangent stresses in the spring cross-section. After some transformations, we obtain:

$$\Phi_0 = 1, \quad \Phi_1 = z, \quad \Phi_2 = z^2 + \tfrac{1}{2}r^2, \quad \Phi_3 = z^3 + \tfrac{3}{2}r^2 z \tag{17a}$$

and

$$\Phi_{j+4} = \sum_{k=0}^{\left[\frac{j}{2}\right]} a_{j\,k} r^{4+2k} z^{j-2k} \qquad j = 0,1,2,... \tag{17b}$$

where $[x]$ is the integer part of real number x and

$$a_{j\,0} = 1, \quad a_{j\,k+1} = -a_{j\,k}\frac{(j-2k)(j-2k-1)}{[4+2(k+1)]2(k+1)}. \tag{17c}$$

The inductive process of derivation ensures the completeness of the system. To present it we consider action of the operator

$$L = \frac{\partial}{\partial r^2} + \frac{\partial}{\partial z^2} - \frac{3}{r}\frac{\partial}{\partial r} \tag{18}$$

resulting from eqn (15), on the monomials $ar^{\alpha}z^{\beta}$, $br^{\alpha+2}z^{\beta-2}$, $cr^{\alpha+4}z^{\beta-4}$:

$$\begin{aligned} L(ar^{\alpha}z^{\beta}) &= a\alpha(\alpha-4)r^{\alpha-2}z^{\beta} + a\beta(\beta-1)r^{\alpha}z^{\beta-2}, \\ L(br^{\alpha+2}z^{\beta-2}) &= b(\alpha+2)(\alpha-2)r^{\alpha}z^{\beta-2} + b(\beta-2)(\beta-3)r^{\alpha+2}z^{\beta-4}, \\ L(cr^{\alpha+4}z^{\beta-4}) &= c(\alpha+4)\alpha r^{\alpha+2}z^{\beta-4} + c(\beta-4)(\beta-5)r^{\alpha+4}z^{\beta-6}. \end{aligned} \tag{19}$$

Now we choose coefficients a, b, c, so that

$$\begin{aligned} a\beta(\beta-1) + b(\alpha+2)(\alpha-2) &= 0 \\ b(\beta-2)(\beta-3) + c(\alpha+4)\alpha &= 0. \end{aligned} \tag{20}$$

Then, summing eqn (19), we obtain

$$\begin{aligned} &L(ar^{\alpha}z^{\beta}) + L(br^{\alpha+2}z^{\beta-2}) + L(cr^{\alpha+4}z^{\beta-4}) \\ &= a\alpha(\alpha-4)r^{\alpha-2}z^{\beta} + c(\beta-4)(\beta-5)r^{\alpha+4}z^{\beta-6}. \end{aligned} \tag{21}$$

Now, if

$$\alpha(\alpha-4) = 0 \quad \text{and} \quad (\beta-4)(\beta-5) = 0, \tag{22}$$

the sum (21) vanishes. Following this procedure, we finally obtain the complete system of polynomials (17). More details of the derivation can be found in [56, 59].

4.2 Preliminary numerical investigations

A problem of a helical spring cross-section (a bounded, simply-connected region) was chosen for the numerical example. Before application of the optimization algorithm, it was necessary to carry out certain preliminary numerical investigations of convergence of the solution proposed. Because of symmetry of the problem with respect to r-axis (see Fig. 5; compare also eqn (27)), only the even functions Φ_j, ($j = 0,2,4,\ldots$) were taken into account. The first calculations were made for a circular cross-section. The T-complete set was supplemented by the particular solution

$$\Phi^{p}(r,z) = \frac{C}{4}\left(r^2 - 2z^2\right) \tag{23}$$

and the boundary equation of the problem

$$\Phi_{\Gamma} = 0 \tag{24}$$

was approximately fulfilled in a system of equidistant collocation points. To decrease the boundary oscillations, number N of collocations points exceeded number n of unknown Trefftz coefficients and the least square procedure was applied.

In case of the Dirichlet problem solved by the boundary collocation approach, the extreme errors always occur on the boundary Γ of investigated area Ω. This fact suggested introduction of the following boundary error, related to, $\hat{\Phi}_8(R,0)$, which was very near to Φ_{max}:

$$\varepsilon = \frac{\max\left|\hat{\Phi}\right|_\Gamma}{\hat{\Phi}_8(R,0)} * 100\% \tag{25}$$

where $\hat{\Phi}$ means the approximated, numerically calculated value of Φ and $\hat{\Phi}_8$ is the result obtained for eight Trefftz functions. The relative error (25) illustrates the quality of the whole approximate solution. Its accuracy is visible in Fig. 6 in which the distribution of the boundary error is presented. We can observe quick decrease of the error with increase of number n of the trial functions.

As it was already mentioned, the number N of the collocation points should be larger than n, but not too much. In Fig. 7, we can observe that the excessive number of the collocation points is not profitable, considering obvious increase of the computer time.

4.3 Optimization procedure of helical spring cross-section

Optimization of the cross-section of helical spring was interesting not only as an exemplary test of the procedure for simply connected regions but also as a solution

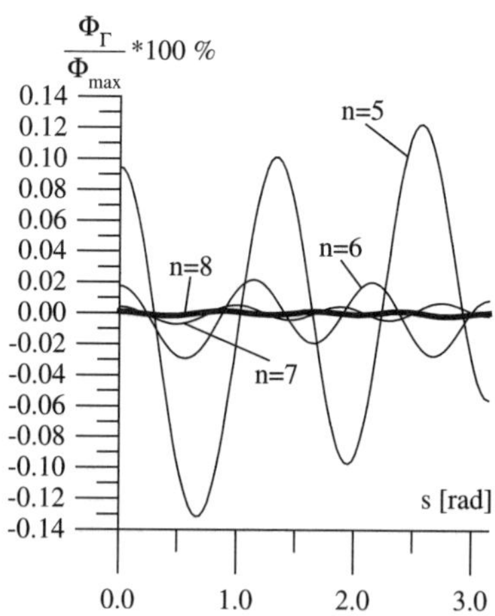

Figure 6: Distribution of boundary solution error for different number n of Trefftz functions (number of collocation points $N = 2n$).

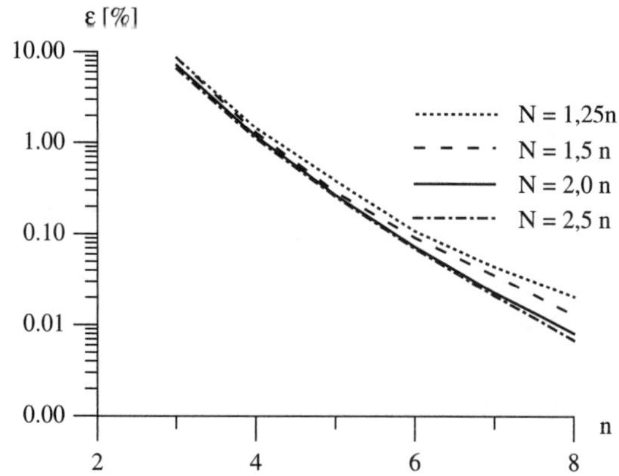

Figure 7: Error measure ε (25) versus number n of trial functions.

of the specific engineering problem. Highly loaded springs in small spaces (e.g. in car making industry) required optimal shapes. This was investigated in a series of Japanese papers [80], also with the help of the boundary element method [51, 52]. The boundary collocation approach occurred to be very convenient in this case.

The circular cross-section boundary described as

$$r = R_0 + \rho_0 \cos\varphi, \quad z = \rho_0 \sin\varphi \tag{26}$$

varied to

$$r = R + \rho_\Gamma(\varphi)\cos\varphi, \quad z = \rho_\Gamma(\varphi)\sin\varphi \tag{27a}$$

where

$$\rho_\Gamma(\varphi) = \rho_0 \sum_{k=0}^{K} C_k \cos k\varphi, \tag{27b}$$

with $C_0, C_1,\ldots, C_k, R$ as optimization (shape) variables. Introducing the condition of the constant volume (constant cross-section area and average radius of the spring), we obtained two constraint equations for the constants C_0 and R [58]:

$$C_0^{\,2} + \frac{1}{2}\sum_{k=1}^{K} C_k^{\,2} = 1 \tag{28a}$$

$$R = R_0 - \frac{\rho_0}{3\pi} \sum_{k,l,m=0}^{K} C_k C_l C_m A_{k\,l\,m} \tag{28b}$$

where

$$A_{k\,l\,m} = \int_0^{2\pi} \cos(k\varphi)\cos(l\varphi)\cos(m\varphi)\cos\varphi \;\mathrm{d}\varphi. \tag{28c}$$

In the example, the main objective function was defined as maximal equivalent stress σ_0 of the structure (σ_{0max}), which was minimized for varying shapes (contours). For comparison, also the uniform contour effort ($\Delta\sigma_0 = \sigma_{0max} - \sigma_{0min} \rightarrow \min$ on Γ) was taken into account. The simple gradient procedure appeared to be sufficient in evaluation of the minimum of the objective functions.

In this numerical example of optimization of the spring cross-section, 10 even Trefftz functions and 20 collocation points were applied. Modification (27b) of the original circular form included up to eight trigonometric terms. Starting from C_3, we already obtained considerable improvement of the spring effort (see Fig. 8). This stands in contrast with the number of 52 shape optimization parameters used in the BEM algorithm [51, 52]. The optimized shape and the distribution of tangent stresses τ along the boundary (for $\sigma_{0max} \rightarrow \min$) are presented in Fig. 9.

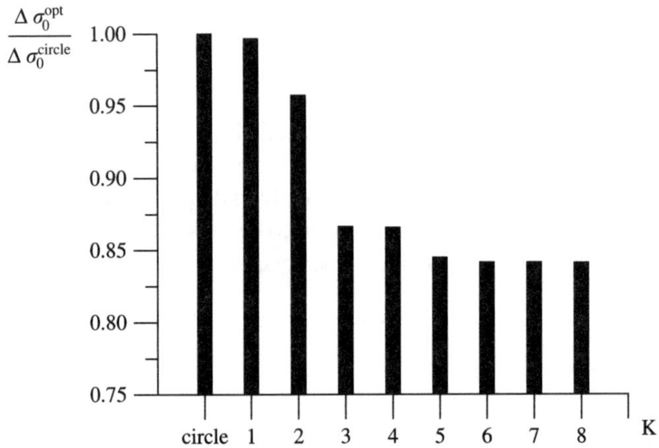

Figure 8: Calculated minimum value of objective function related to $\Delta\sigma_0^{circle}$ (for circular cross-section); example of spring. Influence of number K of optimization parameters.

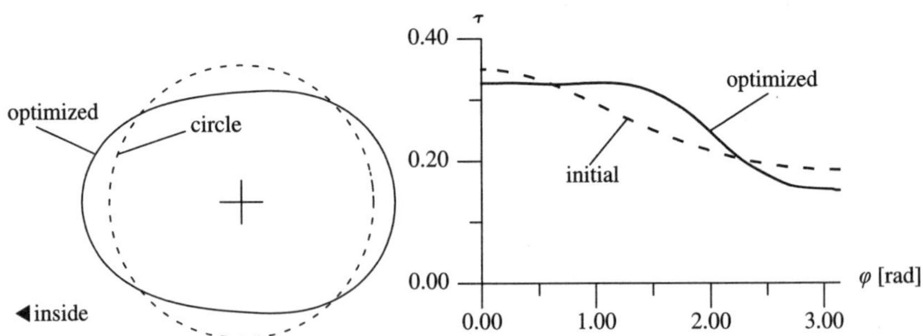

Figure 9: Optimal cross-section of helical spring and distribution of boundary stresses (for minimization of σ_{0max}).

The minimization of $\Delta\sigma_0$ led to different shape and stress distribution in this case [59]. The same approach was also applied to optimization of an extended square plate with a hole [57, 59]. Also in this case only three optimization variables were sufficient to accurately describe modifications of the opening.

The boundary collocation method appeared to be a very convenient tool for the structural optimization. It can be used both in the shape modification [57] and in the optimizing of different structural parameters (e.g. positions of openings [96, 124, 126]). However, certain rules of its application should be observed, which was investigated in the present chapter. The global formulation can be applied to relatively simple structures and objects; however, its investigation is necessary before application of the more complex T-element approach [49, 50, 88].

References

[1] Berger, J.R. & Karageorghis, A., The method of fundamental solutions for layered elastic materials. *Engineering Analysis with Boundary Elements*, **25**, pp. 877–886, 2001.

[2] Berger, J.R., Karageorghis, A. & Martin, P.A., Stress intensity factors computation using the method of fundamental solutions: Mixed-mode problems. *International Journal for Numerical Methods in Engineering*, **67**, pp. 469–483, 2007.

[3] Bowie, O.L. & Conway, H.D., A modified mapping-collocation technique for accurate calculation of stress intensity factors. *International Journal of Fracture Mechanics*, **6**, pp. 199–206, 1970.

[4] Bowie, O.L. & Freese, C.E., Central crack in plane orthotropic rectangular sheet. *International Journal of Fracture Mechanics*, **8**, pp. 49–58, 1972.

[5] Bowie, O.L. & Freese, C.E., Elastic analysis for a radial crack in a circular ring. *Engineering Fracture Mechanics*, **4**, pp. 313–321, 1972.

[6] Bowie, O.L., Application of partitioning to problem-solving in elasticity. In: *International Conference on Fracture Mechanics and Technology*, **2**, March, Hong Kong, pp. 1109–1120, 1977.

[7] Bowie, O.L., Freese, C.E. & Neal, D.M., Solution of plane problems of elasticity utilising partitioning concept. *Journal of Applied Mechanics*, **40**, pp. 767–772, 1973.

[8] Burges, G. & Mahajerin, E., A numerical method for laterally loaded thin plates. *Computer Methods in Applied Mechanics and Engineering*, **49**, pp. 1–15, 1985.

[9] Carperter, W.C., A collocation procedure for determining fracture mechanics parameters at a corner. *International Journal of Fracture*, **24**, pp. 63–74, 1984.

[10] Cartwright, D.J. & Roke, D.P., Evaluation of stress intensity factors. *Journal of Strain Analysis*, **10**, pp. 217–224, 1975.

[11] Chang, C.S. & Conway, H.D., Bond stress in fibre reinforced composite subjected to uniform tension. *Journal of Composite Materials*, **2**, pp. 168–185, 1968.

[12] Chen, C.H. & Cheng, S., Mechanical properties of fiber reinforced composites. *Journal of Composite Materials*, **1**, pp. 30–41, 1967.

[13] Cheng-Xi, L. & Jun, L., A generalised collocation method for stress analysis of cracked plates. *Engineering Fracture Mechanics*, **34**, pp. 1225–1229, 1989.

[14] Cheong, S.K. & Hong, C.S., Analysis of cracks emanating from a circular hole in laminates under various boundary conditions. *Engineering Fracture Mechanics*, **32**, pp. 923–934, 1989.

[15] Cheung, Y.K., Woo, C.W. & Wang, Y.H., Stress intensity factors for a circular arc crack by boundary collocation method. *Engineering Fracture Mechanics*, **34**, pp. 841, 1989.

[16] Cheung, Y.K., Woo, C.W. & Wang, Y.H., The stress intensity factor for a double edge cracked plate by boundary collocation method. *International Journal of Fracture*, **37**, pp. 217–231, 1988.

[17] Cheung-Xi, L., The collocation method for solving two and three dimensional problems of elasticity. *Mechanics Research Communications*, **13**, pp. 325–334, 1986.

[18] Chmielewski, L.H., *Application of the Collocation Method to Calculations of Elastic Plates with Arbitrary Boundary Conditions*. PhD thesis, Warsaw University of Technology, 1985.

[19] Conway, H.D. & Farnham, K.A., Contact stresses between cylindrical shafts and sleeves. *International Journal of Engineering Science*, **5**, pp. 541–546, 1967.

[20] Conway, H.D. & Farnham, K.A., The contact stress problem for indented strips and slabs under conditions of partial slipping. *International Journal of Engineering Science*, **5**, pp. 145, 1967.

[21] Conway, H.D., Vogel, S.M., Farnham, K.A. & So, S., Normal and shearing contact stresses in indented strips and slabs. *International Journal of Engineering Science*, **4**, pp. 342–359, 1966.

[22] Cui, Z.Y., Zhang, Z.Z., Zhang, S.T., Jin, B.S. & Ziong, J.M., The analysis of stress-strain in a finite plate with a blunt crack. *Engineering Fracture Mechanics*, **19**, pp. 465–480, 1984.

[23] De Medeiros, G.C., Partrige, P.W. & Brandao, J.O., The method of fundamental solutions with dual reciprocity for some problems in elasticity. *Engineering Analysis with Boundary Elements*, **28**, pp. 453–461, 2004.

[24] Dong, C.Y. & Lee, K.Y., Stress analysis of an infinite anisotropic elastic medium containing inclusions using the boundary point method. *Engineering Analysis with Boundary Elements*, **28**, pp. 1293–1302, 2004.

[25] Dong, C.Y., Lo, S.H., Cheung, Y.K. & Lee, K.Y., Anisotropic plate bending problems by Trefftz boundary collocation method. *Engineering Analysis with Boundary Elements*, 2**8**, pp. 1017–1024, 2004.

[26] Fett, T. & Munz, D., *Stress intensity factors and weight functions*. Southampton: Computational Mechanics International, 1997.

[27] Fett, T., A boundary collocation analysis of a bimaterial plate with different elastic properties. *Engineering Fracture Mechanics*, **59**, pp. 29–45, 1998.

[28] Fett, T., A boundary collocation study of the edge-bonded plate of dissimilar materials. *FZKA-Report* 5832, Forschungszentrum Karlsruhe, 1996.

[29] Fett, T., A semi-analytical study of the edge-cracked circular disc by use of the boundary collocation method. *Engineering Fracture Mechanics*, **56**, pp. 331–346, 1997.

[30] Gross, B. & Mendelson, A., Plane elastostatic analysis of V-notched plates. *International Journal of Fracture Mechanics*, **8**, pp. 267–276, 1972.

[31] Gross, B. & Srawley, J.E., Stress intensity factors for single edge notch specimen in bending or combined bending tension by boundary collocation of a stress function. *NASA TN D-2609*, 1965.

[32] Gross, B. & Srawley, J.E., Stress intensity factors for three points bend specimens by boundary collocation. *NASA TN D-3092*, 1965.

[33] Gross, B., Roberts, E. & Srawley, J.E., Elastic displacements for various edge cracked plates. *International Journal of Fracture Mechanics*, **4**, pp. 267–276, 1968.

[34] Gross, B., Srawley, J.E. & Brown, W.F. Jr., Stress intensity factors for a single notch tension specimen by boundary collocation of a stress function. *NASA TN D-2395*, 1964.

[35] Hadi, B.K. & Matthews, A., Buckling of anisotropic sandwich panels containing holes: Part I – overall buckling. *Composite Structures*, **52**, pp. 205–216, 2001.

[36] Heaton, M.D., Calculation of the elastic constants of unidirectional fiber-reinforced composite. *British Journal of Applied Physics, Journal of Physics*, sec. D, ser. 2, **1**, pp. 1039–1048, 1968.

[37] Ho, K.C. & Chau, K.T., A finite strip loaded by a bonded-rivet of a different material. *Computers & Structures*, **70**, pp. 203–218, 1999.

[38] Huang, Y. & Hu, K.X., A generalized self-consistent mechanics method for solids containing elliptical inclusions. *Journal of Applied Mechanics*, **62**, pp. 566–572, 1995.

[39] Hufenbach, W. & Zhou, B., Solution for an anisotropic plate with an elastic inclusion and a loaded boundary. *Composite Structures*, **52**, pp. 161–166, 2001.

[40] Hulbert, L.E. & Rybicki, E.F., Boundary point least squares analysis of the free edge effects in some unidirectional fiber composites. *Journal of Composite Materials*, **5**, pp. 164–175, 1971.

[41] Hulbert, L.E., *The Numerical Solution of Two-Dimensional Problems of the Theory of Elasticity*, Ohio State University Engineering Experiment Station Bulletin, no. 198, Reprint of the authors Ph. D. Thesis, No. 63-6186, 1963.

[42] Ishikawa, T. & Kobayashi, S., Elastic properties of unidirectional fibber-reinforced composites II. *Fukago Zairyo Kenkyu (Composite Material Structures)*, **3**, pp. 23–31, 1974.

[43] Isida, M. & Nemat-Nasser, S., A unified analysis of various problems relating to circular holes with edge cracks. *Engineering Fracture Mechanics*, **27**, pp. 571–591, 1987.

[44] Isida, M. & Noguvhi, H., Tension of a plate containing an embedded elliptical crack. *Engineering Fracture Mechanics*, **20**, pp. 387–408, 1984.

[45] Isida, M., Effect of width and length on stress intensity factors of internally cracked plates under various boundary conditions. *International Journal of Fracture Mechanics*, **7**, pp. 301–315, 1971.

[46] Isida, M., Yoshida, T. & Noguvhi, H., Tension and bending of finite thickness plates with a semi-elliptical surface crack. *International Journal of Fracture*, **26**, pp. 157–188, 1984.

[47] Ji, M. & Ishikawa, H., Analysis of an internal central crack with bridging fibers in a finite orthotropic plane. *International Journal of Engineering Science*, **35**, pp. 549–560, 1997.

[48] Jin, W.G., Sheng, N., Sze, K.Y. & Li, J., Trefftz indirect method for plane piezoelectricity. *International Journal for Numerical Methods in Engineering*, **63**, pp. 139–158, 2005.

[49] Jirousek, J. & Wróblewski, A., T-elements: state of the art and future trends. *Archives of Computational Methods in Engineering*, **3–4**, pp. 323–434, 1996.

[50] Jirousek, J. & Zieliński, A.P., Survey of Trefftz-type element formulations. *Computers & Structures*, **2**, pp. 225–242, 1997.

[51] Kamiya, N. & Kita, E., Boundary element method for quasi-harmonic differential equation with application to stress analysis and shape optimization of helical spring. *Computers & Structures*, **37**, pp. 81–86, 1990.

[52] Kamiya, N. & Kita, E., Boundary element method for class of eliptic differential equation and its application: analysis of helical coil spring. *Proc. 3rd Japan/China Symp. BEM*, April, Tokyo, 1990.

[53] Karageorghis, A. & Berger, J.R., The method of fundamental solutions for layered elastic materials. *Department of Mathematics and Statistics, University of Cyprus*, TR/16/1998, 1998.

[54] Karageorghis, A. & Fairweather, G., The Almansi method of fundamental solutions for solving biharmonic problems. *International Journal for Numerical Methods in Engineering*, **26**, pp. 1665–1682, 1988.

[55] Karageorghis, A., Poullikkas, A. & Berger, J.R., Stress intensity factor computation using the method of fundamental solutions. *Computational Mechanics*, **37**, pp. 445–454, 2006.

[56] Karaś, M. & Zieliński, A.P., Application of Trefftz complete functional system to stress analysis in helical spring with an arbitrary wire cross-section. *Journal of Mechanical Engineering (Strojnicki Casopis)*, **49/6**, pp. 426–437, 1998.

[57] Karaś, M. & Zieliński, A.P., Parametric structural shape optimization using the global Trefftz approach. *Journal Theoretical and Applied Mech*anics, **2**, **38**, pp. 285–296, 2000.

[58] Karaś, M. & Zieliński, A.P., Shape optimization of helical spring cross-section by Trefftz aproach. *Proc. VIIth Int. Conf. 'Numerical Methods in Continuum Mechanics'*, ed. V. Kompis, Zilina, VTC, 1998.

[59] Karaś, M., *Application of Generalized Trefftz Method to Optimization of Elastic Structures*. PhD Thesis, supervisor A.P. Zieliński, Cracow University of Technology, Cracow, 2001.

[60] Kobayashi, A.S., Cherepy, R.D. & Kinsel, W.C., A numerical procedure for estimating the stress intensity factor of a crack in a finite plate. *Journal of Basic Engineering*, **86**, pp. 681–684, 1964.

[61] Kobayashi, A.S., Method of collocation applied to edge-notched finite strip subjected to uniaxial tension and pure bending. *Boeing Company*, Seattle, Washington, Document No. D2-23551, 1964.

[62] Kobayashi, S. & Ishikawa, T., Elastic properties of unidirectional fibber-reinforced composites I. *Fukago Zairyo Kenkyu (Composite Material Structures)*, **3**, pp. 12–20, 1974.

[63] Kuang, J.S. & Wang, Y.H., A coupling model of reinforced concrete fracture. *Materials & Structures*, **31**, pp. 602–608, 1998.

[64] Kuang, J.S. & Wang, Y.H., Analysis of interfacial cracks emanating from a hole in a bi-material plate. *European Journal of Mechanics. A – Solids*, **18**, pp. 465–479, 1999.

[65] Lee, K.H., Shukla, A., Parameswaran, V., Chalivera, V. & Hawong, J.S., Static and dynamic fracture analysis for the interface crack of isotropic-orthotropic biomaterial. *KSME International Journal*, **16**, pp. 165–174, 2002.

[66] Lee, L.C. & Conway, H.D., Bond stresses for a composite material reinforced with various arrays of fibers. *International Journal of Solids and Structures*, **14**, pp. 131–139, 1978.

[67] Leissa, A.W. & Clausen, W.E., Application of point matching problem in micromechanics. *Fundamental Aspects of Fiber Reinforced Plastic Composites*, ed. R.T. Schwartz, H.S. Schwartz, Interscience Publishers, New York, 1968, pp. 29–44, Proceedings Conference Held at Dayton, Ohio, Oct. 24–26, May, 1966.

[68] Leissa, A.W., Clausen, W.E. & Agrawal, G.K., Stress and deformation analysis of fibrous composite materials by point matching. *International Journal of Numerical Methods in Engineering*, **3**, pp. 89–101, 1971.

[69] Leitao, V.M.A., Application of multi-region Trefftz method to fracture mechanics. *Engineering with Boundary Elements*, **22**, pp. 251–256, 1998.

[70] Li, Z.C., Lu, T.T. & Hu, H.Y., The collocation Trefftz method for biharmonic equations with crack singularities. *Engineering Analysis with Boundary Elements* **28**, pp. 1293–1302, 2004.

[71] Lo, C.C., Niedenfuhr, F.W. & Leissa, A.W., Further studies in the application of the point matching technique to plate bending and other harmonic and biharmonic boundary value problem. *AFFDT-TR-64-113*, pp. 175, June, 1965.

[72] Madenci, E. & Ileri, L., Analytical determination of contact stresses mechanically fastened composite laminates with finite boundaries. *International Journal Solids and Structures*, **30**, pp. 2469–2484, 1993.

[73] Madenci, E. & Ileri, L., Modified mapping-collocation method revised. *Engineering Fracture Mechanics*, **44**, pp. 127–136, 1993.

[74] Madenci, E., Ileri, L. & Kudva, J.N., Analysis of finite laminates with holes. *International Journal of Solids and Structures*, **30**, pp. 825–834, 1993.

[75] Madenci, E., Sharayev, S., Sergeev, B., Oplinger, D.W. & Shyprykevich, P., Analysis of laminates with multiple fasteners. *International of Solids & Structurs*, **35**, pp. 1793–1811, 1998.

[76] Mangalgalgiri, P.D., Dattaguru, B. & Rao, A.K., Fasteners in composite plates: effects of interfacial friction. *Computers and Mathematics Application*, **11**, pp. 1057–1068, 1985.

[77] Mangalgiri, P.D., Dattaguru, A. & Rumamurthy, T.S., A continuum method for analysis of cracks in anisotropic fields. *International Conference on Fracture Mechanics and Technology*, **2**, March, Hong Kong, pp. 1227–1239, 1977.

[78] Ming, Z.Z., Cheng, S.J. & Xie, H.P., An improved method of collocation for the problem of crack surface subjected to uniform load. *Engineering Fracture Mechanics*, **54**, pp. 731–741, 1996.

[79] Muskhelišvili, N.I., *Certain Basic Problems of Mathematical Theory of Elasticity* (in Russian), Moskva, Izd. AN SSSR, 1949.

[80] Nagaya, K., Stress in coil spring of arbitrary cross-section (1st report, analysis) (in Japanese). *Trans. Japan Soc. Mech. Engrs*, 51a, pp. 2293–2302, 1985.

[81] Newman, J.C. Jr., Crack-opening displacements in centre-crack compact and crack-line wedge-loaded specimens. *NASA TN-D 8268*, 1976.

[82] Newman, J.C. Jr., *Stress analysis of simply and multiply connected regions containing cracks by the method of boundary collocation*. M.Sc. Thesis, Virginia Polytechnic Institute, 1969.

[83] Pao, Y.C. & Maheshwari, M.N., Evaluation on elastic moduli of composite materials by linear programming. *Computer Methods in Applied Mechanics and Engineering*, **3**, pp. 305–318, 1974.

[84] Pei-Oing, G., Stress intensity factors for a rectangular plate with a point-loaded edge crack by a boundary collocation procedure, and an investigation into the convergence of the solution. *Engineering Fracture Mechanics*, **22**, pp. 295–305, 1985.

[85] Persson, E. & Madenci, E., Composite laminates with elliptical pin-loaded holes. *Engineering Fracture Mechanics*, **61**, pp. 279–295, 1998.

[86] Pickett, G., Elastic moduli of fiber reinforced plastic composites. *Fundamental Plastic Composites* (ed. R.T. Schwartz and H.S. Schwartz) Interscience Publishers New York, pp. 13, *Proceedings of Conference held at Dayton*, Ohio, 24–26 May, 1968.

[87] Razi, H., Sergeev, B., Shkaragev, S. & Madenci E., Analysis of sandwich panels with multiple-site damage. *Engineering Fracture Mechanics*, **64**, pp. 255–268, 1999.

[88] Sanecki, H. & Zieliński, A.P., Crack propagation modelled by T-elements. *Engineering Computations*, **23**, pp. 100–123, 2006.

[89] Sergeev, B., Madenci, E. & Ambur, D.R., Stress intensity factors of an arbitrary oriented crack near a hole in longeron web. *Theoretical and Applied Fracture Mechanics*, **31**, pp. 213–222, 1999.

[90] Sergeev, B., Madenci, E. & Ambur, R.D., Influence of bolt spacing and degree of anisotropy in single-lap joints. *Computers & Structures*, **76**, pp. 89–103, 2000.

[91] Sheng, H. & Sze, K.Y., Multi-region Trefftz boundary element method for fracture analysis in plane piezoelectricity. *Computational Mechanics*, **37**, pp. 381–393, 2006.

[92] Sheng, H., Sze, K.Y. & Cheng, Y.K., Trefftz solutions for piezoelectricity by Lekhnitskii's formalism and boundary-collocation method. *International Journal for Numerical Methods in Engineering*, **65**, pp. 2113–2138, 2006.

[93] Shukla, A., Chalivendra, V.B., Parameswaran, V. & Lee, K.H., Photoelastic investigation of interfacial fracture between orthotropic and isotropic materials. *Optics and Lasers in Engineering*, **40**, pp. 307–324, 2003.

[94] Soong, T.C. & Li, C., The rolling contact of two elastic-layer-covered cylinders driving a loading sheet in the nip. *Journal of Applied Mechanics*, **48**, pp. 887–894, 1981.

[95] Soong, T.C. & Li, C., On the unbounded contact between plates and layered cylinders. *Journal of Applied Mechanics*, **47**, pp. 967–984, 1980.

[96] Szybiński, B. & Zieliński, A.P., Folded-plate structures with openings – analysis and optimization, *Computer Assisted Mechanics and Engineering Sciences*, **10/4**, pp. 629–640, 2003.

[97] Timoshenko, S. & Goodier, J.N. *Theory of Elasticity*, 3*rd* ed. McGraw-Hill Book Company, New York, 1982.

[98] Tracy, P., Stress intensity factors for multiple edge cracks in rotating hollow discs. *International Journal of Fracture Mechanics*, **16**, pp. 85–93, 1980.

[99] Tracy, P.G., Elastic analysis of radial cracks emanating from the outer and inner surfaces of a circular ring. *Engineering Fracture Mechanics*, **10**, pp. 291–300, 1979.

[100] Tsamasphyros, G. & Eftaxiopoulos, D.A., Energy considerations in the boundary collocation method applied in fracture. *Engineering Fracture Mechanics*, **54**, pp. 639–652, 1996.

[101] Ugandaoker, A. & Murali, B., Stress intensity factors for finite plate by boundary collocation. *International Journal of Fracture*, **52**, pp. R17–R24, 1991.

[102] Vasilenko, A.T. & Urusova, G.P., Stress state of freely supported multilayered elliptical plates of anisotropic materials. *Mechanics of Composite Materials*, **33**, pp. 349–355, 1997.

[103] Vooren, J.V., Remarks on an existing numerical method to estimate the stress intensity factor of a straight crack in a finite plate. *Journal of Basic Engineering*, **89**, pp. 236–237, 1967.

[104] Wah, T., Stress intensity factors determined by using Westergards's stress functions. *Engineering Fracture Mechanics*, 20, pp. 65–73, 1984.

[105] Wang, J., Andreasen, J.H. & Karihaloo, B.L., The solution of an inhomogeneity a finite plane region and its application to composite materials. *Composite Science and Technology*, **60**, pp. 75–82, 2000.

[106] Wang, T.H. & Cheung, Y.K., Boundary collocation method for cracks emanating from a hole in an anisotropic plate. *Engineering Fracture Mechanics*, **48**, pp. 53–62, 1994.

[107] Wang, Y., Asymmetric crack problems calculated by the boundary collocation method. *Engineering Fracture Mechanics*, **40**, pp. 133–143, 1991.

[108] Wang, Y.H., Cheung, Y.K. & Woo, C.W., The stress intensity factor of a crack emanating from a circular hole in a finite plate by boundary collocation method. *International Journal of Fracture Mechanics*, **43**, pp. 97–108, 1990.

[109] Wilson, H.B. Jr. & Hill, J.L., Plane elastostatics analysis of an infinite plate with a double periodic array of holes or rigid inclusions. *Rohn & Haas Company*, *Report* No. S-50, June, 1965.

[110] Wilson, W.K., Numerical method for determining stress intensity factors of an interior crack in a finite plate. *Journal of Basic Engineering*, **93**, pp. 685–690, 1971.

[111] Wilson, W.K., Stress intensity factors for deep cracks in bending and compact tension specimens. *Engineering Fracture Mechanics*, **2**, pp. 169–171, 1970.

[112] Woo, C.W. & Chan, L.W.S., Composite laminate crack problems using the boundary collocation method. *Engineering Fracture Mechanics*, **45**, pp. 559–574, 1993.

[113] Woo, C.W., Wang, Y.H. & Cheung, Y.K., The mixed mode problems for the cracks emanating from a circular hole in a finite plate. *Engineering Fracture Mechanics*, **32**, pp. 279–288, 1989.

[114] Wu, B. & Rich, T.P., Fracture Mechanics for Die Design and Failure Analysis. *Bucknell University Report*, Lewisburg, PA 17837, 1991.

[115] Wu, B. & Rich, T.P., Stress intensity factors for cracks emanating from pressurised semi-circular cylindrical cavities. *Engineering Fracture Mechanics*, **44**, pp. 383–402, 1993.

[116] Wu, X.J. & Cheng, S.M., A higher-order theory for plane stress conditions of laminates consisting of isotropic layers. *Journal of Applied Mechanics*, **66**, pp. 95–100, 1999.

[117] Xiwu, X., Liangxin, S. & Xuqi, F., Stress concentration of finite composite laminates with elliptical hole. *Computers & Structures*, **57**, pp. 29–34, 1995.

[118] Xu, X.W., Man, H.C. & Yue, T.M., Strength prediction of composite laminates with multiple elliptical holes. *International Journal of Solids and Structures*, **37**, pp. 2887–2900, 2000.

[119] Xu, X.W., Yue, T.M. & Man, H.C., Stress analysis of finite composite laminate with multiple loaded holes. *International of Solids & Structures*, **36**, pp. 919–931, 1999.

[120] Zhan, S.G., Wang, T.C. & Han, X.L., Analysis of two-dimensional finite solids with microcracks. *International of Solids & Structures*, **36**, pp. 3735–3753, 1999.

[121] Zhu, Z., Wang, L., Mohaty, B. & Huang, C., Stress intensity factor for a cracked specimen under compression. *Engineering Fracture Mechanics*, **73**, pp. 482–489, 2006.

[122] Zhu, Z.M., New biaxial failure criterion for brittle, materials in compresion. *Journal of Engineering Mechanics. Transactions of ASCE*, **125**, pp. 1251–1258, 1999.

[123] Zhu, Z.M., Xie, H.P. & Ji, S.C., The mixed boundary problems for a mixed mode crack in a finite plate. *Engineering Fracture Mechanics*, **56**, pp. 647–655, 1997.

[124] Zieliński, A.P. & Sanecki, H., Trefftz method in engineering optimization and sensitivity analysis, *Proc. IVth WCCM*, Buenos Aires, 1998.

[125] Zieliński, A.P., On trial functions applied in the generalized Trefftz method, *Advances in Engineering Software*, **24**, pp. 147–155, 1995.

[126] Zieliński, A.P., Sanecki, H. & Karaś, M.S., Effectiveness of the Trefftz method in different engineering optimization procedures. *Computer Assisted Mechanics and Engineering Sciences*, **8**, pp. 479–493, 2001.

[127] Zong, Z., A complex variable boundary collocation method for plane elastic problems. *Computational Mechanics*, **31**, pp. 284–292, 2003.

CHAPTER 9

Inverse problems in 2D elasticity

1 Introduction

Usually, solutions of inverse problems are connected with minimization of certain functionals, which results in optimization procedures. This kind of formulation is here presented in detail and illustrated by numerical examples. It has first been published in [8] and M. Karaś is a coauthor of this chapter.

The properties of the boundary collocation approach allow to formulate alternative, much more effective, simple direct algorithms, which considerably shorten the time of computer calculations. This is clearly shown in several computational benchmarks for two-dimensional (2D) elastic inverse problems. The proposed algorithms can be applied to any inverse boundary value problem, for which the complete T-function sets are known.

The Herrera-type T-functions were already applied to inverse heat conduction problems, both to stationary [2] and transient [3] formulations, giving very encouraging results. The F-functions (Kupradze functions) were used for steady heat conduction in papers [5, 10, 13, 14, 16], for transient heat conduction in paper [4] and also in the inverse 2D Stokes problems[1].

In [8] the authors focus on a 2D homogeneous elasticity defined in stresses; however, the proposed algorithms can be applied to any boundary value problem for which the complete T-function sets are known. A considered elastic structure occupies region Ω with boundary $\Gamma = \Gamma_g \cup \Gamma_{un}$. In the part Γ_g, the boundary conditions (tractions) are given, while on Γ_{un} they are unknown. Instead, we know from measurements full 2D tensors of stresses $\{\sigma_x, \sigma_y, \sigma_{xy}\}_i$ in fixed control points $\xi_i = (\xi_{ix}, \xi_{iy}),\quad i = 1, 2, ..., K$. Our aim is to recover the unknown boundary conditions on Γ_{un}, i.e. to define the original, governing boundary value problem. Hence, we deal with a typical inverse formulation of this case [9].

2 The Trefftz formulation of the boundary value problem

Consider a 2D elastic problem defined in stresses as

$$\nabla^4 \Phi = 0 \tag{1a}$$

with known, given boundary conditions:

$$\mathbf{t} = \overline{\mathbf{t}}_g \quad \text{on} \quad \Gamma_g \tag{1b}$$

and unknown tractions:

$$\mathbf{t} = \overline{\mathbf{t}}_{un} \quad \text{on} \quad \Gamma_{un}. \tag{1c}$$

The stress function Φ in (1) is related to stress tensor terms by:

$$\sigma_x = \frac{\partial^2 \Phi}{\partial y^2}, \quad \sigma_y = \frac{\partial^2 \Phi}{\partial x^2}, \quad \sigma_{xy} = -\frac{\partial^2 \Phi}{\partial x \partial y}, \tag{2}$$

and to tractions by:

$$\mathbf{t} = \begin{Bmatrix} t_x \\ t_y \end{Bmatrix} = \begin{bmatrix} n_x & 0 & n_y \\ 0 & n_y & n_x \end{bmatrix} \cdot \begin{Bmatrix} \sigma_x \\ \sigma_y \\ \sigma_{xy} \end{Bmatrix} = \begin{bmatrix} n_x & 0 & n_y \\ 0 & n_y & n_x \end{bmatrix} \cdot \begin{Bmatrix} \dfrac{\partial^2 \Phi}{\partial y^2} \\ \dfrac{\partial^2 \Phi}{\partial x^2} \\ -\dfrac{\partial^2 \Phi}{\partial x \partial y} \end{Bmatrix}, \tag{3}$$

where n_x and n_y are coordinates of the outside normal vector to the boundary in a considered point.

In the global Trefftz method, the function Φ is expressed by the T-function series

$$\Phi = \Phi^p + \sum_{i=1}^{M_f} c_i \Phi_i, \tag{4}$$

where Φ^p is the particular solution and $c_1, c_2, \ldots, c_N$ are unknown Trefftz coefficients (in our considerations we assume Φ^p to be zero). The system $\{\Phi_i\}$ can be presented in groups $\{\Psi_i\}$ with four functions Φ_i in one group (for simplicity only cosine terms were taken into account):

$$\begin{aligned} \Psi_0(r,\varphi) &= A_0 + B_0 r^2 + C_0 \ln r + D_0 r^2 \ln r, \\ \Psi_1(r,\varphi) &= \left(A_1 r + B_1 r^3 + C_1 r^{-1} + D_1 r \ln r\right)\cos\varphi, \\ \Psi_n(r,\varphi) &= \left(A_n r^n + B_n r^{n+2} + C_n r^{-n} + D_n r^{-n+2}\right)\cos n\varphi, \quad n \geq 2, \end{aligned} \tag{5}$$

where r,φ are polar coordinates and A_n, B_n, C_n and D_n are unknown coefficients. For bounded domains, we apply only positive powers of r, for unbounded only negative powers and for doubly connected domains both negative and positive (compare numerical example 3).

In collocation formulation, we fix the control points $\boldsymbol{\eta}_1=(\eta_{1x},\eta_{1y}),\dots,\ \boldsymbol{\eta}_r=(\eta_{rx},\eta_{ry})\in\Gamma_{un}$ and $\boldsymbol{\eta}_{r+1}=(\eta_{r+1x},\eta_{r+1y}),\dots,\ \boldsymbol{\eta}_n=(\eta_{nx},\eta_{ny})\in\Gamma_g$. Then, for assumed boundary conditions, we define a linear system of algebraic equations (usually overdetermined)

$$\mathbf{A}\cdot\mathbf{c}=\mathbf{b}, \tag{6}$$

where $\mathbf{b}=\{(t_1^x,t_1^y),\dots,(t_n^x,t_n^y)\}^T$ is a vector consisting of components of the tractions at the collocations points $\boldsymbol{\eta}_1,\dots,\boldsymbol{\eta}_n$. The vector $\mathbf{c}=\{c_1,\dots,c_N\}^T$ gathers unknown coefficients (see eqn (4)) and the matrix $\mathbf{A}$ consists of traction components calculated in manner (3) from single Trefftz functions $\Phi_1,\Phi_2,\dots,\Phi_N$ at the collocation points, i.e. $\mathbf{A}=[A_{ij}]_{\substack{i=1,\dots,2n\\ j=1,\dots,N}}$, where A_{ij} are (see eqn (3))

$$\begin{Bmatrix} A_{2i-1j} \\ A_{2ij} \end{Bmatrix}=\begin{bmatrix} n_x^{\,i} & 0 & n_y^{\,i} \\ 0 & n_y^{\,i} & n_x^{\,i} \end{bmatrix}\cdot\begin{Bmatrix} \dfrac{\partial^2\Phi_j^{(i)}}{\partial y^2} \\ \dfrac{\partial^2\Phi_j^{(i)}}{\partial x^2} \\ -\dfrac{\partial^2\Phi_j^{(i)}}{\partial x\partial y} \end{Bmatrix}, \tag{7}$$

where $n_x^{\,i}$ and $n_y^{\,i}$ are coordinates of the outside normal vector to the boundary at points $\boldsymbol{\eta}_i$.

Equation (6) is an algebraic system that can be solved in the least square sense and binding $\mathbf{c}$ and $\mathbf{b}$ in the linear way:

$$\mathbf{c}=\left(\mathbf{A}^T\mathbf{A}\right)^{-1}\mathbf{A}^T\mathbf{b}. \tag{8}$$

If $\hat{\boldsymbol{\sigma}}$ is a vector of stresses $\hat{\boldsymbol{\sigma}}=\{(\hat{\sigma}_{1x},\hat{\sigma}_{1y},\hat{\sigma}_{1xy}),\dots,(\hat{\sigma}_{kx},\hat{\sigma}_{ky},\hat{\sigma}_{kxy})\}^T$ numerically calculated in the control points $\xi_1=(\xi_{1x},\xi_{1y}),\dots,\xi_k=(\xi_{kx},\xi_{ky})$, it can shortly be expressed as

$$\hat{\boldsymbol{\sigma}}=\mathbf{B}\cdot\mathbf{c}, \tag{9}$$

where $\mathbf{B}$ is a matrix resulting from relations (2) obtained from single Trefftz functions $\Phi_1,\Phi_2,\dots,\Phi_N$ for the above points. From (8) and (9), we obtain

$$\hat{\boldsymbol{\sigma}}=\mathbf{C}\cdot\mathbf{b},\quad\text{where}\quad \mathbf{C}=\mathbf{B}\cdot\left(\mathbf{A}^T\mathbf{A}\right)^{-1}\mathbf{A}^T, \tag{10}$$

which means that also $\hat{\boldsymbol{\sigma}}$ depends linearly on $\mathbf{b}$.

3 Three methods of solving inverse problems

3.1 Optimization of unknown boundary conditions with respect to measurement functional (method 1)

To solve the inverse problem (1), we additionally measure stresses $\bar{\boldsymbol{\sigma}} = \{(\bar{\sigma}_{1x}, \bar{\sigma}_{1y}, \bar{\sigma}_{1xy}), ..., (\bar{\sigma}_{kx}, \bar{\sigma}_{ky}, \bar{\sigma}_{kxy})\}^T$ in the control points $\xi_1, ..., \xi_k$. Then we choose a certain (optimized) distribution of tractions $\mathbf{t}_{un}$ on Γ_{un} and calculate the implicated stresses $\hat{\boldsymbol{\sigma}}$ from (8) to (10). The functional

$$I = \sum_{i=1}^{k} \left[(\hat{\sigma}_{ix} - \bar{\sigma}_{ix})^2 + (\hat{\sigma}_{iy} - \bar{\sigma}_{iy})^2 + (\hat{\sigma}_{ixy} - \bar{\sigma}_{ixy})^2 \right] \tag{11}$$

can now be considered as an optimization measure. Its minimization with respect to the unknown tractions $\mathbf{t}_{un}$ means improvement of the solution of the inverse problem (1) and final calculation of possibly accurate distribution of the above tractions.

To accelerate calculations of $\hat{\boldsymbol{\sigma}}$, we can, instead of (10), express

$$\hat{\boldsymbol{\sigma}} = \mathbf{C_0} + t_1^n \cdot \mathbf{C_1} + t_1^s \cdot \mathbf{C_2} + ... t_r^n \cdot \mathbf{C_{2r-1}} + t_r^s \cdot \mathbf{C_{2r}}, \tag{12}$$

where $\mathbf{C}_0 = \mathbf{C} \cdot \mathbf{b}_0, \mathbf{b}_0 = \{0, ...0, t_{r+1}^n, t_{r+1}^s, ..., t_n^n, t_n^s\}^T$, and the vectors $\mathbf{C}_1, ..., \mathbf{C}_{2r}$ are the particular columns of $\mathbf{C}$. Formula (12) significantly facilitates the whole procedure. Indeed, we calculate the vectors $\mathbf{C}_0$, $\mathbf{C}_1, ..., \mathbf{C}_{2r}$ only once, which considerably decreases the CPU time of the whole searching process.

As an optimization tool, the evolutionary algorithm has been chosen, which is described in the first example of Section 4. The form of the algorithm refers to the specific definition of individuals used in this example (see eqns (16)–(18)).

3.2 Direct comparison of calculated and measured stresses (method 2)

The other idea of solving our problem is to solve, with respect to $t_1^n, t_1^s, ..., t_r^n, t_r^s$ the following equation:

$$\mathbf{C_0} + t_1^n \cdot \mathbf{C_1} + t_1^s \cdot \mathbf{C_2} + ... + t_r^n \cdot \mathbf{C_{2r-1}} + t_r^s \cdot \mathbf{C_{2r}} = \bar{\boldsymbol{\sigma}}, \tag{13}$$

where the vector $\bar{\boldsymbol{\sigma}}$ collects the stresses $\bar{\boldsymbol{\sigma}}_1, ..., \bar{\boldsymbol{\sigma}}_k$ (given by measurement). After solving eqn (13), we extend the discrete information of tractions $t_1^n, t_1^s, ..., t_r^n, t_r^s$ into the approximate distribution of $\hat{\mathbf{t}}_{un}$. This gives us a form of solution of our inverse problem. The illustration of this method is given in Section 4.

3.3 Extrapolation of inside solution (method 3)

The Trefftz approach uses analytical trial functions, which are solutions of governing partial differential equations of a considered problem also in an outside

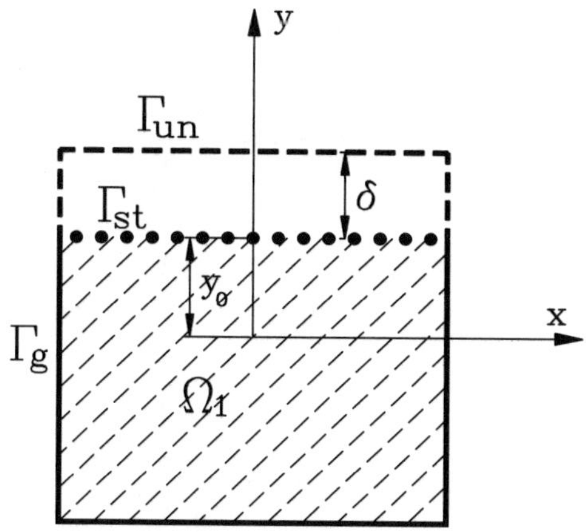

Figure 1: Investigated plate. Positions of control points in Γ_{st}.

neighbourhood of Ω. This observation suggests direct, simultaneous application of the information given in the known boundary conditions and the measurement points $\xi_1,...,\xi_k$ to calculate the unknown Trefftz coefficients **c**. Hence, the next proposed method of solving our inverse problem uses the idea of extrapolation. In our case, we solve the new direct problem defined in the subregion Ω_1 of the region Ω (Fig. 1). This new problem is defined in the following collocation sense.

At the points of part Γ_g of the boundary of subregion Ω_1, we consider traction collocation and at points $\xi_1,...,\xi_k$ situated on part Γ_{st} of the boundary of Ω_1 (Fig. 1) we use stress collocation (that means that we set equations for all components of stress tensor at all points $\xi_1,...,\xi_k$). The resulting equation is following:

$$\tilde{\mathbf{A}} \cdot \mathbf{c} = \tilde{\mathbf{b}}, \tag{14}$$

where $\tilde{\mathbf{b}}$ is a vector consisting of components of the tractions at collocation points situated on Γ_g and all components of the stress tensor at points situated on Γ_{st}. $\tilde{\mathbf{A}}$ is defined similarly as **A** in eqn (6) with this difference that for the points $\xi_1,...,\xi_k$ we collect in $\tilde{\mathbf{A}}$ all components of the entire stress tensor resulting from the single Trefftz functions Φ_i, for $i = 1,...N$.

After solving (14), we obtain a stress field that is defined, theoretically, in the whole space (not only in the region Ω_1). Thus, we can use the information given by this field to recover unknown tractions (boundary conditions) on the boundary Γ_{un}. The illustration of efficiency of the extrapolation method is given in Section 4.

4 Numerical illustration

The methods proposed above were investigated using an example of a rectangular plate. Two different cases of load distribution were considered. The first one was chosen so that the exact, analytical solution was known. This analytical solution

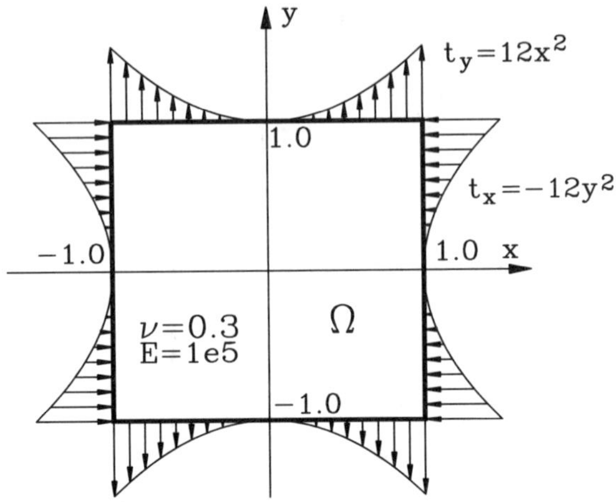

Figure 2: Example 1. Distribution of load in considered plate.

and the known distribution of boundary tractions were considered as references for recovered distributions of the unknown boundary conditions.

In the second and third examples, the analytical solution was unknown. In these examples, numerically obtained results (finite element system ANSYS; very dense mesh) were used, as 'measured' stress tensors at control points $\xi_1,\ldots,\xi_k$, as well as the reference values of tractions.

4.1 First example

Figure 2 presents the considered plate and the given distribution of the normal load (the plate was free of tangential load). Such distribution of the boundary tractions is a result of the stress field defined in Ω by (2) for:

$$\Phi = x^4 - y^4. \tag{15}$$

In our example, we assumed that the whole boundary Γ of the plate was divided into two parts Γ_g and Γ_{un}, as shown in Fig. 1. We used 4×11 collocation points situated equidistantly on all sides of the plate. This gave 88 collocation equations, while the number of T-functions applied was 54. The control points $\xi_1,\ldots,\xi_k$, with $k = 20$, were situated equidistantly along the line $y = y_0$ (Fig. 1).

4.1.1 Method 1

It was assumed that the unknown boundary conditions on Γ_{un} were of the following form:

$$\begin{aligned} t_y &= a_0 + a_1 x + a_2 x^2 + \cdots + a_n x^n, \\ t_x &= b_0 + b_1 x + b_2 x^2 + \cdots + b_n x^n. \end{aligned} \tag{16}$$

Thus, the optimization process (minimization of functional (11)) was developed in the space of parameters $a_0,...,a_n,b_0,...,b_n$. However, not all of these parameters were independent. In our example, the distribution of t_y and t_x should have satisfied the global equilibrium conditions. Thus, we obtained the following relations (see Fig. 2):

$$\int_{-1}^{1} t_x dx = 0, \qquad \int_{-1}^{1} t_y dx = \int_{-1}^{1} 12x^2 dx, \qquad \int_{-1}^{1} x t_y dx = 0. \tag{17}$$

The above conditions gave a relationship between a_0, a_1, b_0 and the others parameters that are independent. The relationships are as follows:

$$a_0 = 4 - \sum_{i=1}^{[n/2]} \frac{a_{2i}}{2i+1}, \quad a_1 = -\sum_{i=1}^{[n/2]} \frac{a_{2i+1}}{2i+3}, \quad b_0 = -\sum_{i=1}^{[n/2]} \frac{b_{2i}}{2i+1}, \tag{18}$$

where $[x]$ means an integer part of the real number x.

To calculate the functional I (11), for given values of parameters $a_0,...,a_n,b_0,...,b_n$, we first evaluate functions (16) in the collocation points $\boldsymbol{\eta}_1,...,\boldsymbol{\eta}_r$ obtaining the values $t_1^{\,n},t_1^{\,s},...,t_r^{\,n},t_r^{\,s}$, which can be substituted into (12). Then we introduce $\hat{\boldsymbol{\sigma}}$ into the functional (11). This operation is repeated very many times in the optimization process; however, the vectors (13) are independent of the parameters $a_0,...,a_n,b_0,...,b_n$ and are calculated only once, which considerably reduces the computational time. It should be underlined that calculation of the vectors $\mathbf{C}_0,\mathbf{C}_1,...,\mathbf{C}_{2r}$ is the most expensive (in sense of CPU time) part of evaluation of the functional I.

We applied an evolutionary algorithm with different number of individuals in the population and different numbers of generations in the searching process. In such an algorithm, the first step is to randomly generate the population of the individuals. In our example, every individual $\mathbf{a} = (a_2,...,a_n,b_1,...,b_n)$ has been connected with a specified distribution of the boundary condition on Γ_{un}. Such individuals in the first step have been generated randomly from a sufficiently large hyper-cube $[a_{2\min}, a_{2\max}] \times ... \times [b_{n\min}, b_{n\max}]$. In our example, the dimensions of the hyper-cube were $[-10^4, 10^4] \times ... \times [-10^4, 10^4]$, while the exact solution was given by $a_2 = 12$ and other parameters equal to zero.

Then the following steps were repeated many times. The whole population was suitably modified by the cross-over (hybridization) operation to generate a new population, from which certain individuals (the worst in sense of functional I (11)) were eliminated. In our example, in this process of passage, we eliminated 75% of population of the older generation.

The hybridization operation has been defined as follows:

$$\bar{\mathbf{a}} = \mu \hat{\mathbf{a}} + (1-\mu)\tilde{\mathbf{a}}. \tag{19}$$

Here the individuals $\hat{\mathbf{a}} = (\hat{a}_2,...,\hat{a}_n,\hat{b}_1,...,\hat{b}_n)$ and $\tilde{\mathbf{a}} = (\tilde{a}_2,...,\tilde{a}_n,\tilde{b}_1,...,\tilde{b}_n)$ are parents of a new one $\bar{\mathbf{a}} = (\bar{a}_2,...,\bar{a}_n,\bar{b}_1,...,\bar{b}_n)$ and, in the first version, $\mu \in (0,1)$ is a randomly chosen real number. In the second version, we initially optimized the constant μ by minimization of a square function approximating the function $\mu \mapsto I(\mu\hat{\mathbf{a}} + (1-\mu)\tilde{\mathbf{a}})$.

To obtain a larger variety of the population, a kind of mutation has been applied, which can be defined as a random perturbation:

$$\overline{\mathbf{a}} = \hat{\mathbf{a}} + \boldsymbol{\varepsilon}. \tag{20}$$

Here $\hat{\mathbf{a}}$ and $\overline{\mathbf{a}}$ are the individuals before and after mutation, respectively, and a vector $\boldsymbol{\varepsilon} = (\varepsilon_2,...,\varepsilon_n,\varepsilon'_1,...,\varepsilon'_n)$ is randomly chosen from the hyper-cube $q^m\left([a_{2\min}, a_{2\max}]\times...\times[b_{n\min}, b_{n\max}]\right)$, where m is a current number of generation, and q is a number very close to 1.0 (in our example $q = 0.9995$). Such a mutation was applied with the probability 15% for each individual of the current population.

The results obtained by the above optimization process are presented in Fig. 3. It can be observed in Fig. 3a–c that the larger number n in (16) cannot improve the solution in this case because of parabolic form of the analytical solution (Fig. 2). To illustrate the convergence process of the solution, the simple Tschebyshev norm:

$$\boldsymbol{\varepsilon} = \max_{\eta\in\Gamma_{un}} \frac{\left|\hat{t}_{un}(\boldsymbol{\eta}) - \overline{t}_{un}(\boldsymbol{\eta})\right|}{\max \overline{t}_{un}} * 100\% \tag{21}$$

versus the number of generations is shown in Fig. 4.

4.1.2 Method 2

The results obtained by the method of direct comparison of numerically calculated and measured stresses are presented in Fig. 5a–d. The number of applied Trefftz functions M_f should not be too small (see Fig. 5a) and should satisfy the condition:

$$M_f < N_b, \tag{22}$$

where N_b is the number of equations in (6). In our example, we had two traction conditions in 44 collocation points, but four equations given by tangential conditions at the corner points were doubled. Thus, the number of independent equations was $N_b = 84$, and $M_f = 87$ appeared to be too large (see Fig. 5c). Figure 5d shows that in (13) the number $2r$ of unknowns (in our case $2r = 22$) must be smaller than the number $3k$ of equations. It should, however, be emphasised that in our investigations the results were very accurate (like in Fig. 5b) in the wide range of values of M_f ($M_f \in [43, 83]$).

4.1.3 Method 3

Investigation of the extrapolation method concerned two aspects. The first one was dependence of results on the distance of control points $\xi_1,...,\xi_k$ from the boundary Γ_{un}. This distance was represented by number y_0 or equivalently by number δ (see Fig. 1). The second aspect was dependence of the results with respect to the number k of control points $\xi_1,...,\xi_k$). In every numerical calculation, the control points were placed equidistantly on the segment with ends at the points $\xi_1 = (-0.95, y_0)$ and

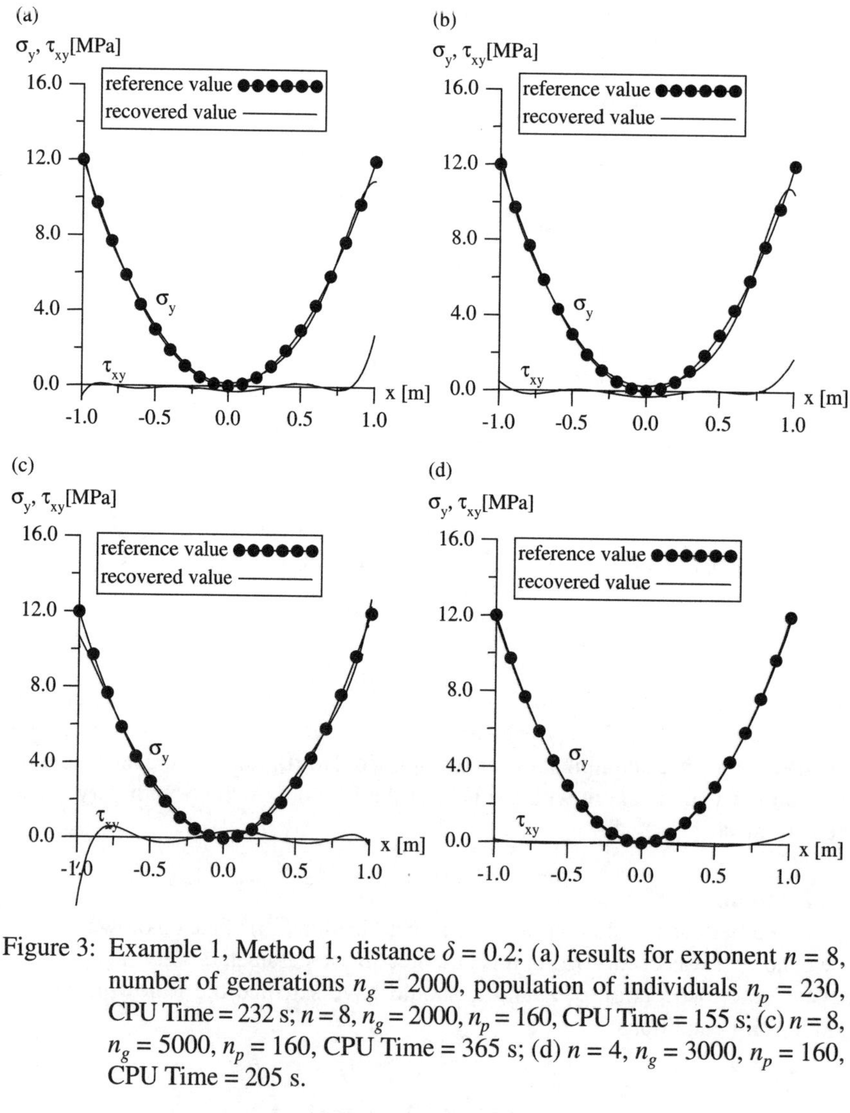

Figure 3: Example 1, Method 1, distance $\delta = 0.2$; (a) results for exponent $n = 8$, number of generations $n_g = 2000$, population of individuals $n_p = 230$, CPU Time = 232 s; $n = 8$, $n_g = 2000$, $n_p = 160$, CPU Time = 155 s; (c) $n = 8$, $n_g = 5000$, $n_p = 160$, CPU Time = 365 s; (d) $n = 4$, $n_g = 3000$, $n_p = 160$, CPU Time = 205 s.

$\xi_k = (-0.95, y_0)$. The choice of these points is justified by the condition $\xi_1,\ldots,\xi_k \in \Omega\backslash\Gamma$ (we want to measure the entire tensor at these points). Figure 6 shows the obtained results represented by the fit error ε (see eqn (21)). In this example, they are very accurate because the exact solution can be obtained as a linear combination of the applied Trefftz functions.

4.2 Second example

Figure 7 presents the considered plate and the distribution of normal compression in this case. The upper distribution is assumed to be unknown, and the

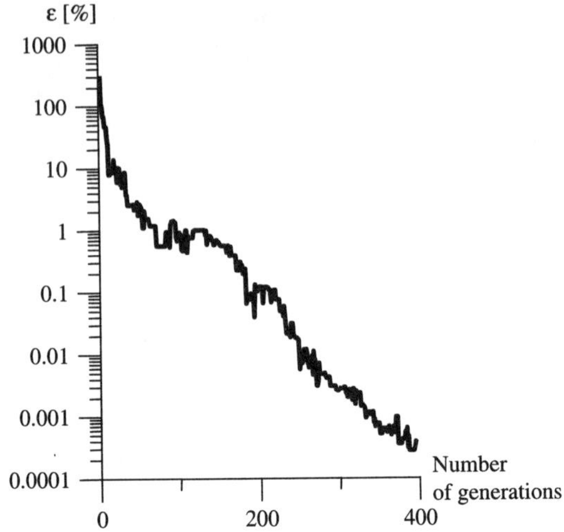

Figure 4: Example 1, Method 1. Evolution of error measure ε for recovered distribution of $\hat{\mathbf{t}}_{un}; n = 4, \quad \delta = 0.2, \quad n_p = 200$, CPU Time = 7 s.

division of the boundary Γ into Γ_g and Γ_{un} is defined as in Fig. 1. This example is more natural because there is not any exact analytical solution, which can be obtained as a linear combination of the applied Trefftz functions. The reference solution is calculated here with the help of the FE code ANSYS with very dense element mesh.

4.2.1 Method 1

For the numerical calculations, again parameterization (16) of the unknown boundary condition was used. Also in this case not all the parameters were independent because the distribution of t_y and t_x should have satisfied the following global equilibrium conditions:

$$\int_{-1}^{1} t_x dx = 0, \quad \int_{-1}^{1} t_y dx = \int_{-1}^{1} 100(1 - x^2)dx, \quad \int_{-1}^{1} x t_y dx = 0. \tag{23}$$

The above conditions give, like in (18), the relationship between a_0, a_1, b_0 and the other parameters that are independent. Figure 8a and b presents the recovered tractions on the boundary Γ_{un}, while Fig. 9a and b shows the evolution of the error ε defined in (21). The obtained results confirm efficiency of the proposed method. In Fig. 9b, however, we observe stabilization of the error at a level of approximately 3%. This phenomenon was also observed in Fig. 3a–c because the larger number n in (15) cannot improve the parabolic solution in both cases.

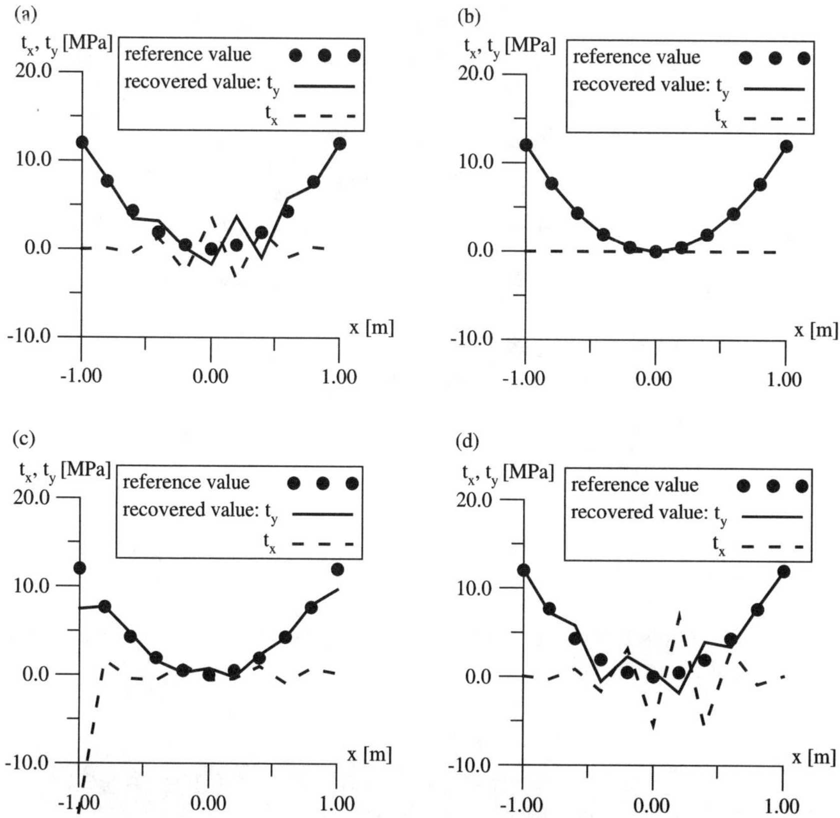

Figure 5: Example 1, Method 2. Recovered distribution of t_y and t_x (solid and dashed line) for $y_0 = 0.8$; $N_b = 84$ and $k = 20$; (a) number of T-functions $M_f = 31$, (b) $M_f = 43$, (c) $M_f = 87$, (d) $M_f = 43$ but with $k = 6$.

4.2.2 Method 2

The method of direct comparison was also applied to the second numerical example. The number of equations and collocation points were the same as in the previous example. Hence, in (22) we had again $N_b = 80$. The obtained results were very accurate for small δ (Fig. 10a and b).

For larger δ, the solution is less stable. In Fig. 10c for $M_f = 70$, we observe surprisingly good results, while for $M_f = 74$ (Fig. 10d) the recovered distribution of tractions is much less accurate.

4.2.3 Method 3

The results shown in Fig. 11 illustrate the extrapolation method, which works very well for smaller distance δ (up to $\delta = 0.3$). For larger δ, the information on Γ_{un} from measurements on Γ_{st} is less accurate and the results considerably worsen.

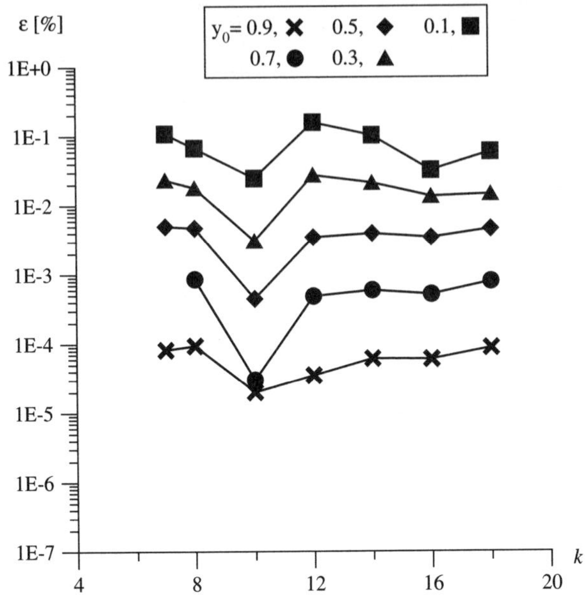

Figure 6: Example 1, Method 2. Fit error ε (eqn (21)) with respect to y_0 and number of control points k.

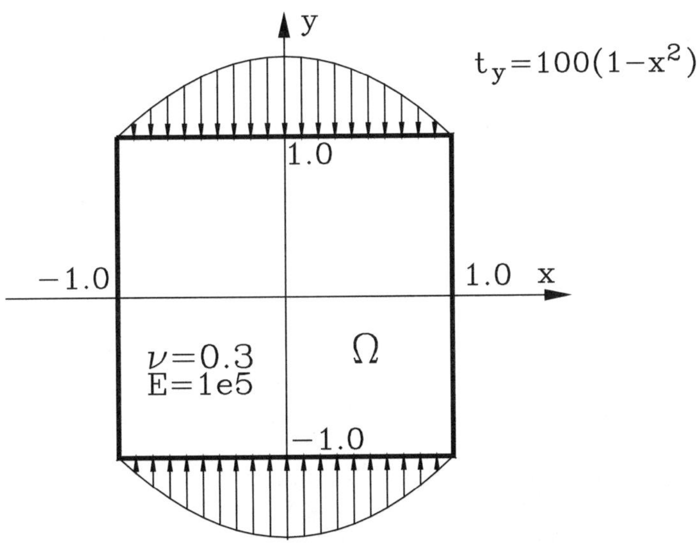

Figure 7: Example 2. Distribution of normal load in considered plate.

It is characteristic that starting from a certain minimal number k_{min}=10, the variable k has rather small influence on ε.

In this example, the authors tried to discover nature of behaviour of the results with an increase of the distance δ. Deterioration of conditioning of the solution

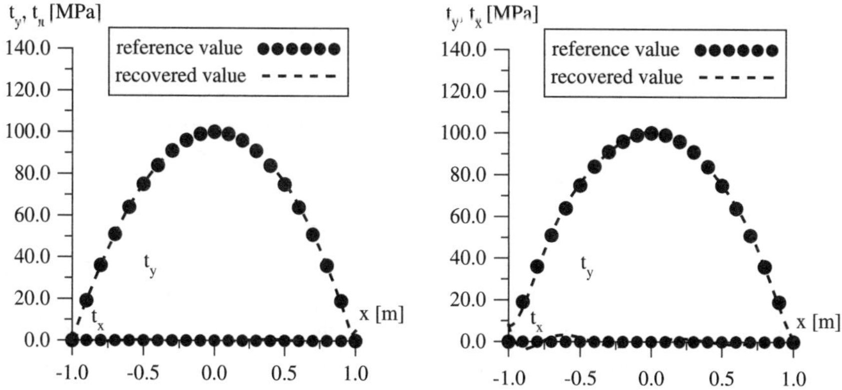

Figure 8: Example 2, Method 1; $k = 20$, $\delta = 0.2$, $n_g = 2000$, $n_p = 160$; (a) $n = 4$, CPU Time = 155 s; (b) $n = 6$, CPU Time = 155 s.

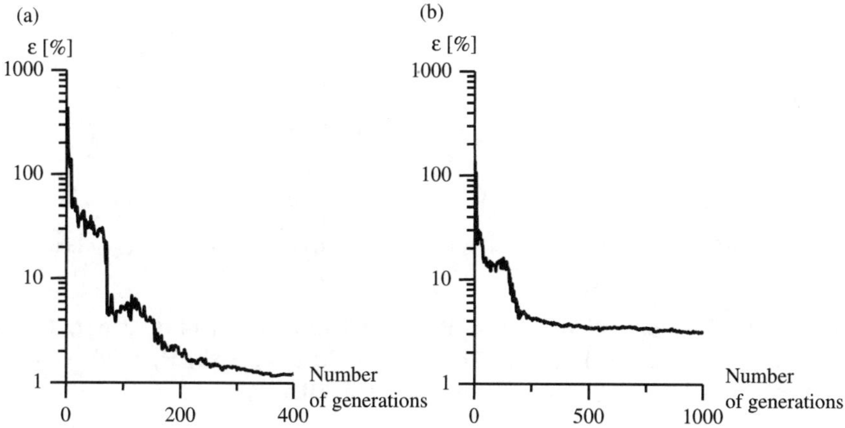

Figure 9: Example 2, Method 1. Evolution of error ε for recovered distribution of $\hat{\mathbf{t}}_{un}$, $n_p = 160$, $y_0 = 0.8$, $k = 20$; (a) $n = 4$, CPU Time = 7 s, (b) $n = 6$, CPU Time = 7 s.

matrices was supposed to be a reason of the error. However, the investigations denied this hypothesis. The following conditioning measures were taken into account:

$$\gamma_1 = \left\| \mathbf{D} \cdot \mathbf{D}^{-1} - \mathbf{I} \right\| \quad \text{and} \quad \gamma_2 = \left\| \mathbf{D} \right\| \cdot \left\| \mathbf{D}^{-1} \right\|, \tag{24}$$

where the norm of the square matrix $\mathbf{D}=\tilde{\mathbf{A}}^T \cdot \tilde{\mathbf{A}}$ (see eqn (14)) was given by $\left\| \mathbf{D} \right\| = \Sigma_{i,j} \, |d_{ij}|$.

Also the Tikhonov regularization procedure [12] has been applied (compare Section 1 in Chapter 5). In this procedure, the ill-conditioned matrix $\mathbf{D}$ is replaced by $\mathbf{D} + a\mathbf{I}$, where $\mathbf{I}$ means the identity matrix and a is an independent parameter of

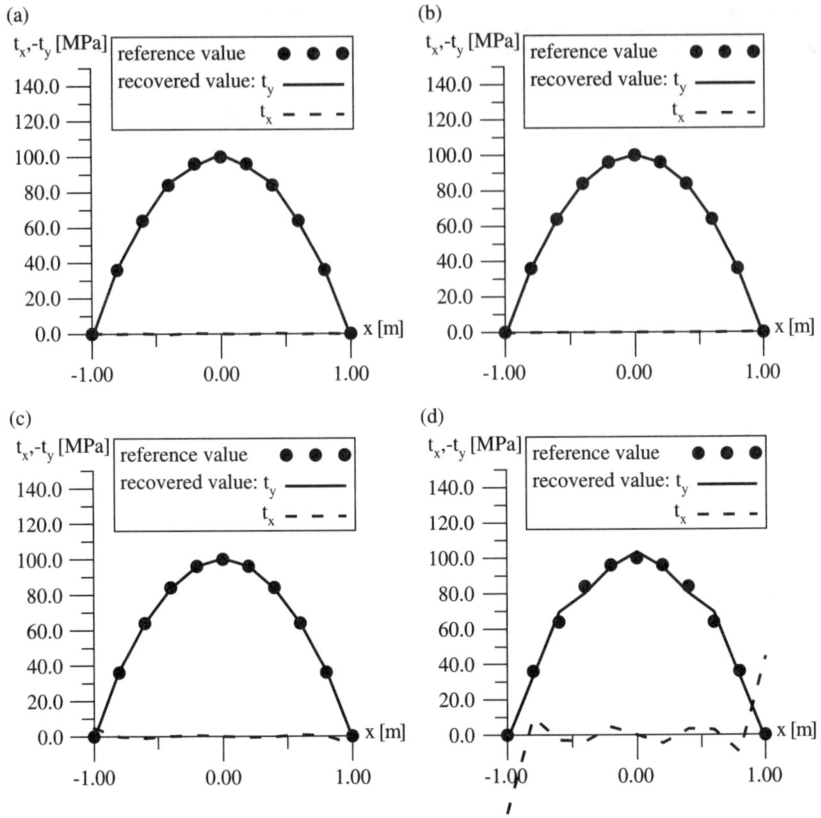

Figure 10: Example 2, Method 2. Recovered distribution of t_y and t_x (solid and dashed line), $k = 20$; (a) $M_f = 70, \delta = 0.2$, (b) $M_f = 74, \delta = 0.2$, (c) $M_f = 70, \delta = 0.5$, (d) $M_f = 74, \delta = 0.5$.

the regularization. This corrected matrix with the suitably chosen coefficient a can considerably improve the conditioning, without substantial deterioration of the searched solution [10].

The influence of the parameter a of the Tikhonov regularization on conditioning indices γ_1, γ_2 and on the fit error ε (21) is shown in Fig. 12a and b. In addition, the investigated matrix was normalized by division of its terms by the smallest term on its diagonal. The lack of positive reaction of the error ε on the increase of a proves good conditioning of the investigated matrix.

4.3 Third example

As the third example, a rectangular plate with a rivet inclusion was considered (Fig. 13). In the example, the exterior square boundary of the plate was load free, but the internal boundary was subjected to a normal non-uniform compression

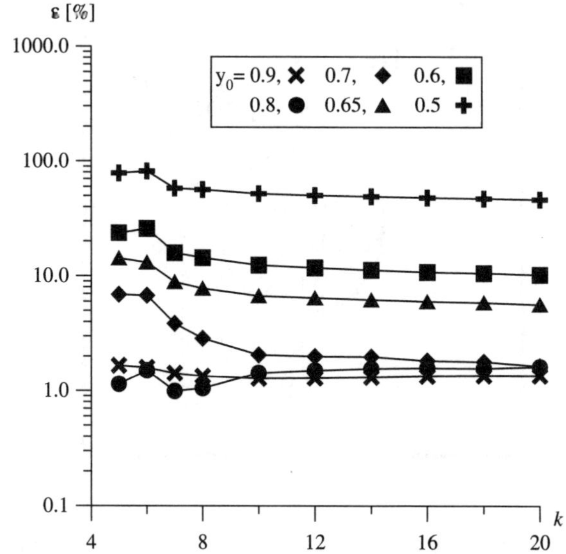

Figure 11: Example 2, Method 3. Fit error ε (21) versus coordinate $y_0 = 1\text{–}\delta$ and number of control points k.

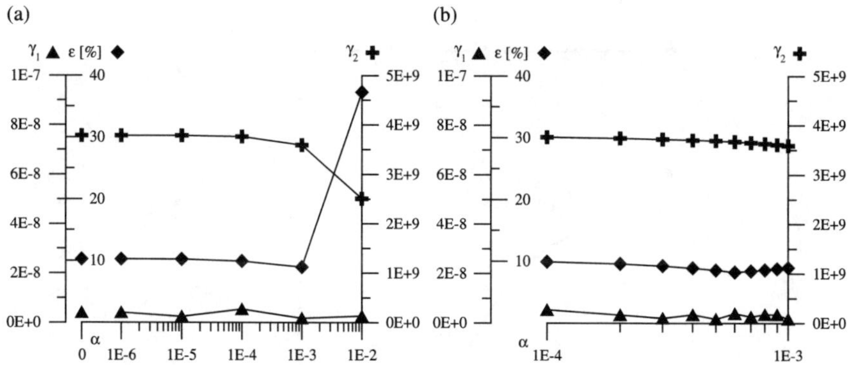

Figure 12: Example 2, Method 3. Conditioning indices γ_1, γ_2 and fit error ε (21) versus parameter a (for $y_0 = 0.6$); $a = 0$ means lack of regularization; (a) full diagram, (b) detail.

caused by the imposed uniform radial displacement $\overline{u}_r$. Because of the symmetry with respect to both *OX* and *OY* axes, only one quarter of the considered plate was investigated and the suitable, symmetric Trefftz functions were applied. It should be noted that as a reference value of the unknown boundary conditions as well as measurements in the control points, the numerical results from the finite element system ANSYS (with very dense mesh) were used.

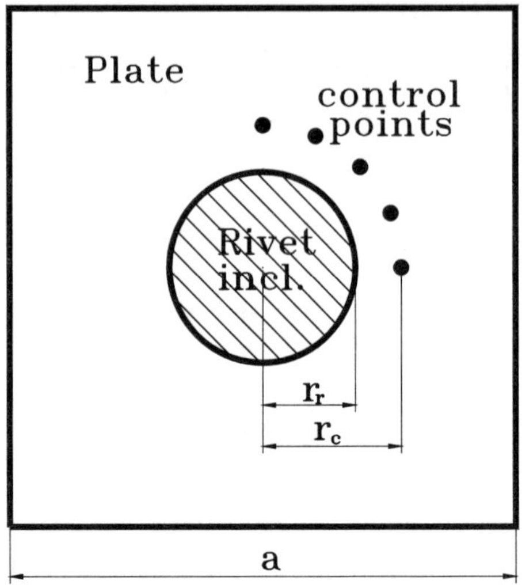

Figure 13: Plate with rivet inclusion. Positions of control points.

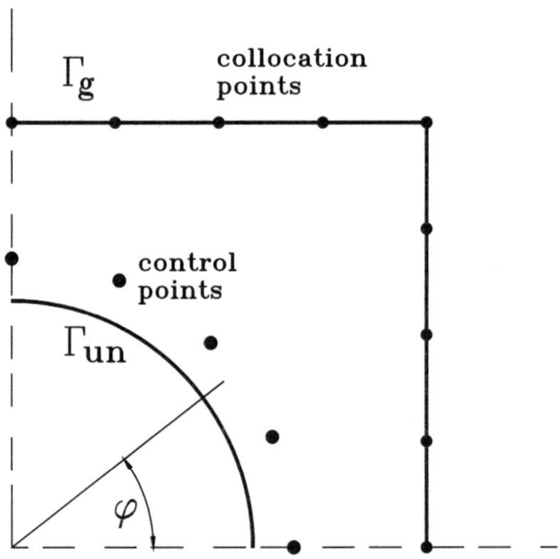

Figure 14: Division of the boundary: Γ_g – with known boundary conditions and Γ_{un} – with unknown ones. Positions of collocation and control points.

As in the previous examples, we divided the whole boundary Γ of the plate into two parts Γ_g and Γ_{un} (Fig. 14). We used 2×8 collocations points on the boundary Γ_g and 11 points on Γ_{un} (for method 1 and method 2) or 9 points with tensor collocation situated at the control points (for method 3).

4.3.1 Method 1

In this example, the unknown boundary conditions on Γ_{un} was assumed to be of the form:

$$t_n = a_0 + a_1 \cos 2\varphi + a_2 \cos 4\varphi + ... + a_n \cos 2n\varphi,$$
$$t_t = b_1 \sin 2\varphi + b_2 \sin 4\varphi + ... + b_n \sin 2n\varphi, \tag{25}$$

where t_n means normal tractions, t_t the tangential tractions, a_0, a_1,..., a_n, b_1,...,b_n- unknown parameters of the solution and φ is a variable explained in Fig. 14. The above form has been chosen because of symmetry of the example with respect to both *OX* and *OY* axes.

The results obtained by the evolutionary algorithm (as in the previous examples) are presented in Fig. 15, while convergence of the solution measured by the error ε (21) is shown in Fig. 16. In Fig. 16b, it is visible that the terms $a_3 \cos 6\varphi$, $a_4 \cos 8\varphi$, $b_3 \sin 6\varphi$ and $b_4 \sin 8\varphi$ have small influence on the solutions. On the other hand, their presence considerably enlarges dimension of the optimization problem, which results in worse convergence of ε (21).

4.3.2 Method 2

In the method of direct comparison of the measured and numerically calculated stresses, we assumed that the distribution of tractions on Γ_{un} was given by step-wise linear functions of the variable φ. The other details of the method were identical as in the first and second examples. We used 11 collocation points on the boundary Γ_{un}, 2×8 points on Γ_g, 11 control points and $M_f = 45$ T-functions. The obtained results are presented in Fig. 17.

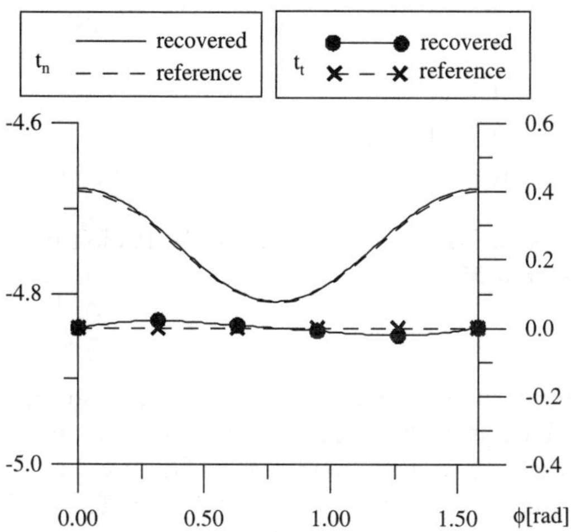

Figure 15: Example 3, Method 1. Recovered distribution of tractions on Γ_{un}, $r_r = 30$, $r_c = 35$, $n_p = 150$, $n = 4$, CPU Time = 31 s.

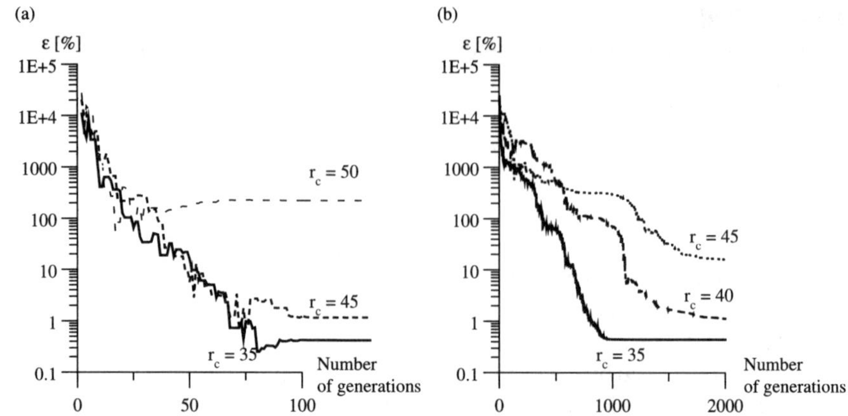

Figure 16: Example 3, Method 1. Evolution of error ε (21) for recovered distribution $\hat{\mathbf{t}}_{un}$, $n_p = 150$, $r_r = 30$; (a) $n = 2$, CPU Time = 4 s, (b) $n = 4$, CPU Time = 31 s.

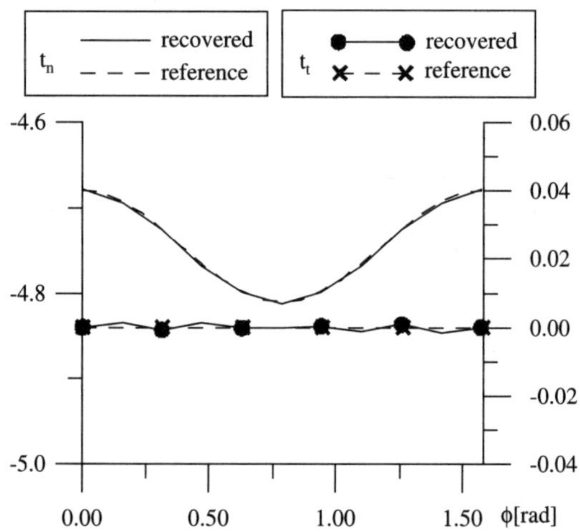

Figure 17: Example 3, Method 2. Recovered distribution of tractions on Γ_{un}, $r_r = 30$, $r_c = 35$, $a = 200$.

4.3.3 Method 3

The extrapolation method was investigated for different radii r_r of the hole (rivet inclusion) and different positions of the control points measured by the radius r_c of the circle along which these points were situated (see Fig. 13).

Accuracy of the obtained results is visible in Fig. 18. The results are very good for the difference $\delta = r_c - r_r$ smaller or equal to 10, good for $\delta = 15$ and a little worse for δ=20 (especially for $r = 30$). The reason of such behaviour of the recovered distribution is the higher stress concentration in the case of smaller r_r.

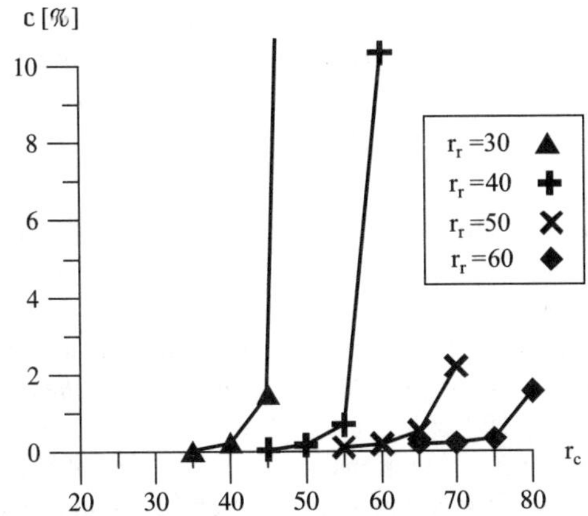

Figure 18: Example 3, Method 3. The fit error (21) versus different values of hole radius r_r and positioning (r_c) of control points (stress collocation points); $a = 200$.

5 Conclusions

Three different methods of solving inverse structural boundary value problems with the boundary collocation approach are presented in this chapter. The first method – optimization – works very well; however, this solution takes substantial amount of computer time (see CPU Time in the examples). This can be a considerable disadvantage in larger problems. The second method is very promising in spite of its limitations with respect to the applied number M_f of the T-functions. Its results for the larger distance δ were better than in case of the other formulations.

The first two methods result in estimation of the unknown tractions along the boundary Γ_{un}. To obtain the full unknown stress state, it is necessary to solve the direct problem with these estimated tractions along this boundary. In the first method, additionally, it is possible to compare the measured and estimated stresses in the control points (error measure (11)).

In the third method, the full boundary collocation solution is extrapolated outside the region of the solved boundary value problem. This approach is more sensitive to the distance δ, but its simplicity and natural benefit from T-functions properties suggest further more intensive investigations of the method.

In the presented examples, the conditioning of the solution matrices was investigated in detail, showing good stability of the obtained results. However, shapes of the investigated regions were rather simple. In case of more complex shapes of structures, division of an investigated region into finite subregions (elements) is necessary because the global Trefftz formulation would result in ill-conditioned matrices [17]. For such cases, the present investigation can be treated as the first step in application of the proposed formulations to the Trefftz finite element approach [6, 7, 11, 18, 19]. The structural inverse problems solved by the T-element approach are included into current directions of the authors' research [15].

References

[1] Chen, C.W., Young, D.L., Tsai, C.C. & Murugesan, K., The method of fundamental solutions for inverse 2D Stokes problems. *Computational Mechanics*, **37**, pp. 2–14, 2005.

[2] Ciałkowski, M. *et al.*, Method of heat polynomials in solving the inverse heat conduction problems, *Zeitschrift fűr Angewandte Mathematik und Mechanik*, **79**, pp. T709–710, 1999.

[3] Ciałkowski, M., Trefftz functions as basic functions of FEM in application to solution of inverse heat conductions problem. *Computer Assisted Mechanics and Engineering Sciences*, **8**, pp. 247–260, 2001.

[4] Dong, C.F., Sun, F.Y. & Meng, B.Q., A method of fundamental solutions for inverse heat conduction problems in anisotropic medium equations. *Engineering Analysis with Boundary Elements*, **31**, pp. 75–82, 2007.

[5] Jin, B., Zheng, Y. & Marin, L., The method of fundamental solutions for inverse boundary value problems associated with the steady-state heat conduction in anisotropic media. *International Journal for Numerical Methods in Engineering*, **65**, pp. 1865–1891, 2006.

[6] Jirousek, J. & Wróblewski, A., T-elements: State of the art and future trends. *Archives of Computational Methods in Engineering* **3(4)**, pp. 323–434, 1996.

[7] Jirousek, J. & Zieliński, A.P., Survey of Trefftz-type element formulation. *Computers and Structures*, **63**, pp. 225–242, 1997.

[8] Karaś, M.S. & Zieliński, A.P., Boundary-value recovery by the Trefftz approach in structural inverse problems. *Communications in Numerical Methods in Engineering* , **24**, pp. 605–625, 2008.

[9] Liu, G.R. & Han, X., *Computational Inverse Techniques in Non-Destructive Evaluation*, CRC Press: Boca Raton and London, 2003.

[10] Marin, L., Numerical solution of Cauchy problem for steady-heat transfer in two-dimensional functionally graded materials. *International Journal of solids and Structures*, **42**, pp. 4338–4351, 2005.

[11] Qin, H.Q., *The Trefftz Finite and Boundary Element Method*, WITpress: Southampton and Boston, 2000.

[12] Tikhonov, A., On certain problems of linear algebra and a stable method of their solution (in Russian). *Doklady Akademii Nauk SSSR*, **163**, pp. 591–594, 1965.

[13] Wang, Y. & Rudy, Y., Application of the method of fundamental solutions to potential-based inverse electrocardiography. *Annals of Biomedical Engineering*, **34**, pp. 1272–1288, 2006.

[14] Wei, T., Hon, Y.C. & Ling, L., Method of fundamental solutions with regularization techniques for Cauchy problems of elliptic operators. *Engineering Analysis with Boundary Elements*, **31**, pp. 373–385, 2007.

[15] Wróblewski, A. & Zieliński, A.P., Structural inverse problems solved by the T-element approach. *Computer Assisted Mechanics and Engineering Sciences*, **13**, pp. 473–480, 2006.

[16] Young, D.L., Tsai, C.C., Chen, C.W. & Fan, C.M., The method of fundamental solutions and condition number analysis for inverse problems of Laplace equation. *Computers and Mathematics with Applications*, **55**, pp. 1189–1200, 2008.

[17] Zieliński, A.P. & Herrera, I., Trefftz method: fitting boundary conditions. *International Journal for Numerical Methods in Engineering*, **24**, pp. 871–891, 1987.

[18] Zieliński, A.P., Sanecki, H. & Karaś, M.S., Effectiveness of the Trefftz method in different engineering optimization procedures. *Computer Assisted Mechanics and Engineering Sciences*, **8**, pp. 479–493, 2001.

[19] Zieliński, A.P., Special Trefftz elements and improvement of their conditioning. *Communications Numerical Methods in Engineering*, **13**, pp. 765–775, 1997.

CHAPTER 10

Three-dimensional problem described by elliptic equations

1 General information

The number of papers, in which the boundary collocation method applied to solve three-dimensional (3D) boundary problems has been used, is much smaller than those describing two-dimensional formulations. It is so mainly because the collocation method, especially its simplest version, shows in this case its disadvantages. Fulfilling boundary conditions by the collocation, that is in a finite number of boundary points, in case of 3D problems not always guarantees accurate solutions. However, there are branches of mechanics, in which the boundary collocation method has been successfully used to solve 3D problems, and solutions obtained by the most popular methods were more time- and work consuming. For example, to such problems belongs creeping flow through a system of rigid particles.

2 3D potential problems

The boundary collocation method in 3D potential problems was used mainly to determine temperature distribution and also in certain electrical field formulations. The respective trial functions were usually connected with general solutions of the Laplace equation in a spherical coordinate system.

To illustrate a frequent algorithm of the solution, let us consider a temperature field in solid matrix with N inclusions of another material [80]. A convenient, global Cartesian coordinate system (x,y,z) should be chosen in such a way that its origin is situated in the first inclusion centre (Fig. 1). Besides, we introduce local spherical coordinate systems (r_i,θ_i,ϕ_i) with origins in centres of all the inclusions. From the equation of energy balance and Fourier's law of heat conduction, we obtain the following set of governing equations:

$$\nabla_s^2 T = 0 \tag{1}$$

$$\nabla_s^2 T_i = 0, \quad i = 1,2,...,N, \tag{2}$$

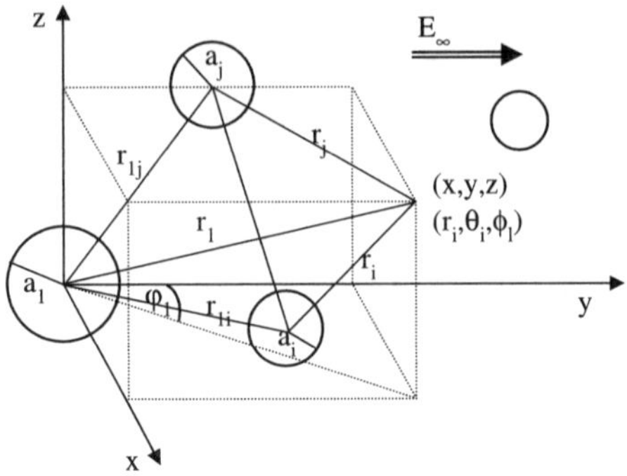

Figure 1: Composite area.

where T is a temperature field in the matrix, T_i is a temperature field in ith inclusion, and

$$\nabla_s^2 = \frac{1}{r^2}\frac{\partial}{\partial r}\left(r^2\frac{\partial}{\partial r}\right) + \frac{1}{r^2 \sin\theta}\frac{\partial}{\partial\theta}\left(\sin\theta\frac{\partial}{\partial\theta}\right) + \frac{1}{r^2\sin^2\theta}\frac{\partial}{\partial\phi} \tag{3}$$

is the Laplace operator in spherical coordinate system (r,θ,ϕ).

The boundary conditions required for determination of the temperature field result from continuity of temperature and heat stream on the inclusion–matrix interfaces and from homogeneity of the heat stream far from inclusions (equal E_∞). If we assume that inclusions are spheres of diameters a_i, those conditions take the form:

$$T = T_i \quad \text{for } r_i = a_i \tag{4a}$$

$$k\frac{\partial T}{\partial r_i} = k_i\frac{\partial T_i}{\partial r_i} \quad \text{for } r_i = a_i \tag{4b}$$

$$T \to T_\infty = E_\infty x \quad \text{when} \left(x^2 + y^2 + z^2\right) \to \infty, \tag{4c}$$

where k and k_i are heat conduction coefficients of the matrix and ith inclusion, respectively.

For this formulation of the boundary value problem, it is useful to divide the considered area into $N+1$ subareas (large elements) in such a way that one of them is a complex area occupied by the matrix and the rest are spheres occupied by particular inclusions. Then, in each subarea we deal with the different solutions:

- in matrix area

$$T = E_\infty \sum_{j=1}^{N} \sum_{n=0}^{\infty} \sum_{m=0}^{n} r_j^{-n-1} P_n^m\left(\mu_j\right)\left[R_{jmn}\cos\left(m\phi_j\right) + S_{imn}\sin\left(m\phi_j\right)\right] + E_\infty x \tag{5a}$$

- in area of *i*th inclusion

$$T_i = E_\infty \sum_{n=0}^{\infty} \sum_{m=0}^{n} r_i^n P_n^m (\cos\theta_i) \left[\overline{R}_{imn} \cos(m\phi_i) + \overline{S}_{imn} \sin(m\phi_i) \right] \quad (5b)$$
$$i = 1, 2, ..., N,$$

where P_n^m are associated Legendre functions and R_{jmn}, S_{jmn}, $\overline{R}_{imn}$, $\overline{S}_{imn}$ are unknown constants, which should be determined from the remaining boundary conditions.

Solutions (5) have been chosen in such a way that boundary condition (4c) is fulfilled exactly. After cutting the infinite series in solutions (5) to a finite number of terms, assigning an adequate number of collocation points on the inclusion surface and fulfilling boundary conditions (4a and b) strictly in those points – a set of linear equations on the unknown series coefficients can be obtained. The described procedure refers to the case, where inclusions have any convex, not far from spherical, shape. When the inclusions are exactly spherical, the procedure is much simpler.

Other applications of the BCM to solutions of 3D potential problems are presented in papers [127, 155] and to the 3D Poisson equation in [97].

3 3D creeping flow

The set of equations governing incompressible, viscous liquid flow (called the Navier–Stokes equations) is non-linear because of the convection term. Neglecting this term is rational in case when the Reynolds number is very small. Then, the Stokes set of linear equations is obtained. They govern relatively slow liquid flow, having rather high viscosity. This flow is called creeping and its governing equations in general case are

$$\nabla \cdot \mathbf{v} = 0 \quad (6a)$$

$$\nabla p = \mu \nabla^2 \mathbf{v}, \quad (6b)$$

where $\mathbf{v}$ is a velocity vector, p means pressure and μ is a liquid viscosity coefficient.

To solve 3D problems of creeping flow, the boundary collocation method has been used by the following authors: Ganatos, Pfeffer & Weinbaum [42–44], Dagan, Pfeffer & Weinbaum [33], Dagan, Weinbaum & Pfeffer [34], Ganatos, Weinbaum & Pfeffer [45], Tozeren [168], Kucaba-Piętal [110, 111], Yan, Weinbaum, Ganatos & Pfeffer [186], Hassonjee, Ganatos & Pfeffer [48], Larson & Higdon [113], Keh & Yang [92], Chapman & Higdon [16], Keh & Tseng [91], Wang [182], Keh & Chen [83], Shankar [158, 159], Kuo & Keh [112], Chen & Keh [17, 18], Chen & Ye [21], Hsu & Keh [56], Keh & Chang [76, 77], Keh & Chen [81], Lu & Lee [116, 117], Keh & Chen [82], Keh & Hsu [85], Keh & Huang [86], Lu & Yen [118], Lu, Yen, Tseng & Tsao [119], Young, Jane, Fan, Murugesan & Tsai [189] and Young, Chiu, Fan, Tsai & Lin [188].

In order to observe complexity of the problem, let us consider flow through a N-body system. Similar to the case of temperature field in the former section, we assume N local spherical coordinate systems with origins in centres of those bodies. The assumed solution is superposition of single solutions in one spherical system:

$$\mathbf{v} = \mathbf{U}_\infty + \sum_{j=1}^{N}\sum_{n=1}^{\infty}\begin{bmatrix} \nabla\times\left(\mathbf{r}_j \chi_{-n-1}^{(j)}\right) \\ +\nabla\phi_{-n-1}^{(j)} - \dfrac{n-2}{2\eta(2n-1)} r_j^2 \nabla p_{-n-1}^{(j)} + \dfrac{n+1}{\eta n(2n-1)} \mathbf{r}_j p_{-n-1}^{(j)} \end{bmatrix}, \tag{7}$$

where $\chi_n^{(j)}$, $\phi_n^{(j)}$, $p_n^{(j)}$ are spherical harmonic functions expressed by:

$$\begin{bmatrix} \chi_{-n-1}^{(j)} \\ \phi_{-n-1}^{(j)} \\ p_{-n-1}^{(j)} \end{bmatrix} = r_j^{-n-1} \sum_{m=0}^{n} P_n^m\left(\cos\theta_j\right)\left\{\begin{bmatrix} A_{jmn} \\ C_{imn} \\ E_{imn} \end{bmatrix}\cos\left(m\phi_j\right) + \begin{bmatrix} B_{jmn} \\ D_{imn} \\ F_{imn} \end{bmatrix}\sin\left(m\phi_j\right)\right\}, \tag{8}$$

$\mathbf{U}_\infty$ is a known constant velocity far from the round flow bodies and A_{jmn}, B_{jmn}, C_{jmn}, D_{jmn}, E_{jmn}, F_{jmn} are unknown coefficients.

The unknown constants are determined by fulfilment of boundary conditions resulting from lack of the liquid slip on surface of the overflowed bodies (flow velocity equals here to zero). Then, again after cutting of the infinite series to certain number of its first expressions and choosing an adequate number of collocation points on surfaces of the bodies, a linear equation set for the unknown coefficients can be obtained.

In case of plane and axisymmetric flows, there is a possibility to introduce the stream function, which fulfils continuity equation (6a). Then equation set (6b) leads to a single equation for this function.

4 Spatial (3D) elastostatic problems

In linear elasticity theory, boundary value problems can be formulated in two ways. In the first case, taking equilibrium equations as a starting point, we can express them by displacements. In the second case, we start from continuity equations and express them by stresses. In the first case, we deal with the Lame–Navier displacement equations:

$$\mu\nabla^2\mathbf{u} + (\lambda+\mu)\nabla(\nabla\cdot\mathbf{u}) + \mathbf{X} = \mathbf{0} \tag{9}$$

and in the second case, with the Beltrami–Michell stress system:

$$\nabla^2\sigma_{xx} + \frac{3}{1+v}\frac{\partial^2 s}{\partial x^2} = -2\frac{\partial X_1}{\partial x} - \frac{v}{1-v}\nabla\cdot\mathbf{X} \tag{10a}$$

$$\nabla^2\sigma_{yy} + \frac{3}{1+v}\frac{\partial^2 s}{\partial y^2} = -2\frac{\partial X_2}{\partial x_2} - \frac{v}{1-v}\nabla\cdot\mathbf{X} \tag{10b}$$

$$\nabla^2\sigma_{zz} + \frac{3}{1+v}\frac{\partial^2 s}{\partial z^2} = -2\frac{\partial X_z}{\partial x_z} - \frac{v}{1-v}\nabla\cdot\mathbf{X} \tag{10c}$$

$$\nabla^2\sigma_{xy} + \frac{3}{1+v}\frac{\partial^2 s}{\partial x\partial y} = -\left(\frac{\partial X_x}{\partial y} + \frac{\partial X_y}{\partial x}\right) \tag{10d}$$

$$\nabla^2\sigma_{xz} + \frac{3}{1+v}\frac{\partial^2 s}{\partial x\partial z} = -\left(\frac{\partial X_x}{\partial z} + \frac{\partial X_z}{\partial x}\right) \tag{10e}$$

$$\nabla^2\sigma_{yz} + \frac{3}{1+v}\frac{\partial^2 s}{\partial y\partial z} = -\left(\frac{\partial X_y}{\partial z} + \frac{\partial X_z}{\partial y}\right), \tag{10f}$$

where **u** is a displacement vector, $\mathbf{X} = X_x\mathbf{i} + X_y\mathbf{j} + X_z\mathbf{k}$ is a mass forces vector, σ_{xx}, σ_{yy}, σ_{zz}, σ_{xy}, σ_{yz}, σ_{xz} are stress components, $s = \sigma_{xx} + \sigma_{yy} + \sigma_{zz}$ and λ, μ, v are material constants of elastic medium.

The boundary collocation method was often used to solve spatial problems of elastostatics. They were expressed both by the displacement as well as the stress equations and applied by the following authors: Cheung [25], Hulbert [60], Hooke & Demushi [53], Redekop [151, 152], Isida & Noguvhi [67], Isida, Yoshida & Noguvhi [68], Patterson & Sheikh [138], Cheung-Xi [25], Redekop & Cheung [149], Hyer & Cohen [65], Lucking, Hoa & Sankar [120], Marin [123], Poullikkas, Karageorghis & Georgiou [139, 142], Tsai [170].

For example, the paper by Redekop & Cheung [149] presents the discussed method basing on the fundamental Kelvin solution, i.e. a stress state in the given point $P = (x,y,z)$ of elastic body caused by concentrated force $F_{xk}\mathbf{i} + F_{yk}\mathbf{j} + F_{zk}\mathbf{k}$ acting in source point $Q_k = (x_k,y_k,z_k)$. The solution is there given by the explicit formulas:

$$\begin{aligned}\sigma_{xxk} &= \left[-(1-2v)\Delta xr^{-3} - 3\Delta x^3r^{-5}\right]F_{xk} + \left[(1-2v)yxr^{-3} - 3\Delta x^2\Delta yr^{-5}\right]F_{yk} \\ &+ \left[(1-2v)\Delta zr^{-3} - 3\Delta x^2\Delta zr^{-5}\right]F_{zk}\end{aligned} \tag{11a}$$

$$\sigma_{yyk} = \left[(1-2\nu)\Delta xr^{-3} - 3\Delta y^3\Delta xr^{-5}\right]F_{xk} + \left[-(1-2\nu)\Delta yr^{-3} - 3\Delta y^3 r^{-5}\right]F_{yk} + \left[(1-2\nu)\Delta zr^{-3} - 3\Delta y^2\Delta zr^{-5}\right]F_{zk} \tag{11b}$$

$$\sigma_{xxk} = \left[(1-2\nu)\Delta xr^{-3} - 3\Delta z^2\Delta xr^{-5}\right]F_{xk} + \left[(1-2\nu)\Delta yr^{-3} - 3\Delta z^2\Delta yr^{-5}\right]F_{yk} + \left[-(1-2\nu)\Delta zr^{-3} - 3\Delta z^3 r^{-5}\right]F_{zk} \tag{11c}$$

$$\sigma_{xyk} = \left[-(1-2\nu)\Delta yr^{-3} - 3\Delta x^2\Delta xr^{-5}\right]F_{xk} + \left[-(1-2\nu)\Delta xr^{-3} - 3\Delta x\Delta y^2 r^{-5}\right]F_{yk} + \left[-3\Delta x\Delta y\Delta zr^{-5}\right]F_{zk} \tag{11d}$$

$$\sigma_{yzk} = \left[-3\Delta x\Delta y\Delta zr^{-5}\right]F_{xk} + \left[-(1-2\nu)\Delta zr^{-3} - 3\Delta z\Delta y^2 r^{-5}\right]F_{yk} + \left[-(1-2\nu)\Delta yr^{-3} - 3\Delta y\Delta z^2 r^{-5}\right]F_{zk} \tag{11e}$$

$$\sigma_{xzk} = \left[-(1-2\nu)\Delta zr^{-3} - 3\Delta z\Delta x^2 r^{-5}\right]F_{xk} + \left[-3\Delta x\Delta y\Delta zr^{-5}\right]F_{yk} + \left[-(1-2\nu)\Delta xr^{-3} - 3\Delta x\Delta z^2 r^{-5}\right]F_{zk}, \tag{11f}$$

where $\Delta x = x - x_k$, $\Delta y = y - y_k$, $\Delta z = z - z_k$, $r^2 = \Delta x^2 + \Delta y^2 + \Delta z^2$ and F_{xk}, F_{yk}, F_{zk} are force intensity coefficients at point Q_k.

The resulting stress state at any point P is expressed as superposition of K stress states caused by forces placed in the source points located outside the investigated object. Thus, it takes the form:

$$\sigma^{(n)} = \sum_{k=1}^{K}\left[G_{xk}^{(n)}F_{xk} + G_{yk}^{(n)}F_{yk} + G_{zk}^{(n)}F_{zk}\right] \quad n = 1,2,...,6, \tag{12a}$$

where

$$\sigma^{(1)} = \sigma_{xx}, \qquad \sigma^{(2)} = \sigma_{yy}, \qquad \sigma^{(3)} = \sigma_{zz} \tag{12b}$$

$$\sigma^{(4)} = \sigma_{xy}, \qquad \sigma^{(6)} = \sigma_{yz}, \qquad \sigma^{(6)} = \sigma_{xz}, \tag{12c}$$

where $G_{xk}^{(n)}$, $G_{yk}^{(n)}$, $G_{zk}^{(n)}$ are functions expressed by means of formulas (11).

The source points Q_k are located not very far from the surface of the body (Fig. 2). The unknown intensity coefficients of source forces are determined by fulfilling boundary conditions resulting from known tractions on a surface of the considered body in finite number of points $P_m = (x_m, y_m, z_m)$. The method described in the paper published by Redekop & Cheung [149] has been successfully tested for three differently loaded objects.

5 Other applications of boundary collocation method to 3D problems

Other worth mentioning cases of application of the boundary collocation method in 3D and plane problems are (see Table 1):

- axisymmetric solutions in elastostatics,
- stress and displacement state of shells,
- steady thermo-elasticity problem,
- problems of wave diffraction and reflection.

It is not possible to mention all single applications of the boundary collocation method applied to solve 3D problems. Except examples gathered in Table 1 and other discussed in Chapters 5 and 6, we may signalize interesting papers by the following authors: Slater [160], Balcerzak & Raynor [8], Hocney [49], Harrington [47], Mullin, Sadburg & Velline [129], Dally & Erisman [35], Bates [9], Bobone [14], Read [145–148], Scholtz & Trass [156], Symm [164], Bolle & Fye [15], Wilton & Mitra [185], Davies [36], Zijl [190], Ishikawa, Koyama & Kobayashi [66], Joosten, Zijl & VanStralen [69], Khila & Wolf [95], Sultanian & Sastri [163], Zijl, Ramakers & VanStralen [191], Yang [187], Sluyter, VanStralen & Zijl [161], Mazur [126], Kin & Russel [96], Mahajerin [122], Reichel [153, 154], Kołodziej [103, 104], Moore, Saffman & Tanveer [128], Vanden-Broeck [174], Kołodziej & Kleiber [101], Christiansen [26–28], Keh & Yang [92, 93], Dias & Christodoulides

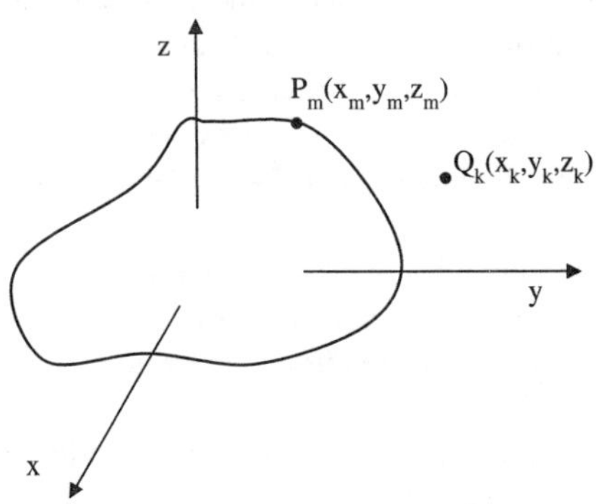

Figure 2: Position of exemplary source point Q_k and collocation points P_m.

Table 1: Papers in which the boundary collocation method was applied to problems not discussed in present monograph.

Problem	References
Axisymmetric problems of elastostatics	[29, 54, 59, 72–74, 125, 150, 184]
Stress and displacement state of shells	[30–32, 39, 50, 52, 58, 60, 61, 108, 131, 132, 134, 151, 152]
Steady thermo-elasticity problem	[10, 11, 51, 52, 57, 62–64, 115, 121, 133, 134, 136, 137, 143, 144, 157, 162, 163, 165–167, 176–178]
Problems of wave diffraction and reflection	[1–6, 12, 13, 15, 23, 24, 38, 40, 70, 71, 98–100, 105–107, 109, 130, 169, 171, 172, 183, 185]

[37], Keh & Lien [89], Kołodziej & Musielak [102], Vanden-Broeck [173], Chen & Keh [20], Karageorghis [75], Keh & Tseng [91], Vanden-Broeck [175], Wang [180], Keh, Chen & Kim [94], Keh & Chen [78], Lee & Vanden-Broeck [114], Fransear, Bouet, Celis, Gabrielli, Huet & Maurin [41], Wang [179], Glav [46], Keh & Chiou [84], Keh & Jan [87], Keh & Luo [90], Matsuyama, Moriyasu & Nishiguchi [124], Wang [181], Hsu & Tseng [55], Keh & Li [88], Chen [22], Poullikkas, Karageorghis & Georgiou [140, 141], Chen & Cai [19], Balakrishnam & Ramachandran [7] and Nishida [135].

References

[1] Albertsen, N.C., Chesneaux, J.M., Christiansen, S. & Wirgin, A., Evaluation of round-off error by interval and stochastic arithmetic method in numerical application of the Rayleigh theory to the study of scattering from an uneven boundary. In: *Proceedings 3rd International Conference on Mathematical and Numerical Aspects Wave Propagation Phenomena,* Mandelieu – La Napoule, France, Apr. 24–28, 1995, ed. G. Cohen, Philadelphia, PA, SIAM, pp. 338–346, 1995.

[2] Al-Rizzo, H.M. & Tranquilla, J.M., Electromagnetic scattering from dielectrically coated axisymmetric objects using the generalized point-matching technique. I. Theoretical formulation. *Journal of Computational Physics,* **119**, pp. 356–373, 1995.

[3] Al-Rizzo, H.M. & Tranquilla, J.M., Electromagnetic scattering from dielectrically coated axisymmetric objects using the generalized point-matching technique. II. Numerical results and comparisons. *Journal of Computational Physics*, **119**, pp. 342–355, 1995.

[4] Al-Rizzo, H.M. & Tranquilla, J.M., Electromagnetic wave scattering by highly elongated and geometrically composite objects of large size parameters: the generalized multipole technique. *Applied Optics*, **34**, pp. 3502–3521, 1995.

[5] Al-Rizzo, H.M., Tranquilla, J.M., Al-Amri, S.M. & Alhafid, H.T., Application of the generalized multipole technique (GMTP) to high-frequency electromagnetic scattering from perfectly conducting and dielectric bodies of revolution. *Journal of Computational Physics*, **136**, pp. 1–18, 1997.

[6] Alves, C.J.S. & Valtchev, S.S., Numerical comparison of two meshfree methods for acoustic wave scattering. *Engineering Analysis with Boundary Elements*, **29**, pp. 371–382, 2005.

[7] Balakrishnam, K. & Ramachandran, P.A., A particular solution Trefftz method for non-linear Poisson problem in heat and mass transfer. *Journal of Computational Physics*, **150**, pp. 239–267, 1999.

[8] Balcerzak, E.M. & Raynor, S., Solution for several shapes of externally pressurised hydrostatic thrust bearings. *Applied Scientific Research*, Section A, **11**, pp. 189–217, 1962.

[9] Bates, R.H.T., The point matching method for interior and exterior two-dimensional boundary value problems. *IEEE Transactions on Microwave Theory and Techniques*, MTT-**15**, pp. 185–187, 1967.

[10] Blanchard, J.P. & Ghoniem, N.M., An eigenfunction approach to singular thermal stresses in bonded strips. *Journal of Thermal Stresses*, **12**, pp. 501–527, 1989.

[11] Blanchard, J.P. & Ghoniem, N.M., Analysis of singular stress fields in duplex components. *Journal of Nuclear Materials*, **172**, pp. 54–70, 1990.

[12] Boag A., Leviatan Y. & Boag, A., Analysis of acoustic scattering from fluid bodies using a multipoint source model. *IEEE Transactions on Ultrasonic, Ferroelectrics and Frequency Control*, **36**, pp. 119–128, 1989.

[13] Boag, A., Leviatan, Y. & Boag, A., Analysis of acoustic scattering from fluid cylinders using multifilament source model. *Journal of Acoustical Society of America*, **83**, pp. 1–8, 1988.

[14] Bobone, R., The method of solution functions: a computer-oriented solution of boundary value problems as applied to nuclear reactors – Part 1, Cylindrical Reactors in $R - \theta$ geometry. *Nuclear Engineering Science*, **29**, pp. 337–353, 1967.

[15] Bolle, D.M. & Fye, D.M., Application of point matching method to scattering from quadrilateral cylinders. *Electronic Letters*, **7**, pp. 577–579, 1971.

[16] Chapman, A.M. & Higdon, J.J.L., Oscillatory Stokes flow in periodic porous media. *Physics of Fluids A*, **4**, pp. 2099–2116, 1992.

[17] Chen, P.Y. & Keh, H.J., Diffusiophoresis of a colloidal sphere in nonelectrolyte gradients parallel to one or two plane walls. *Chemical Engineering Science*, **57**, pp. 2883–2899, 2002.

[18] Chen, P.Y. & Keh, H.J., Boundary effects on osmophoresis: motion of a spherical vesicle parallel to two plane walls. *Chemical Engineering Science*, **58**, pp. 4449–4464, 2003.

[19] Chen, S.B. & Cai, A.P., Hydrodynamic interactions and mean settling velocity of porous particles in a dilute suspension. *Journal of Colloid and Interface Science*, **217**, pp. 328–340, 1999.

[20] Chen, S.B. & Keh, H.J., Axisymmetric electrophoresis of multiple colloidal spheres. *Journal of Fluid Mechanics*, **238**, pp. 251–276, 1992.

[21] Chen, S.B. & Ye, X., Boundary effect on slow motion of a composite sphere perpendicular to two parallel impermeable plates. *Chemical Engineering Science*, **55**, pp. 2441–2453, 2000.

[22] Chen, S.B., Axisymmetric motion of multiple composite spheres: Solid core with permeable shell, under creeping flow conditions. *Physics of Fluids*, **10**, pp. 1550–1563, 1998.

[23] Chen, W. & Hon, Y.C., Numerical investigation on convergence of boundary knot method in the analysis of homogeneous Helmholtz, modified Helmholtz, and convection-diffusion problems. *Computer Methods in Applied Mechanics Engineering*, **192**, pp. 1859–1875, 2003.

[24] Chen, W., Meshfree boundary particle method applied to Helmholtz problems. *Engineering Analysis with Boundary Elements*, **26**, pp. 577–581, 2002.

[25] Cheung-Xi, L., The collocation method for solving two and three dimensional problems of elasticity. *Mechanics Research Communications*, **13**, pp. 325–334, 1986.

[26] Christiansen, S., A brief mathematical analysis of electrochemical machining problems. In: *Proceedings of the Fourth European Conference on Mathematics in Industry*, B.G. Teubner Stuttgart and Kluwer Academic Publishers, 1991.

[27] Christiansen, S., A stability analysis of Eulerian solution method for moving boundary problems in electrochemical machining. *Journal of Computational and Applied Mathematics*, **33**, pp. 269–296, 1990.

[28] Christiansen, S., An analysis of a moving boundary problem in porous flow. *Zeitschrift für Angewandte Mathematik und Mechanik (ZAMM)*, **71**, pp. T813–T814, 1991.

[29] Chu, W.L. & Conway, H.D., A numerical method for computing the stresses around an axisymmetrical inclusion. *International Journal of Mechanical Science*, **12**, pp. 575–588, 1970.

[30] Clausen, W.E. & Leissa, A.W., Stress coupling between holes in shallow spherical shells. In: *Symposium of the International Association for Shell Structures*, Leningrad, September, 1960.

[31] Clausen, W.E., *The approximate analysis of shallow spherical shells with holes*. Ph. D. Thesis, No. 66-1764, Ohio State University, 1965.

[32] Conway, H.D. & Leissa, A.W., Application of the point-matching method to shallow-spherical-shell theory. *Journal of Applied Mechanics*, **29**, pp. 745–747, 1962.

[33] Dagan, Z., Pfeffer, R. & Weinbaum, S., Axisymmetric stagnation flow of a spherical particle near a finite planar surface at zero Reynolds number. *Journal of Fluid Mechanics*, **122**, pp. 273–294, 1982.

[34] Dagan, Z., Weinbaum, S. & Pfeffer, R., General theory for the creeping motion of a finite sphere along the axis of circular orifice. *Journal of Fluid Mechanics*, **127**, pp. 143–170, 1982.

[35] Dally, J.W. & Erisman, E.R., An analytic separation method for photoelasticity. *Experimental Mechanics*, **6**, pp. 493–499, 1966.

[36] Davies, J.B., A least-squares boundary residual method for the numerical solution of scattering problems. *IEEE Transactions MTT*, MTT-**21**, pp. 99–104, 1973.

[37] Dias, F. & Christodoulides, P., Ideal jets falling under gravity. *Physics of Fluids A*, **3**, pp. 1711–1717, 1991.

[38] Dugan, J.P., A variational method for scattering by an arbitrary shaped obstacles. *Variational Methods in Engineering*, **11**, pp. 11/49–11/58, Conference at Southampton, England, 25 September 1972, 1973.

[39] Eringen, A.C., Naghdi, A.K. & Thiel, C.C., State of stress in a circular cylindrical shell with a circular hole. *Welding Research Council Bulletin*, No. 102, 1965.

[40] Fairweather, G.F., Karageorghis, A. & Martin, P.A., The method of fundamental solutions for scattering and radiation problems. *Engineering Analysis with Boundary Elements*, **27**, pp. 759–769, 2003.

[41] Fransear, J., Bouet, V., Celis, J.P., Gabrielli, C., Huet, F. & Maurin, G., Perturbation of the flow of current to a disk electrode by an insulating sphere. *Journal of the Electrochemical Society*, **142**, pp. 4181–4189, 1995.

[42] Ganatos, P., Pfeffer, R. & Weinbaum, S., A strong interaction theory for the creeping motion of a sphere between plane parallel boundaries. Part 1. Perpendicular motion. *Journal of Fluid Mechanics*, **99**, pp. 739–753, 1980.

[43] Ganatos, P., Pfeffer, R. & Weinbaum, S., A numerical solution technique for three-dimensional Stokes flows with application to the motion of strongly interacting sphere in a plane. *Journal of Fluid Mechanics*, **84**, pp. 79–111, 1978.

[44] Ganatos, P., Pfeffer, R. & Weinbaum, S., A strong interaction theory for the creeping motion of a sphere between plane parallel boundaries. Part 2. Parallel motion. *Journal of Fluid Mechanics*, **99**, pp. 755–785, 1980.

[45] Ganatos, P., Weinbaum, S. & Pfeffer, R., Gravitational and zero-drag motion of a sphere of arbitrary size in an inclined channel at low Reynolds number. *Journal of Fluid Mechanics*, **124**, pp. 27–43, 1982.

[46] Glav, R., The point-matching method on dissipative silencers of arbitrary cross-section. *Journal of Sound and Vibration*, **189**, pp. 123–135, 1996.

[47] Harrington, R.F., On the calculation of scattering conducting cylinders. *IEEE Transactions on Antennas and Propagation*, **AP-13**, pp. 812–813, 1965.

[48] Hassonjee, Q., Ganatos, P. & Pfeffer R., A strong-interaction theory for the motion of arbitrary three-dimensional clusters of spherical particles at low Reynolds number. *Journal of Fluid Mechanics*, **197**, pp. 1–37, 1988.

[49] Hocney, R.W., A solution of Laplace equation for a round hole in square peg. *Journal of Society of Industrial Applied Mathematics*, **12**, pp. 1–14, 1964.

[50] Hoffman, R.E. & Ariman, T., Thermal and mechanical stresses in nuclear reactor vessels. *Nuclear Engineering and Design*, **20**, pp. 31–55, 1972.

[51] Hoffman, R.E. & Ariman, T., Thermal bending of plates with circular holes. *Nuclear Engineering and Design*, **12**, pp. 231–238, 1970.

[52] Holy, Z.J., Three dimensional temperature and thermoelastic stress field in a heat producing sphere due to arbitrary surface heat transfer. *Nuclear Engineering and Design*, **6**, pp. 395–420, 1967.

[53] Hooke, C.J. & Demushi, G., Stress concentration factors, around the intersection of cylindrical holes in an infinite elastic body. *Journal of Strain Analysis*, **12**, pp. 217–222, 1977.

[54] Hooke, C.J., Numerical solution of axisymmetric-stress problems by point matching. *Journal of Strain Analysis*, **5**, pp. 25–37, 1970.

[55] Hsu, J.P. & Tseng, M.T., Electrostatic interactions between two identical thin discs of arbitrary orientation in an electrolyte solution. *Langmuir*, **13**, pp. 1810–1819, 1997.

[56] Hsu, Y.S. & Keh, H.J., Boundary effects on osmophoresis: Motion of a spherical vesicle perpendicular to two plane walls. *Chemical Engineering Sciences*, **61**, pp. 434–448, 2006.

[57] Hulbert, L.E. & Sampath, S.G., Thermal stress analysis of multiholed composite plates and cylinders by the boundary points-least-squares method. *Journal of Pressure Vessel and Technology*, **96**, pp. 214–219, 1974.

[58] Hulbert, L.E. & Simonsen, F.A., Analysis of stresses in shallow spherical shells with periodically spaced holes. *Journal of Engineering for Industry*, **92**, pp. 834–840, 1970.

[59] Hulbert, L.E. & Simonsen, F.A., On the solution of axisymmetric elastic stress problems by boundary point least squares technique. In: *Proceedings of First International Conference on Pressure Vessel Technology, Part I – Design Analysis*, Delft, The Netherlands, pp. 415–422, 1969.

[60] Hulbert, L.E., Heat conduction in multiconnected composite region with radiation boundary conditions. In: *Proceedings of the First International Conference on Structural Mechanics in Reactor Technology*, **6**, paper 6L, pp. 11–26, Berlin, West Gramany, September, 1972.

[61] Hulbert, L.E., Hopper, A.T. & Rybicki, E.F., The stress analysis of plates with single and clustered nozzles by the boundary point least square method. *Journal of Engineering Industry*, **93**, pp. 929–934, 1971.

[62] Hulbert, L.E., Solution of thermal stress problems in tube sheets by the boundary point least square method. *Journal of Engineering for Industry*, **92**, pp. 339–349, 1970.

[63] Hulbert, L.E., Solution of thermoelastic problems for plane multiplyconnected regions by boundary point least squares. In: *Proceedings of International Conference 'Thermal Stresses and Thermal Fatigue'*, Berkeley Castle, Gloucestershire, England, pp. 268–283, 1969.

[64] Hulbert, L.E., Verification of computer program based on assumed mode methods. In: *Symposium on Engineering Computer Software*, ed. I. Berman, New York, American Society of Mechanical Engineers, pp. 17–25, 1971.

[65] Hyer, M.W. & Cohen, D., Calculation of stresses in stiffened composite panels. *AIAA Journal*, **26**, pp. 852–857, 1988.

[66] Ishikawa, T., Koyama, K. & Kobayashi, S., Thermal expansion coefficients of unidirectional composites. *Journal of Composite Materials*, **12**, pp. 153–168, 1978.

[67] Isida, M. & Noguvhi, H., Tension of a plate containing an embedded elliptical crack. *Engineering Fracture Mechanics*, **20**, pp. 387–408, 1984.

[68] Isida, M., Yoshida, T. & Noguvhi, H., Tension and bending of finite thickness plates with a semi-elliptical surface crack. *International Journal of Fracture*, **26**, pp. 157–188, 1984.

[69] Joosten, J.G.H., Zijl, W. & Stralen, S.J.D. Van, Growth of a vapour bubble under combined gravitational and non-uniform temperature field. *International Journal of Heat and Mass Transfer*, **26**, pp. 1143–1149, 1978.

[70] Kaknert, F.M., Numerical methods in electromagnetic scattering theory. *Journal of Quantitative Spectroscopy & Radiative Transfer*, **79–80**, pp. 775–824, 2003.

[71] Karageorghis, A. & Fairweather, G., The method of fundamental solutions for axisymmetric acoustic scattering and radiation problems. *Journal the Acoustical Society of America*, **104**, pp. 3212–3218, 1998.

[72] Karageorghis, A. & Fairweather, G., The method of fundamental solutions for axisymmetric elasticity problems. *Department of Mathematics and Statistics, University of Cyprus*, TR/3/1999, 1999.

[73] Karageorghis, A. & Fairweather, G., The method of fundamental solutions for axisymmetric elasticity problems. *Computational Mechanics*, **25**, pp. 524–532, 2000.

[74] Karageorghis, A. & Smyrlis, Y.S., Matrix decomposition MFS algorithms for elasticity and thermo-elasticity problems in axisymmetric domains. *Journal of Computational and Applied Mathematics*, **206**, pp. 774–795, 2007.

[75] Karageorghis, A., Modified method of fundamental solutions for harmonic and biharmonic problems with boundary singularities. *Numerical Methods for Partial Differential Equations*, **8**, pp. 1–19, 1992.

[76] Keh, H.J. & Chang, Y.C., Creeping motion of a slip spherical particle in a circular cylindrical pore. *International Journal of Multiphase Flow*, **33**, pp. 726–741, 2007.

[77] Keh, H.J. & Chang, Y.C., Thermophoresis of an aerosol sphere perpendicular to two plane walls. *AIChE Journal*, **52**, pp. 1690–1704, 2006.

[78] Keh, H.J. & Chen, J.B., Particle interactions in electrophoresis. V. Motion of multiple spheres with thin but finite electrical double layers. *Journal of Colloid and Interface Science*, **158**, pp. 199–222, 1993.

[79] Keh, H.J. & Chen, L.S., Droplet interactions in axisymmetric thermocapillary motion. *Journal of Colloidal and Interface Science*, **151**, pp. 1–16, 1992.

[80] Keh, H.J. & Chen, L.S., Droplet interactions in thermocapilary migration. *Chemical Engineering Science*, **48**, pp. 3565–3582, 1993.

[81] Keh, H.J. & Chen, L.S., Effective viscosity of a suspension of fluid droplets. *Journal of the Chinese Institute of Chemical Engineers*, **31**, pp. 295–307, 2000.

[82] Keh, H.J. & Chen, P.Y., Thermophoresis of an aerosol sphere parallel to one or two plane walls. *AIChE Journal*, **49**, pp. 2283–2299, 2003.

[83] Keh, H.J. & Chen, S.H., Thermophoresis of an arbitrary three-dimensional array of interacting arbitrary spheres. *Journal of Aerosol Science*, **27**, pp. 1035–1061, 1996.

[84] Keh, H.J. & Chiou, J.Y., Electrophoresis of a colloidal sphere in a circular cylindrical pore. *AIChE Journal*, **42**, pp. 1397–1406, 1996.

[85] Keh, H.J. & Hsu, F.C., Photophoresis of an aerosol sphere normal to a plane wall. *Journal of Colloid and Interface Science*, **289**, pp. 94–103, 2005.

[86] Keh, H.J. & Huang, C.H., Slow motion of axisymmetric particles along their axes of revolution. *International Journal of Engineering Sciences*, **42**, pp. 1621–1644, 2004.

[87] Keh, H.J. & Jan, J.S., Boundary effects on diffusiophoresis and electrophoresis: motion of a colloidal sphere normal to plane wall. *Journal of Colloidal and Interface Science*, **183**, pp. 458–475, 1996.

[88] Keh, H.J. & Li, W.J., A study of bipolar spheroids in an electrolytic cell. *Journal of the Electromechanical Society*, **144**, pp. 1323–1331, 1997.

[89] Keh, H.J. & Lien, L.C., Electrophoresis of a colloidal sphere along the axis of a circular orifice or circular disk. *Journal of Fluid Mechanics*, **224**, pp. 305–333, 1991.

[90] Keh, H.J. & Luo, S.C., Particle interactions in diffusiophoresis: Axisymmetric motion of multiple spheres in electrolyte gradients. *Langmuir*, **12**, pp. 657–667, 1996.

[91] Keh, H.J. & Tseng, Y.K., Slow motion of multiple droplets in arbitrary three-dimensional configurations. *AIChE Journal*, **38**, pp. 1881–1904, 1992.

[92] Keh, H.J. & Yang, F.R., Particle interaction in electrophoresis. IV. Motion of arbitrary three-dimensional clusters of spheres. *Journal of Colloidal and Interface Science*, **145**, pp. 362–389, 1991.

[93] Keh, H.J. & Yang, F.R., Particle interactions in electrophoresis. III. Axisymmetric motion of multipole spheres. *Journal of Colloidal and Interface Science*, **139**, pp. 105–116, 1990.

[94] Keh, H.J., Chen, J.B. & Kim, S., Axisymmetric electrophoresis of coaxial spheroids. *Colloids and Surfaces A: Physicochemical and Engineering Aspects*, **75**, pp. 147–162, 1993.

[95] Khila, A.M. & Wolff, I., The point-matching solution for magnetically tunable cylindrical cavities and ferrite planar resonators. *IEEE Transactions MTT*, MTT-**26**, pp. 592–598, 1979.

[96] Kin, S. & Russel, E.C., The hydrodynamic interaction between two sphere in a Brinkman medium. *Journal of Fluid Mechanics*, **154**, pp. 235–268, 1985.

[97] Kita, K., Ikeda, Y. & Kamiya, N., Trefftz solution for boundary value problem of three-dimensional Poisson equation. *Engineering Analysis with Boundary Elements*, **29**, pp. 383–390, 2005.

[98] Kleev, A.I. & Manenkov, A.B., Method of adaptive collocation in two-dimensional diffraction problems. *Radiophysics and Quantum Electronics*, **29**, pp. 414–420, 1986.

[99] Kleev, A.I. & Manenkov, A.B., The convergence of point-matching techniques. *IEEE Transaction on Antennas and Propagation*, **37**, pp. 50–54, 1989.

[100] Kleev, A.I., Diffraction by a periodic surface and the adaptive collocation method. *Radiophysics and Quantum Electronics*, **33**, pp. 181–187, 1990.

[101] Kołodziej, J.A. & Kleiber, M., Boundary collocation method vs FEM for some harmonic 2-D problems. *Computers & Structures*, **33**, pp. 155–168, 1989.

[102] Kołodziej, J.A. & Musielak, G., Domain decomposition in boundary collocation method. In: *Parallel Algorithms for Partial Differential Equations. Proceedings of the Sixth GAMM-Seminar Kiel, 1990, Notes on Numerical Fluid Mechanics*, **31**, pp. 140–156, 1991.

[103] Kołodziej, J.A., Determination of the interaction coefficient in mixture equations for heat conduction in unidirectional composites. *Archives of Thermodynamics*, **8**, pp. 87–107, 1987.

[104] Kołodziej, J.A., Influence of the porosity of porous medium on the effective viscosity in Brinkman's filtration equation. *Acta Mechanika*, **75**, pp. 241–254, 1988.

[105] Kondapalli, P.S., Shipy, D.J. & Fairweather, G., The method of fundamental solutions for scattering and transmission of waves in fluids and solids. *Technical Report* CCS-90-5, University of Kentucky, 1990.

[106] Kondapalli, P.S., Shipy, D.S. & Fairwather, G., The method of fundamental solutions for transmission and scattering of elastic waves. *Computer Methods in Applied Mechanics and Engineering*, **96**, pp. 255–269, 1992.

[107] Koopman, G.H., Song, L. & Fahnline, J.B., A method for computing acoustic fields based on the principle of wave superposition. *Journal of Acoustical Society of America*, **86**, pp. 2433–2438, 1989.

[108] Korda, P.E., *Critical evaluation of the theory of shallow shells and application of digital computers to mixed boundary value problems of shells*. Ph. D. Thesis, No. 64-9574, Ohio State University, 1964.

[109] Kress, R. & Mohsen, A., On the simulation source technique for exterior problems in acoustics. *Mathematical Methods in the Applied Science*, **8**, pp. 585–597, 1986.

[110] Kucaba-Piętal, A., Application of boundary collocation methods in fluid mechanics to Stokes flow problems. *Lecture Notes in Computer Science*, 1988/2001, pp. 498–505, 2001.

[111] Kucaba-Piętal, A., Nonaximmetric Stokes flow past a torus in the presence of a wall. *Archives of Mechanics*, **38**, pp. 647–663, 1986.

[112] Kuo, J. & Keh, H.J., Motion of a colloidal sphere covered by a layer of adsorbed polymers normal to a plane surface. *Journal of Colloid and Interface Science*, **210**, pp. 296–308, 1999.

[113] Larson, R.E. & Higdon, J.J.L., A periodic grain consolidation model of porous media. *Physics of Fluids A*, **1**, pp. 38–46, 1989.

[114] Lee, J. & Vanden-Broeck, J.M., Two-dimensional jets falling from funnels and nozzles. *Physics of Fluids A*, **5**, pp. 2454–2460, 1993.

[115] Lo, C.C., *The Solution of Plane Harmonic and Biharmonic Boundary Value Problems in the Theory of Elasticity*. Ph. D. Thesis, No. 65-2633, Purdue University, 1964.

[116] Lu, S.Y. & Lee, C.T., Thermophoretic motion of a spherical aerosol particle in a cylindrical pore. *Aerosol Science and Technology*, **37**, pp. 455–459, 2003.

[117] Lu, S.Y. & Lee, C.T., Thermophoretic motion of an aerosol particle in a non-concentric pore. *Journal of Aerosol Science*, **32**, pp. 1341–1358, 2001.

[118] Lu, S.Y. & Yen, Y.M., A first-pasage scheme for determination of overall rate constants for non-diffusion-limited suspensions. *Journal of Chemical Physics*, **116**, pp. 3128–3133, 2002.

[119] Lu, S.Y., Yen, Y.M., Tseng, C.Y. & Tsao, H.K., Overal rate constants for diffusion and incorporation in clusters of spheres. *Journal of Chemical Physics*, **117**, pp. 3431–3439, 2002.

[120] Lucking, W.M., Hoa, S.V. & Sankar, T.S., A boundary collocation method for stress analysis of laminate edges. *Computers & Structures*, **34**. pp. 655–661, 1990.

[121] Mahajerin, E. & Burgess, G., Fundamental collocation method applied to plane thermoelasticity problems. *Computers & Structures*, **57**, pp. 795–797, 1995.

[122] Mahajerin, E., On the approximation of Helmholtz's differential equation. *Mechanics Research Communications*, **12**, pp. 283–287, 1985.

[123] Marin, L., A meshless method for solving the Cauchy problem in three-dimensional elastostatics. *Computers and Mathematics with Applications*, **50**, pp. 73–92, 2005.

[124] Matsuyama, K.I., Moriyasu, M. & Nishiguchi, K., Estimation of radial heat flux distribution in plasma arc cutting with a least-squares-collocation method on the boundary. *Welding in the World*, **37**, pp. 301–307, 1996.

[125] Matthews, G.J. & Hooke, C.J., Solution of axisymmetric torsion problem by point matching. *Journal of Strain Analysis*, **6**, pp. 217–222, 1971.

[126] Mazur, J., Analysis of quasicyclic ferrite junction by point-matching technique. *Electronic Letters*, **19**, pp. 140–142, 1983.

[127] Mera, N.S., Elliot, L. & Ingham, D.B., Boundary identification for a 3D Laplace equation using a genetic algorithm. *Computer Methods in Applied Mechanics and Engineering*, **194**, pp. 4411–4424, 2005.

[128] Moore, D.W., Saffman, P.G. & Tanveer, S., The calculation of some Batchelor flow: The Sadovskii vortex and rational corner flow. *Physics of Fluids*, **31**, pp. 978–990, 1988.

[129] Mullin, C.R., Sadburg, R. & Velline, C.D., A numerical technique for the determination of scattering cross section of infinite cylinders of arbitrary geometrical cross section. *IEEE Transactions on Antennas and Propagation*, AP-**13**, pp. 142–149, 1965.

[130] Na, H.G. & Kim, H.T., Scattering analysis of conducting bodies of revolution using fictitious currents and point-matching. *IEEE Transactions on Antennas and Propagation*, **43**, pp. 426–430, 1995.

[131] Naghdi, A.K. & Eringen, A.C., Stress distribution in a circular cylindrical shell with a circular cutout. *Ingenieur-Archiv*, **34**, pp. 161–172, 1965.

[132] Naghdi, A.K. & Gersting, J.M. Jr., The effect of transverse shear acting on the edge of a circular cutout in a simply supported circular cylindrical shell. *Ingenieur-Archiv*, **42**, pp. 141–150, 1973.

[133] Naghdi, A.K., Solution of Poisson equation in a semicircular region with a hole. *Journal of Applied Mechanics*, **47**, pp. 677–679, 1980.

[134] Neidenfuhr, F.W., Leissa, A.W. & Lo, C.C., A study of the point matching method as applied to thermally and transversally loaded plate and other boundary value problems. *Technical Report* No. AFFDL-TR-159, AD-610146, *U.S. Air Force Flight Dynamics Lab.*, 1964.

[135] Nishida, K., Numerical methods for Oseen's linearized equations in three-dimensional exterior domains. *Journal of Computational and Applied Mathematics*, **152**, pp. 405–409, 2003.

[136] Ondris, L., Remarks concerning the solution of stationary plane thermal problems by the boundary point least squares method. *Computer Methods in Applied Mechanics and Engineering*, **31**, pp. 331–336, 1982.

[137] Ondris, L., Solution of some symmetrical plane thermal problems by the boundary point least squares method. *Computer Methods in Applied Mechanics and Engineering*, **28**, pp. 309–325, 1981.

[138] Patterson, C. & Sheikh, M.A., A modified Trefftz method for three-dimensional elasticity. *Boundary elements, Proceedings of 5th International Conference,* Hiroshima 1983, eds. C.A. Brebbia, T. Futagami & M. Tanaka, Springer-Verlag: New York, 1983.

[139] Poullikkas, A., Karageorghis, A. & Georgiou, G., The method of fundamental solutions for three-dimensional elastostatics problems. *Computer & Structures*, **80**, pp. 365–370, 2002.

[140] Poullikkas, A., Karageorghis, A. & Georgiou, G., The numerical solution of three dimensional Signorini problems with the method of fundamental solutions. *Department of Mathematics and Statistics, University of Cyprus*, TR/6/1999, 1999.

[141] Poullikkas, A., Karageorghis, A. & Georgiou, G., The method of fundamental solutions for Singorini problems. *IMA Journal of Numerical Analysis*, **18**, pp. 273–285, 1998.

[142] Poullikkas, A., Karageorghis, A. & Georgiou, G., The numerical solution of three-dimensional Signorini problems with the method of fundamental solutions. *Engineering Analysis with Boundary Elements*, **25**, pp. 221–227, 2001.

[143] Rao, K.S., Bapu Rao, M.N. & Ariman, T., Thermal stresses in elastic plates with elliptic holes. *Journal of Engineering Industry*, **96**, pp. 827–832, 1974.

[144] Rao, K.S., Bapu Rao, M.N. & Ariman, T., Thermal stresses in plates with circular holes. *Nuclear Engineering and Design*, **15**, pp. 97–112, 1971.

[145] Read, F.H., A collocation method for solving Laplace's equation. *Journal of Computational Physics*, **6**, pp. 527–532, 1970.

[146] Read, F.H., Accurate calculations of double-aperture electrostatic immersion lenses. *Journal of Scientific Instruments, Journal Physics Section E,* ser. 2, **2**, pp. 165–169, 1969.

[147] Read, F.H., Asymmetric electrostatic lenses of three apertures. *Journal of Scientific Instruments, Journal of Physics, Section E*, **3**, pp. 127–131, 1970.

[148] Read, F.H., Calculated properties of electrostatic Einzel lenses of three apertures. *Journal of Scientific Instruments, Journal Physics Section E,* ser. 2, **2**, pp. 679–684, 1969.

[149] Redekop, D. & Cheung, R.S.W., Fundamental solutions for the collocation method in three-dimensional elastostatics. *Computers & Structures*, **26**, pp. 703–707, 1987.

[150] Redekop, D. & Thompson, J.C., Use of fundamental solutions in the collocation method in axisymmetric elastostatics. *Computers & Structures*, **17**, pp. 485–490, 1983.

[151] Redekop, D., Three-dimensional analysis of an elastic plate-cylinder intersection by the boundary-point-least-squares technique. *Journal of Pressure Vessel and Technology*, **99**, pp. 17–25, 1977.

[152] Redekop, D., Three-dimensional analysis of an intersection of a spherical shell with a cylindrical shell. In: *Developments in Stress Analysis for Pressurised Components*, ed. R.W. Nichols, London, Applied Science Publishers LTD, 1977, also in Russian: *Mechanika, Novoe v Zarubežnoj Nauke*, **24**, pp. 151–171, 1980.

[153] Reichel, L., On the determination of boundary collocation points for solving some problems for the Laplace operator. *Journal of Computational and Applied Mathematics*, **11**, pp. 175–196, 1984.

[154] Reichel, L., On the numerical solution of some 2-D electromagnetic interface problems by the boundary collocation method. *Computer Methods in Applied Mechanics and Engineering*, **53**, pp. 1–11, 1985.

[155] Saavedra, I. & Power, H., Adaptive refinement scheme for the least-squares approach of the method of fundamental solution for three-dimensional harmonic problems. *Engineering Analysis with Boundary Elements*, **28**, pp. 1123–1133, 2004.

[156] Scholtz, M.T. & Trass, O., Mass transfer in a nonuniform impinging jet, Part I, Stagnation flow-velocity and pressure distribution. *AIChE Journal*, **16**, pp. 82–90, 1970.

[157] Sekiya, T. & Takeuti, Y., Approximate solution of thermal problems in plane elasticity. In: *Proceedings 8th Japan National Congress on Applied Mechanics*, pp. 119–122, 1958.

[158] Shankar, P.N., Three-dimensional eddy structure in a cylindrical container. *Journal of Fluid Mechanics*, **342**, pp. 97–118, 1997.

[159] Shankar, P.N., Three-dimensional Stokes flow in a cylindrical container. *Physics of Fluids*, **10**, pp. 540–549, 1998.

[160] Slater, J.C., Electron energy bands in metals. *Physics*, **45**, pp. 794–801, 1934.

[161] Sluyter, W.M., Stralen, S.J.D. Van & Zijl, W., The method of characteristics applied to numerical solution of gas bubble implosion and fragmentation. *International Journal of Heat and Mass Transfer*, **25**, pp. 1103–1111, 1982.

[162] Sultanian, B.K. & Sastri, V.M.K., Effect of geometry on heat conduction in coolant channels of a liquid rocket engine. *Warme – und Stoffubertragung*, **14**, pp. 254–251, 1980.

[163] Sultanian, B.K. & Sastri, V.M.K., Steady heat conduction in a circular duct with circumferentially varying heat transfer coefficients. *Regional Journal Energy, Heat and Mass Transfer*, pp. 101–110, 1979.

[164] Symm, G.T., The longitudinal shear modulus of an unidirectional fibrous composite. *Journal of Composite Materials*, **4**, pp. 426–428, 1970.

[165] Takeuti, Y. & Sekiya, T., Thermal stresses in a polygonal cylinder with a circular hole under internal heat generation. *Zeitschrift für Angewandte Mathematik und Mechanik (ZAMM)*, **48**, pp. 237–246, 1968.

[166] Takeuti, Y. & Sumi, N., Steady-state thermoelastic problem for a triangular prism with a circular hole under heat generation. *International Journal of Engineering Science*, **5**, pp. 145, 1968.

[167] Takeuti, Y., Thermal stresses in heat-generating multibore square or circular region. *Nuclear Engineering and Design*, **14**, pp. 201–210, 1970.

[168] Tozeren, H., Torque on eccentric spheres flowing in tubes. *Journal of Applied Mechanics*, **49**, pp. 279–283, 1982.

[169] Tranquilla, J.M. & Al-Rizzo, H.M., Electromagnetic scattering from dielectric-coated axisymmetric objects using the generalized point-matching technique (GPMT). *IEEE Transactions on Antennas and Propagation*, **43**, pp. 63–71, 1995.

[170] Tsai, C.C., The method of fundamental solutions for three-dimensional elastostatic problems of transversely isotropic solids. *Engineering Analysis with Boundary Elements*, **31**, pp. 586–594, 2007.

[171] Ushijima, T. & Chiba, F., A fundamental solution method for the reduced wave problem in a domain exterior to a disc. *Journal of Computational and Applied Mathematics*, **152**, pp. 545–557, 2003.

[172] Ushijima, T. & Chiba, F., Error estimates for a fundamental solution method applied to reduced wave problem in a domain exterior to a disc, *Journal of Computational and Applied Mathematics*, **159**, pp. 137–148, 2003.

[173] Vanden-Broeck, J.M., Cavity flow of a fluid with surface tension past a circular cylinder. *Physics of Fluids A*, **3**, pp. 263–266, 1991.

[174] Vanden-Broeck, J.M., Joukowskii's model for a rising bubble. *Physics of Fluids*, **31**, pp. 974–977, 1988.

[175] Vanden-Broeck, J.M., Rising bubble in a two-dimensional tube: Asymptotic behaviour for small values of the surface tension. *Physics of Fluids A*, **4**, pp. 2332–2334, 1992.

[176] Wah, T., In-plane stresses in polygonal plates due to thermal inclusions. *Mechanics Research Communications*, **9**, pp. 301–310, 1982.

[177] Wah, T., In-plane thermal stresses in elastic plates. *Journal of Strain Analysis*, **19**, pp. 97–101, 1984.

[178] Wah, T., Thermal stresses in elastic plates. In-plane thermal stresses in elastic plates. *Journal of Structural Mechanics*, **12**, pp. 79–85, 1984.

[179] Wang, C.Y., Flow and heat transfer between plates with longitudinal fins. *Applied Scientific Research*, **54**, pp. 23–38, 1995.

[180] Wang, C.Y., Potential field between finned boundaries. *Mechanics Research Communications*, **19**, pp. 29–36, 1992.

[181] Wang, C.Y., Rotation of a slotted cylinder in a fluid at low Reynolds number – implication on the no-slip condition for grates. *Mechanics Research Communications*, **23**, pp. 305–312, 1996.

[182] Wang, C.Y., Stokes flows through a thin screen with patterned holes. *AIChE Journal*, **40**, pp. 419–423, 1994.

[183] Williams, W., Parke, N.G., Morgan, D.A. & Sherman, C.H., Acoustic radiation from a finite cylinder. *Journal of Acoustical Society of America*, **36**, pp. 2316–2322, 1964.

[184] Wilson, H.B. Jr. & Goree, J.C., Axisymmetric contact stress about a smooth elastic sphere stressed uniformly at infinity. *Journal of Applied Mechanics*, **34**, pp. 960–966, 1967.

[185] Wilton, D.R. & Mitra, R., A new numerical approach to the calculation of electromagnetic scattering properties of two-dimensional bodies of arbitrary cross section. *IEEE Transactions AP*, AP-**20**, pp. 310–317, 1972.

[186] Yan, Z., Weinbaum, S., Ganatos, P. & Pfeffer, R., The three-dimensional hydrodynamic interaction of a finite sphere with a circular orifice at low Reynolds number. *Journal of Fluid Mechanics*, **174**, pp. 39–68, 1987.

[187] Yang, J.W., Heat transfer analysis of internally-cooled fuel element. *Nuclear Engineering and Design*, **61**, pp. 123–129, 1980.

[188] Young, D.L., Chiu, C.L., Fan, C.M., Tsai, C.C. & Lin, Y.C., Method of fundamental solutions for multidimensional Stokes equations by the dual-potential formulation. *European Journal of Mechanics B/Fluids*, **25**, pp. 877–893, 2006.

[189] Young, D.L., Jane, S.S., Fan, C.M., Murugesan, K. & Tsai, C.C., The method of fundamental solutions for 2D and 3D Stokes problems. *Journal of Computational Physics*, **211**, pp. 1–8, 2006.

[190] Zijl, W., Global collocation approximations of the flow and temperature fields around a gas and vapour bubble. *International Journal of Heat and Mass Transfer*, **20**, pp. 487–498, 1977.

[191] Zijl, W., Ramakers, F.J.M. & Stralen, S.J.D. Van, Global numerical solutions of growth and departure of a vapour bubble at a horizontal superheated wall in a pure liquid and a binary mixture. *International Journal of Heat and Mass Transfer*, **22**, pp. 401–420, 1979.

CHAPTER 11

Non-homogenous equations and nonlinear boundary conditions

1 Non-homogenous harmonic problems

1.1 Integral form solution

Problems governed by the Laplace equation or Poisson equation with constant right side (which could easily be transformed to Laplace equation) have been already discussed in Chapter 6. In general, however, a particular integral of the non-homogenous Laplace equation is not always easy to determine. Therefore, there are not many papers concerning application of the boundary collocation method to solve the non-homogenous harmonic or biharmonic problems. Below, we present a method to determine this particular solution, which have been used, for example, by Poullikkas, Karageorghis & Georgiou in their paper [32].

Let us consider the two-dimensional (2D) Poisson equation in the form:

$$\nabla^2 u(\mathbf{x}) = f(\mathbf{x}) \quad \text{in } \Omega \tag{1}$$

with boundary conditions

$$u(\mathbf{x}) = g(\mathbf{x}) \quad \text{on } \Gamma_1 \tag{2a}$$

$$\frac{\partial u(\mathbf{x})}{\partial n} = h(\mathbf{x}) \quad \text{on } \Gamma_2, \tag{2b}$$

where Ω is a plane, bounded area, ∇^2 is the 2D Laplace operator, $u(\mathbf{x})$ is a function we search, $f(\mathbf{x})$, $g(\mathbf{x})$ and $h(\mathbf{x})$ are known, $\partial u/\partial n$ means derivative of u normal to Γ_2 and $\Gamma = \Gamma_1 \cup \Gamma_2$ means the boundary of Ω.

Let $u_0(\mathbf{x})$ be a particular solution of eqn (1), that is

$$\nabla^2 u_0(\mathbf{x}) = f(\mathbf{x}) \quad \text{in } \Omega. \tag{3}$$

Then function $\upsilon(\mathbf{x}) = u(\mathbf{x}) - u_0(\mathbf{x})$ fulfils Laplace equation:

$$\nabla^2 \upsilon(\mathbf{x}) = 0 \quad \text{in } \Omega, \tag{4}$$

with modified boundary conditions in the form:

$$\upsilon(\mathbf{x}) = g(\mathbf{x}) - u_0(\mathbf{x}) \quad \text{on } \Gamma_1 \tag{5a}$$

and

$$\frac{\partial \upsilon(\mathbf{x})}{\partial n} = h(\mathbf{x}) - \frac{\partial u_0(\mathbf{x})}{\partial n} \quad \text{on } \Gamma_2. \tag{5b}$$

Now, the above boundary value problem can be solved by means of the standard collocation method, provided we know particular solution u_0 ($\mathbf{x}$). This solution can be based on the Newton potential, which in case of plane area leads to the integral:

$$u_0(\mathbf{x}) = \frac{1}{2\pi} \int_{\Omega} \ln|\mathbf{x} - \mathbf{q}| f(\mathbf{q}) d\Omega(\mathbf{q}), \quad \mathbf{x} \equiv (x, y) \in \Omega, \ \mathbf{q} \equiv (\xi, \eta) \in \Omega. \tag{6}$$

Some difficulties concerning calculation of that integral can be overcome by using a method suggested by Atkinson [2]. Let Ω_0 be an area which contains $\bar{\Omega} = \Omega \cup \Gamma$, that is $\Omega_0 \supset \bar{\Omega}$. Then $f(\mathbf{q})$ can be smoothly extended to Ω_0 and

$$u_0(\mathbf{x}) = \frac{1}{2\pi} \int_{\Omega_0} \ln|\mathbf{x} - \mathbf{q}| f(\mathbf{q}) d\Omega(\mathbf{q}), \quad \mathbf{x} \in \bar{\Omega} \tag{7}$$

is another particular solution of eqn (1). Area Ω_0 is chosen so that calculation of integral (7) is possibly easy. In order to avoid problems of integration with the singularity in point $\mathbf{x} = \mathbf{q}$, the following transformation of variables is applied:

$$\xi = x + r\left[b_x(\theta) - x\right], \quad \eta = y + r\left[b_y(\theta) - y\right], \tag{8}$$

where $0 \le r \le 1$, $0 \le \theta \le 2\pi$, $(\xi,\eta) = \mathbf{q} \in \Omega_0$, $(b_x(\theta),b_y(\theta)) \in \Gamma_0$, that means we continue integration with respect to new variables (r, θ). Solution (7) can then be transformed to:

$$u_0(\mathbf{x}) = \frac{1}{2\pi} \int_{\Omega_0} \ln\left(r\sqrt{\left[b_x(\theta) - x\right]^2 + \left[b_y(\theta) - y\right]^2} \right) rH(\mathbf{x},\theta) f_0(\mathbf{x},r,\theta) dr d\theta, \tag{9}$$

where $|J| = rH(\mathbf{x},\theta)$ is the transformation Jacobian and

$$H(\mathbf{x},\theta) = \frac{db_y(\theta)}{d\theta}\left[b_x(\theta) - x\right] - \frac{db_x(\theta)}{d\theta}\left[b_y(\theta) - y\right]. \tag{10}$$

For simplicity, integral area Ω_0 can be chosen in form of an ellipse defined as

$$b_x = a\cos\theta, \quad b_y = b\sin\theta \tag{11}$$

where a and b are its semi-axes. In this case

$$H(\mathbf{x},\theta) = ab\left(1 - \frac{x}{a}\cos\theta - \frac{y}{b}\sin\theta\right) \tag{12}$$

and

$$\begin{aligned} u_0(\mathbf{x}) = \frac{1}{2\pi}\int_0^{2\pi}\int_0^1 \ln\left(r\sqrt{(a\cos\theta - x)^2 + (b\sin\theta - y)^2} \right) \\ \times r\,ab\left(1 - \frac{x}{a}\cos\theta - \frac{y}{b}\sin\theta\right) f_0(\mathbf{x},r,\theta)\,dr\,d\theta, \end{aligned} \tag{13}$$

where $f_0(\mathbf{x},r,\theta) = f(\mathbf{q}), \quad \mathbf{x} \in \bar{\Omega}$.

The normal derivative appearing in boundary condition (5b) is then presented in the form:

$$\begin{aligned} \frac{\partial u_0(\mathbf{x})}{\partial n} = -\frac{1}{2\pi}\int_0^{2\pi}\int_0^1 \left[(a\cos\theta - x)n_x + (b\sin\theta - y)n_y\right] \\ \times \frac{H(\mathbf{x},\theta) f_0(\mathbf{x},R,\theta)}{(a\cos\theta - x)^2 + (b\sin\theta - y)^2}\,dR\,d\theta, \end{aligned} \tag{14}$$

where n_x and n_y are components of a unit vector perpendicular to Γ_2 in point $\mathbf{x}$.

The integral in formula (14) does not have any singularity and can be calculated numerically with the help of standard procedures. Function (13) can be expressed in the form:

$$\begin{aligned} u_0(\mathbf{x}) = \frac{1}{2\pi}\int_0^{2\pi} H(\mathbf{x},\theta)\int_0^1 r\ln(r) f_0(\mathbf{x},r\theta)\,dr\,d\theta \\ + \frac{1}{4\pi}\int_0^{2\pi}\int_0^1 \ln\left\{\left[b_x(\theta) - x\right]^2 + \left[b_y(\theta) - y\right]^2\right\} \\ \times rH(\mathbf{x},\theta) f_0(\mathbf{x},r\theta)\,dr\,d\theta \equiv I_1 + I_2, \end{aligned} \tag{15}$$

where inner integral in expression I_1 has a weak singularity $r \ln r$, which can be also calculated using standard procedures.

After determination of function $u_0(\mathbf{x})$ (15) and its normal derivative (14), we insert them into boundary conditions (5) and then solve the boundary value problem (4 and 5) by means of the method described in Chapter 6.

1.2 Solutions with radial basis functions

Consider a problem described by 2D Poisson equation (1) with inhomogeneous boundary condition

$$Bu = \overline{g}(x,y) \quad \text{on } \Gamma \tag{16}$$

where $\overline{g}(x,y)$ is a known function and B is an operator of boundary conditions. In physical interpretation for case of heat transfer, u is temperature and $f(x,y)$ is a function of heat generation.

Let $\{P_i = (x_i, y_i)\}_{i=1}^{M}$ denote the set of M interpolation points in Ω, of which $\{(x_i, y_i)\}_{i=1}^{M_1}$ are interior points and $\{(x_i, y_i)\}_{i=M_1+1}^{M}$ are boundary points (see Fig. 1).

The right-hand side function in (1) is approximated by radial basis functions (RBFs) as

$$f(x,y) = \sum_{m=1}^{M} \alpha_m \hat{\varphi}_m(r_m) + \sum_{k=1}^{K} \beta_k \tilde{\varphi}_k(x,y), \tag{17}$$

where $\hat{\varphi}_m(r_m) = \hat{\varphi}_m\left(\sqrt{(x-x_m)^2 + (y-y_m)^2}\right)$ is a radial basis function, $\{\tilde{\varphi}_k(x,y)\}_{k=1}^{K}$ is a basis for Φ, the set of d-variate polynomials of degree $n \le m-1$ and $K = \binom{m+d-1}{d}$ is the dimension of Φ.

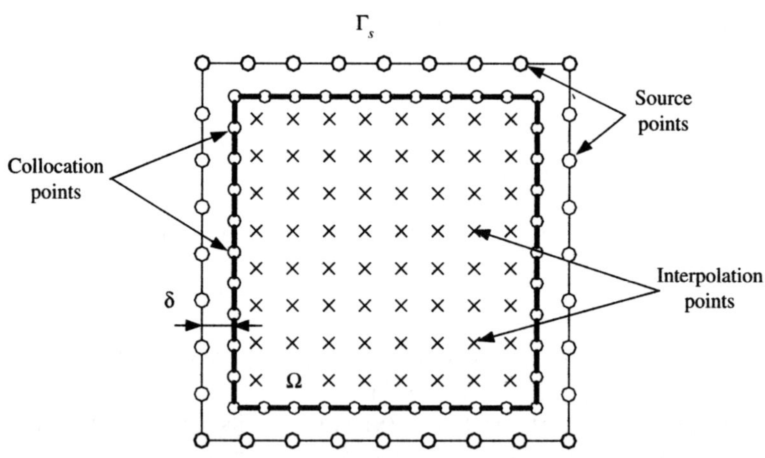

Figure 1: Distribution of collocation, source and interpolation points.

The coefficients a_m and β_k can be found by solving the system of linear equations

$$\sum_{m=1}^{M} a_m \hat{\varphi}_m(r_{mi}) + \sum_{k=1}^{K} \beta_k \tilde{\varphi}_k(x_i, y_i) = b(x_i, y_i), \quad 1 \le i \le M \tag{18a}$$

$$\sum_{m=1}^{M} a_m \tilde{\varphi}_k(x_m, y_m) = 0, \quad 1 \le k \le K, \tag{18b}$$

where $r_{ij} = \sqrt{(x_i - x_j)^2 + (y_i - y_j)^2}$ and $\{(x_i, y_i)\}_{i=1}^{M}$ are the interpolation points in $\Omega \cup \Gamma$.

The approximate particular solutions u_p of eqn (1) can be obtained using coefficients a_m and β_k

$$u_p(x, y) = \sum_{m=1}^{M} a_m \hat{u}_m(r_m) + \sum_{k=1}^{K} \beta_k \tilde{u}_k(x, y), \tag{19a}$$

where

$$\nabla^2 \hat{u}_j(x, y) = \hat{\varphi}_j(x, y) \quad \text{for } j = 1, 2, \ldots, M \tag{19b}$$

$$\nabla^2 \tilde{u}_j(x, y) = \tilde{\varphi}_j(x, y) \quad \text{for } j = 1, 2, .., K. \tag{19c}$$

The general solution of differential eqn (1) now can be given as

$$u = u_h + u_p, \tag{20}$$

where u_h is the solution of the boundary value problem in the form:

$$\nabla^2 u_h = 0 \quad \text{in } \Omega \tag{21a}$$

and

$$Bu_h = \bar{g}(x, y) - Bu_p \quad \text{on } \Gamma. \tag{21b}$$

The boundary collocation method with fundamental solutions is used to solve problem (21a), meaning that

$$u_h(x, y) = \sum_{n=1}^{N} c_n \ln\left((x - x_n)^2 + (y - y_n)^2\right), \tag{22}$$

where x_n, y_n are coordinates of source points. These points are situated on a certain contour Γ_s geometrically similar to the boundary Γ (see Fig. 1). A distance between the boundary and the line of sources is defined by method parameter δ.

Enforcement of the boundary conditions yields:

$$\sum_{n=1}^{N} c_n B \ln\left((x_i - x_n)^2 + (y_i - y_n)^2\right) = \bar{g}(x_i, y_i) - Bu_p(x_i, y_i) \qquad \text{for } i = 1, 2, \ldots, N_c \tag{23}$$

where N_c is a number of collocation points on the boundary Γ. If $N_c > N$, the system of eqn (23) can be solved in the least square sense.

During the past two decades, the radial basis functions have been studied and applied to many areas of science and engineering. The following authors should here be mentioned: Schaback [38], Hickernell & Hon [17], Rippa [36], Cheng, Young & Tsai [7], Kansa & Hon [19], Wang & Liu [48], Trahan & Wyatt [41], Xiao & McCarthy [49], Xie, Nogami & Wang [51] and Xiao, Gama, Gillespie & Kansa [50].

Various types of RBFs can be used for interpolation of right-hand side function in eqn (1). All of them are also dependent on their shape parameters (see Table1), which considerably influence quality of the approximation. The radial basis functions presented in Table 1 have been used by numerous authors. Also in paper [24], several numerical experiments with the RBFs have been presented.

Table 1: Exemplary radial basis functions.

	Name of radial basis function	$\hat{\varphi}(r_m)$	Shape parameters
1	Polynomial	$a_0 + a_1 r_m + a_2 r_m^2 + a_3$	a_0, a_1, a_2, a_3
2	Polyharmonic spline function	$\begin{cases} 0 & \text{for } r_m = 0 \\ r_m^n \ln r_m & \text{for } r_m \neq 0 \end{cases}$	n
3	Multiquadrics	$\sqrt{r_m^2 + c^2}$	c
4	Wendland's functions	$\begin{cases} \left(1 - \frac{r_m}{a_0}\right)^4 \left(1 + \frac{4r_m}{a_0}\right) & \text{for } r_m \le a_0 \\ 0 & \text{for } r_m > a_0 \end{cases}$	a_0
5	Inverted multiquadrics	$\frac{1}{\sqrt{r_m{}^2 + c^2}}$	c

If the particular radial basis function $\hat{\varphi}(r_m)$ has been chosen, one has to determine $\hat{u}(r_m)$, which is evaluated by integrating eqn (19b). The governing equation for $\hat{u}(r_m)$ is thus given by:

$$\frac{1}{r_m}\frac{d}{dr_m}\left(r_m\frac{d\hat{u}}{dr_m}\right)=\hat{\varphi}(r_m). \tag{24}$$

Expressions of the particular solutions for various radial basis functions from Table 1 are listed in Table 2. Supplementing Table 3 contains polynomial functions $\tilde{\varphi}_k$ (see eqn (19c)) and also their respective particular solutions.

An optimization problem for suitable selection of parameters for the method of fundamental solutions and the RBFs has been considered in [24]. It can be written as

$$\max \rightarrow F(\mathbf{X}) \text{ Subject to: } g_j(\mathbf{X})=0 \quad \text{for } j=1,2,...,J, \tag{25}$$

where $F(\mathbf{X})$ is an objective function, $\mathbf{X}$ is a design variables vector and $g_j(\mathbf{X})$ are constraints defining the searching space in which J is the number of constraints. The optimization procedure using the genetic algorithm is presented in detail in paper [24] and illustrated by well convergent numerical examples.

Table 2: Particular solutions for radial basis functions from Table 1.

	Name of radial basis function	$\hat{u}(r)$
1	Polynomial	$\hat{u}(r)=\dfrac{r^2\left(900a_0+400a_1r+225a_2r^2+144a_3r^3\right)}{3600}$
2	Polyharmonic spline function	$\hat{u}(r)=\begin{cases}\dfrac{r^{n+2}\left((n+2)\ln(r)-2\right)}{(n+2)^3} & \text{for} \quad r\neq 0\\ 0 & \text{for} \quad r=0\end{cases}$
3	Multiquadrics	$\hat{u}(r)=-\frac{1}{3}c^3\ln\left(c\sqrt{c^2+r^2}+c\right)+\frac{1}{9}\left(4c^2+r^2\right)\sqrt{c^2+r^2}$
4	Wendland's functions	$\hat{u}(r)=\begin{cases}\dfrac{r^2}{4}-\dfrac{5r^4}{8a_0^2}+\dfrac{4r^5}{5a_0^3}-\dfrac{5r^6}{12a_0^4}+\dfrac{4r^7}{49a_0^5} & \text{dla} \quad r\leq a_0\\ \dfrac{259a_0^2}{5880}+\dfrac{a_0^2\ln\left(r/a_0\right)}{14} & \text{dla} \quad r>a_0\end{cases}$
5	Inverted multiquadrics	$\hat{u}(r)=-c\ln\left(\sqrt{c^2+r^2}+c\right)+\sqrt{c^2+r^2}$

Table 3: Polynomial functions $\tilde{\varphi}$ and their particular solutions.

k	$\tilde{\varphi}_k(x,y)$	$\tilde{u}_k$
1	1	$\frac{1}{4}\left(x^2+y^2\right)$
2	x	$\frac{1}{8}x^3+\frac{1}{8}xy^2$
3	y	$\frac{1}{8}y^3+\frac{1}{8}yx^2$
4	xy	$\frac{1}{12}(x^3y)+\frac{1}{12}(y^3x)$
5	x^2	$\frac{1}{14}\left(x^4+x^2y^2-\frac{y^4}{6}\right)$
6	y^2	$\frac{1}{14}\left(y^4+x^2y^2-\frac{x^4}{6}\right)$

In the last years, the method of radial basis functions for determination of particular solutions in the Poisson-type equations has been intensively developed. One can find different aspects of this method in papers [1, 3–6, 13, 18, 23, 25–27, 30, 31, 42, 47].

2 Non-homogenous biharmonic problems

Let us now consider a 2D, non-homogenous biharmonic equation:

$$\nabla^4 u = f(\mathbf{x}) \quad \text{in}\,\Omega \tag{26}$$

with boundary conditions

$$u(\mathbf{x}) = g_1(\mathbf{x}), \qquad \frac{\partial u(\mathbf{x})}{\partial n} = h_1(\mathbf{x}) \quad \text{on}\,\Gamma \tag{27a}$$

or

$$u(\mathbf{x}) = g_2(\mathbf{x}), \qquad \nabla_1^2 u(\mathbf{x}) = h_2(\mathbf{x}) \quad \text{on}\,\Gamma \tag{27b}$$

where $g_1(\mathbf{x})$, $g_2(\mathbf{x})$, $h_1(\mathbf{x})$ and $h_2(\mathbf{x})$ are known functions.

If $u_0(\mathbf{x})$ is a particular solution of eqn (26), then $\upsilon(\mathbf{x}) = u(\mathbf{x}) - u_0(\mathbf{x})$ fulfils homogenous biharmonic equation:

$$\nabla^4 \upsilon = 0 \quad \text{in}\ \Omega, \tag{28}$$

with boundary conditions

$$\upsilon(\mathbf{x}) = g_1(\mathbf{x}) - u_0, \quad \frac{\partial \upsilon(\mathbf{x})}{\partial n} = h_1(\mathbf{x}) - \frac{\partial u_0(\mathbf{x})}{\partial n} \quad \text{on } \Gamma \tag{29a}$$

or

$$\upsilon(\mathbf{x}) = g_2(\mathbf{x}) - u_0(\mathbf{x}), \quad \nabla_1^2 \upsilon(\mathbf{x}) = h_2(\mathbf{x}) - \nabla_1^2 u_0(\mathbf{x}) \quad \text{on } \Gamma. \tag{29b}$$

A particular solution of eqn (26) is presented in the form:

$$u_0(\mathbf{x}) = \frac{1}{8\pi} \int_\Omega |\mathbf{x} - \mathbf{q}|^2 \ln|\mathbf{x} - \mathbf{q}| f(\mathbf{q}) d\Omega(\mathbf{q}) \quad \text{where } \mathbf{x} \in \Omega. \tag{30}$$

If $f(\mathbf{q})$ can be smoothly extended to Ω_0 which contains $\bar{\Omega}$, then

$$u_0(\mathbf{x}) = \frac{1}{8\pi} \int_\Omega |\mathbf{x} - \mathbf{q}|^2 \ln|\mathbf{x} - \mathbf{q}| f(\mathbf{q}) d\Omega(\mathbf{q}) \quad \text{where } \mathbf{x} \in \Omega. \tag{31}$$

After variable transformation (8) function (31) becomes

$$\begin{aligned} u_0(\mathbf{x}) = & \frac{1}{8\pi} \int_0^{2\pi} \left([b_x(\theta) - x]^2 + [b_y(\theta) - y]^2 \right) H(\mathbf{x}, \theta) \\ & \times \int_0^1 R^3 \ln(R f_0(\mathbf{x}, R, \theta)) \, dR d\theta \\ & + \frac{1}{16\pi} \int_0^{2\pi} \int_0^1 \left([b_x(\theta) - x]^2 + [b_y(\theta) - y]^2 \right) \\ & \times \ln\left([\mathrm{b}_x(\theta) - x]^2 + [b_y(\theta) - y]^2 \right) R^3 H(\mathbf{x}, \theta) f_0(\mathbf{x}, R, \theta) dR d\theta \end{aligned} \tag{32}$$

The normal derivative and the Laplacian of function $u_0(\mathbf{x}), \mathbf{x} \in \bar{\Omega}$ are then expressed by formulas:

$$\begin{aligned} \frac{\partial u_0(\mathbf{x})}{\partial n} = & -\frac{1}{2\pi} \int_0^{2\pi} \left[(a\cos\theta - x) n_x + (b\sin\theta - y) n_y \right] H(\mathbf{x}, \theta) \\ & \times \int_0^1 R^2 \ln(R f_0(\mathbf{x}, R, \theta)) \, dR d\theta \\ & - \frac{1}{8\pi} \int_0^{2\pi} \int_0^1 \left[(a\cos\theta - x) n_x + (b\sin\theta - y) n_y \right] \\ & \times \left(1 + \ln\left[(a\cos\theta - x)^2 + (b\sin\theta - y)^2 \right] \right) \\ & \times R^2 H(\mathbf{x}, \theta) f_0(\mathbf{x}, R, \theta) dR d\theta \end{aligned} \tag{33a}$$

$$\nabla_1^2 u_0(\mathbf{x}) = \frac{1}{2\pi}\int_0^{2\pi} H(\mathbf{x},\theta)\int_0^1 R\ln(R f_0(\mathbf{x},R,\theta))\,dRd\theta$$
$$+\frac{1}{4\pi}\int_0^{2\pi}\int_0^1 \left(2+\ln\left[(a\cos\theta - x)^2 + (b\sin\theta - y)^2\right]\right). \tag{33b}$$
$$\times RH(\mathbf{x},\theta) f_0(\mathbf{x},R,\theta)\,dRd\theta.$$

After determining solution $u_0(\mathbf{x})$ from expression (32), its normal derivative and Laplacian (33), we insert into boundary conditions (29) and then solve homogenous boundary problem (28 and 29) by means of the method described in Chapter 7.

3 Boundary collocation method applied to non-linear conditions

Four main cases, in which the boundary collocation method has so far been applied to non-linear conditions, are described in literature as

- non-linear conditions in formulating boundary value problems
- determination of positions of source or collocation points
- steady boundary value problems with variable, free boundary
- surface wave problems

Nearly all present applications of the boundary collocation method in non-linear context concern 2D harmonic problems. The above applications have been discussed as examples of these problems.

3.1 Non-linear boundary conditions

Non-linear boundary conditions result from physical relations of a problem discussed. For example, in the paper of Karageorghis & Fairweather [20], a steady heat exchange (as a result of convection and radiation) in a rectangular rib has been considered. For the 2D Laplace equation defining the temperature field on the cooled body surface, the following condition has been applied:

$$\frac{\partial \phi}{\partial n} = B\phi + \chi\phi^2, \tag{34}$$

where ϕ is a non-dimensional temperature, B is the Biot number and χ is a non-dimensional constant being a combination of environment temperature, the Steffan–Boltzman constant and the heat conductivity coefficient.

If the approximate solution (fulfilling the Laplace equation) in the form:

$$\phi = \sum_{i=1}^{N} a_i \varphi_i(X,Y) \tag{35}$$

is inserted into condition (34) to fulfil it in a finite number of points on the boundary Γ, then a non-linear algebraic set of equations for unknown constants a_i is obtained. In paper [20], the fundamental solutions of the Laplace equation have been used as trial functions $\varphi_i(X,Y)$, whereas constants a_i have been determined from a non-linear functional (built with help of the collocation method), which has been minimized in sense of the least square approach.

3.2 Unknown position of source points or collocation points

Another example of non-linearity in the boundary collocation context is determination of unknown positions of the source or collocation points during the calculational procedure. Let us consider the 2D Laplace equation:

$$\nabla^2 \phi = 0, \quad (X,Y) \in \Omega, \tag{36}$$

with boundary condition

$$B\phi(X,Y) = g(X,Y), \quad (X,Y) \in \Gamma, \tag{37}$$

where B is a linear operator.

Solution of the boundary problem solution (36 and 37) is presented as:

$$\phi = \sum_{i=1}^{N} a_i \ln\left[(X-\xi_i)^2 + (Y-\zeta_i)^2\right], \tag{38}$$

where both, coefficients a_i and source point positions (ξ_i, ζ_i) outside area Ω, are considered as unknown.

The source point location and values of coefficients a_i are determined by the nonlinear functional:

$$\begin{aligned} &F(a_1, a_2, \dots a_N, \xi_1, \xi_2, \dots, \xi_N, \zeta_1, \zeta_2, \dots, \zeta_N) \\ &= \sum_{k=1}^{M} \left| \sum_{i=1}^{N} a_i \ln\left[(X_k-\xi_i)^2 + (Y_k-\zeta_i)^2\right] \right|^2, \end{aligned} \tag{39}$$

where (X_k, Y_k) are coordinates of collocation points on the boundary Γ and $M \geq 3N$. Minimization of functional (39) leads to the set of $3N$ non-linear algebraic equations resulting from zeroing of the partial derivatives:

$$\frac{\partial F}{\partial a_i} = 0, \qquad \frac{\partial F}{\partial \xi_i} = 0, \qquad \frac{\partial F}{\partial \zeta_i} = 0, \quad i = 1, 2, \dots, N. \tag{40}$$

That non-linear collocation method with unknown position of source points has been applied in many papers by Karageorghis, Fairweather and their co-workers [21], whereas Golik & Kołodziej [16] and Mendes & Kołodziej [29] have presented that approach with unknown positions of collocation points.

3.3 Steady problem with variable boundary

The free (variable) boundary problem takes place when we solve differential equation with partial derivatives in an area with boundary, which before solution has partly unknown shape. This part is called the free boundary [15]. We have additional boundary condition on that boundary, which enables determination of its position during the solution. In the steady state, the free boundary problem for 2D Laplace's equation is defined in the following way. The equation:

$$\nabla^2 \phi(\mathbf{x}) = 0 \quad \text{in } \Omega \tag{41}$$

should be solved with the following boundary conditions

$$B_1\phi(\mathbf{x}) = g_1(\mathbf{x}) \quad \text{on } \Gamma_u \tag{42a}$$

$$\left.\begin{aligned} B_2\phi(\mathbf{x}) &= g_2(\mathbf{x}) \\ B_3\phi(\mathbf{x}) &= g_3(\mathbf{x}) \end{aligned}\right\} \quad \text{on } \Gamma_s, \tag{42b}$$

where $g_1(\mathbf{x})$, $g_2(\mathbf{x})$, $g_3(\mathbf{x})$ are known functions. The complete boundary $\Gamma = \Gamma_u \cup \Gamma_s$ consists of Γ_u, which is the motionless boundary part and its location is known and Γ_s being a boundary part, the position of which must be determined during the solution process (hence, we have two conditions on Γ_s).

To solve the above problem, Tanveer [39], Dias, Keller & Vanden-Broeck [12], Vanden-Broeck & Keller [43], Vanden-Broeck [45], Dias & Vanden-Broeck [11], Karageorghis [22], Poullikkas, Karageorghis & Georgiou [33] and Vanden-Broeck [44] have used the boundary collocation method. In numerical implementation, it leads to a solution of non-linear algebraic equations.

3.4 Surface wave problems

Let us consider a bi-periodical non-linear problem of surface waves with 2π-period in liquid running parallelly to horizontal axis X. The bottom of the area, where the liquid (water) is waving, is flat and defined by the straight line $Y = 0$ (axis Y is upward). The unknown wave shape is described by function $Y = \eta(X,t)$, where t is time. In the 2π-period area, velocity potential $\phi(X,Y,t)$ fulfils the 2D Laplace equation:

$$\nabla^2 \phi(X,Y,t) = 0, \tag{43}$$

with periodic conditions

$$\phi(0,Y,t) = \phi(2\pi,Y,t) \tag{44a}$$

$$\frac{\partial}{\partial X}\phi(0,Y,t) = \frac{\partial}{\partial X}\phi(2\pi,Y,t) \tag{44b}$$

boundary condition at the bottom

$$\frac{\partial \phi(X,Y,t)}{\partial Y} = 0 \quad \text{for } Y = 0 \tag{45a}$$

and the free surface conditions

$$\frac{\partial \eta(X,t)}{\partial t} = \frac{\partial \phi(X,Y,t)}{\partial Y} - \frac{\partial \eta(X,t)}{\partial X}\frac{\partial \phi(X,Y,t)}{\partial Y} \quad \text{for } Y = \eta(X,t) \tag{45b}$$

$$\begin{aligned}\frac{\partial \phi(X,Y,t)}{\partial t} = H - \eta(X,t) - \frac{1}{2}\left[\frac{\partial \phi(X,Y,t)}{\partial X}\right]^2 \\ -\frac{1}{2}\left[\frac{\partial \phi(X,Y,t)}{\partial Y}\right]^2 - \frac{\partial \eta(X,t)}{\partial X}\frac{\partial \phi(X,Y,t)}{\partial X}\frac{\partial \phi(X,Y,t)}{\partial Y} \\ \text{for } Y = \eta(X,t),\end{aligned} \tag{45c}$$

where H is a known constant representing non-dimensional pressure on the free surface.

The presented formulation is a non-linear initial boundary problem requiring sometimes a two-step solution method. It consists in alternating boundary solution (boundary collocation method with non-linear conditions on the free surface) and initial problem, which serves to define the wave shape in the next time step. In certain specific cases, the above approach becomes the boundary value problem. The boundary collocation method was used in this case by Tong, Currie & Dugan [40], Rienecker & Fenton [35], Fenton & Rienecker [14], Meiron, Saffman & Yuen [28], Saffman & Yuen [37], Renardy [34], Vanden-Broeck [46], Christiansen [8, 9] and Dias & Tuck [10].

References

[1] Alves, C.J.S. & Chen, C.S., A new method of fundamental solutions applied to nonhomogeneous elliptic problems. *Advances in Computational Mathematics*, **23**, pp. 125–142, 2001.

[2] Atkinson, K.E., The numerical evaluation of partial solution for Poisson's equation. *IMA Journal of Numerical Analysis*, **5**, pp. 319–338, 1985.

[3] Balakrishnam, K. & Ramachandran, A., Oscillatory interpolation in the method of fundamental solution for nonlinear Poisson problems. *Journal of Computational Physics*, **172**, pp. 1–18, 2001.

[4] Chen, C.S., Muleshkov, A.S., Golberg, M.A. & Mattheij, R.M.M., A mesh-free approach to solving the axisymmetric Poisson's equation. *Numerical Methods for Partial Differential Equations*, **21**, pp. 349–367, 2005.

[5] Chen, W. & Tanaka, M., A meshless, integration-free, and boundary-only RBF technique. *Computers and Mathematics with Applications*, **43**, pp. 379–391, 2002.

[6] Chen, W., Shen, L.J., Shen, Z.J. & Yuan, G.W., Boundary knot method for Poisson equations. *Engineering Analysis with Boundary Elements*, **29**, pp. 756–760, 2005.

[7] Cheng, H.D., Young, D.L. & Tsai, C.C., Solution of Poisson's equation by iterative DRBEM using compactly supported, positive definite radial basic function. *Engineering Analysis with Boundary Elements*, **24**, pp. 549–557, 2000.

[8] Christiansen, S., A stability analysis of a Eulerian method for some surface gravity wave problems. In: *Nonlinear Hyperbolic Equations – Theory, Computation Methods, and Applications*, eds. J. Ballman, & R. Jeltsch, *Notes Numer. Fluid Mech.*, **24**, (Vieweg, Braunschweig), pp. 75–84, 1989.

[9] Christiansen, S., An elementary analysis of saw-tooth instabilities on a moving boundary for surface gravity water waves. *Zeitschrift für Angewandte Mathematik und Mechanik (ZAMM)*, **68**, pp. T284–T286, 1988.

[10] Dias, F. & Tuck, E.O., A steady breaking wave. *Physics of Fluids A*, **5**, pp. 277–279, 1993.

[11] Dias, F. & Vanden-Broeck, J.M., Flow emerging from a nozzle and falling under gravity. *Journal of Fluid Mechanics*, **213**, pp. 465–477, 1990.

[12] Dias, F., Keller, J.B. & Vanden-Broeck, J.M., Flow over rectangular weirs. *Physics of Fluids*, **31**, pp. 2071–2076, 1988.

[13] Elansari, M., Ouazar, D. & Cheng, A.H.D., Boundary solution of Poisson's equation using radial basis function collocated on Gaussian quadrature nodes. *Communications in Numerical Methods in Engineering*, **17**, pp. 455–464, 2001.

[14] Fenton, J.D. & Rienecker, M.M., A Fourier method for solving non-linear water-wave problems: application to solitary-wave interaction. *Journal of Fluid Mechanics*, **118**, pp. 411–443, 1982.

[15] Friedman, A., Free boundary problems in science and technology. *Notices of the American Mathematical Society*, **47**, pp. 854–861, 2000.

[16] Golik, W.L. & Kołodziej, J.A., An adaptive boundary collocation method for linear PDEs. *Numerical Methods for Partial Differential Equations*, **11**, pp. 555–560, 1995.

[17] Hickernell, F.J. & Hon, Y.C., Radial basis function approximation of the surface wind field from scattered data. *International Journal of Applied Scientific Computation*, **4**, pp. 221–247, 1998.

[18] Jin, B. & Zheng, Y., Boundary knot method for some inverse problems associated with the Helmholtz equation. *International Journal for Numerical Methods in Engineering*, **65**, pp. 1636–1651, 2005.

[19] Kansa, E.J. & Hon, Y.C., Circumventing the ill-conditioning problem with multiquadric radial basis functions: applications to elliptic partial differential equations. *Computers and Mathematics with Applications*, **39**, pp. 123–137, 2000.

[20] Karageorghis, A. & Fairweather G., The method of fundamental solutions for the solutions of nonlinear plane potential problems. *IMA Journal of Numerical Analysis*, **9**, pp. 231–242, 1989.

[21] Karageorghis, A. & Fairweather, G., The method of fundamental solutions for axisymmetric acoustic scattering and radiation problems. *Journal of Acoustical Society of America*, **104**, pp. 3212–3218, 1998.

[22] Karageorghis, A., The method of fundamental solution for steady-state free boundary problems. *Journal of Computational Physics*, **98**, pp. 119–128, 1992.

[23] Kita, E., Ikeda, Y. & Kamiya, N., Sensitivity analysis scheme of boundary value problem of 2D Poisson equation by using Trefftz method. *Engineering Analysis with Boundary Elements*, **29**, pp. 738–748, 2005.

[24] Kołodziej, J.A. & Klekiel, T., Optimal parameters of method of fundamental solutions for Poisson problems in heat transfer by means of genetic algorithms. *Computer Assisted Mechanics and Engineering Sciences*, **15**, pp. 99–112, 2008.

[25] Kołodziej, J.A. & Uściłowska-Gajda, A., Comparison of different radial basis function in the Poisson equation solutions by the method of fundamental solution. *Studium Vilnense*, **1**, pp. 240–244, 2003.

[26] Li, J., Mathematical justification for RBF-MFS. *Engineering Analysis with Boundary Elements*, **25**, pp. 897–901, 2001.

[27] Li, X., On convergence of the method of fundamental solutions for solving the Dirichlet problem of Poisson's equation. *Advances in Computational Mechanics*, **23**, pp. 265–277, 2005.

[28] Meiron, D.I., Saffman, P.G. & Yuen, H.C., Calculation of steady three-dimensional deep-water waves. *Journal of Fluid Mechanics*, **124**, pp. 109–121, 1982.

[29] Mendes, A.C. & Kołodziej, J.A., An adaptive boundary collocation method for creeping flow between eccentric cylinders. *Advances in Fluid Mechanics*, **9**, pp. 85–91, *First International Conference on Advances in Fluid Mechanics*, AFM 96, 1996.

[30] Palmisano, E., Ramachandran, P.A., Balakrishnan, K. & Al-Dahhan, M., Computation of effectiveness factors for partially wetted catalyst pellets using the method of fundamental solution. *Computers & Chemical Engineering*, **27**, pp. 1431–1444, 2003.

[31] Partridge, P.W. & Sensale, B., The method of fundamental solutions with dual reciprocity for diffusion and diffusion-convection using subdomains. *Engineering Analysis with Boundary Elements*, **24**, pp. 633–641, 2000.

[32] Poullikkas, A., Karageorghis, A. & Georgiou, G., The method of fundamental solutions for inhomogeneous elliptic problems. *Computational Mechanics*, **22**, pp. 100–107, 1998.

[33] Poullikkas, A., Karageorghis, A. & Georgiou, G., The method of fundamental solutions for Singorini problems. *IMA Journal of Numerical Analysis*, **18**, pp. 273–285, 1998.

[34] Renardy, Y., Trapping of water waves above a round sill. *Journal of Fluid Mechanics*, **132**, pp. 105–118, 1983.

[35] Rienecker, M.M. & Fenton, J.D., Fourier approximation method for steady water waves. *Journal of Fluid Mechanics*, **104**, pp. 119–137, 1981.

[36] Rippa, S., An algorithm for selecting a good parameter c in radial basis function. *Advances in Computational Mathematics*, **11**, pp. 193–210, 1999.

[37] Saffman, P.G. & Yuen, H.C., Finite-amplitude interfacial waves in the presence of a current. *Journal of Fluid Mechanics*, **123**, pp. 459–476, 1982.

[38] Schaback, R., Error estimates and condition numbers for radial basis function interpolation. *Advances in Computational Mathematics*, **3**, pp. 251–264, 1995.

[39] Tanveer, S., The effect of surface tension on the shape of Hele-Shaw cell bubble. *Physics of Fluids*, **29**, pp. 3537–3548, 1986.

[40] Tong, J.S.C., Currie, I.G. & Dugan, J.P., Surface waves in the lee of obstacles. In: *Fourth Canadian Congress of Applied Mechanics. Proceedings of Cancam*, 13 May 1973, Montreal, Ecole Polytechnoque, pp. 695–696, 1973.

[41] Trahan, J. & Wyatt, R.E., Radial basis function interpolation in the quantum trajectory method: optimization of the multi-quadric shape parameter. *Journal of Computational Physics*, **185**, pp. 27–49, 2003.

[42] Uściłowska-Gajda, A., Kołodziej, J.A., Ciałkowski, M. & Frączkowiak, A., Comparison of two types of Trefftz method for the solution of inhomogeneous elliptic problems. *Computer Assisted Mechanics and Engineering Sciences*, **10**, pp. 375–389, 2003.

[43] Vanden-Broeck, J.M. & Keller, J.B., Pouring flows with separation. *Physics of Fluids A*, **1**, pp. 156–158, 1989.

[44] Vanden-Broeck, J.M., A model for the free-surface flow due to a submatged source in water of infinite depth. *Journal of the Australian Mathematical Society, Series B – Applied Mathematics*, **39**, pp. 528–539, 1998.

[45] Vanden-Broeck, J.M., Bow flows in water of finite depth. *Physics of Fluids A*, **1**, pp. 1328–1330, 1989.

[46] Vanden-Broeck, J.M., Steep gravity waves: Havelock's method revised. *Physics of Fluids*, **29**, pp. 3084–3085, 1986.

[47] Wang, H., Qin, Q.H. & Kang, Y.L., A new meshless method for steady-state heat conduction problems in anisotropic and inhomogeneous media. *Archive of Applied Mechanics*, **74**, pp. 563–579, 2005.

[48] Wang, J.G. & Liu, G.R., The optimal shape parameters of radial basis functions used for 2-D meshless methods. *Computer Methods in Applied Mechanics and Engineering*, **191**, pp. 2611–2630, 2002.

[49] Xiao, J.R. & McCarthy, M.A., A local Heaviside weighted meshless method for two-dimensional solids using radial basis functions. *Computational Mechanics*, **31**, pp. 301–315, 2003.

[50] Xiao, J.R., Gama, B.A., Gillespie, J.W. Jr. & Kansa, E.J., Meshless solution of 2D contact problems by subdomain variational inequality and MLPG method with radial basis functions. *Engineering Analysis with Boundary Elements*, **29**, pp. 95–106, 2005.

[51] Xie, H., Nogami, T. & Wang, J., A radial boundary node method for two-dimensional elastic analysis. *Engineering Analysis with Boundary Elements*, **27**, pp. 853–862, 2003.

CHAPTER 12

Physically non-linear, elastic–plastic problems

1 Scheme of solution

The T-complete systems applied in this chapter for solving elastic and elastoplastic problems are fundamental solutions with singularities outside the investigated area Ω (the Kupradze functions). For the linear displacement Navier equation:

$$\frac{1}{1-2v}\frac{\partial^2 u_j}{\partial x_i \partial x_j}+\frac{\partial^2 u_i}{\partial x_j \partial x_j}+\frac{1}{G}f_i=0, \tag{1}$$

where u_i are displacements, v is a Poisson's ratio, G is Kirchhoff's modulus and f_i are components of body forces per unit volume. These solutions have the form [2, 3] (two-dimensional problem, a point number μ)

$$U_{ij}^{\mu}=C_1\left(C_2\delta_{ij}\ln r-\frac{y_i y_j}{r^2}\right)\Bigg|_{\mu} \tag{2}$$

where

$$U_{ij}^{\mu}=C_1\left(C_2\delta_{ij}\ln r-\frac{y_i y_j}{r^2}\right)\Bigg|_{\mu},\quad C_2=3-4v,$$

$$y_i=x_i-\xi_{i\mu},\ r^2=y_i y_i$$

and δ_{ij} is Kronecker's delta. The formula (2) describes a displacement into 'i' direction in the point (x_1,x_2) caused by a unit force concentrated in point $(\xi_{1\mu},\xi_{2\mu})$ and acting into 'j' direction. This formula is symmetric, that is we can exchange indices i and j.

Beside the Kelvin functions (2), we use also influence functions (non-symmetric) describing tractions acting into i direction at the point (x_1,x_2) caused by the same unit force [2, 3, 14]:

$$P_{ij}^{\mu} = C_3\left(\frac{1}{r^2}\right)\left[C_4\left(n_j y_i - n_i y_j\right) + \left(C_4\delta_{ij} + \frac{2y_i y_j}{r^2}\right) y_k n_k\right]\Bigg|_{\mu}, \tag{3}$$

where

$$C_3 = 2GC_1,\ C_4 = 1 - 2v$$

$n_i = \cos(n,x_i)$, n is the outside normal to the contour.

Hence, the general integral will be sought in the form

$$\begin{Bmatrix} u_i^g \\ p_i^g \end{Bmatrix} = \sum_{j,\mu} a_{j\mu} \begin{Bmatrix} U_{ij}^{\mu} \\ P_{ij}^{\mu} \end{Bmatrix} \tag{4}$$

and the final solution, as the sum

$$u_i = u_i^g + u_i^{\text{prt}}, \tag{5}$$

where u_i^{prt} means a particular integral of the problem.

In the case of purely displacement boundary conditions, the system u_i^{prt} is complete. For mixed or stress-type conditions, it will sometimes be necessary to supplement the system by functions describing rigid body motions [7] (see also the numerical examples in the next section).

In the Trefftz approach, the applied trial functions fulfil differential equations (1). Hence, it has only been necessary to fulfil the boundary conditions in a possibly accurate way. We present this in an example of the integral fit, but there exists full analogy between this type of the fit in the numerical sense and the boundary collocation [12] (see also Section 2 in Chapter 2).

There are two kinds of the weighted residuals in this case. The first one is the direct fit, which results from the least-squares condition:

$$\int_{\Gamma_1} U_{ik}^{v}\left(u_i^g + u_i^{\text{prt}} - \overline{u}_i\right) d\Gamma_1 + W\int_{\Gamma_2} P_{ik}^{v}\left(p_i^g + p_i^{\text{prt}} - \overline{p}_i\right) d\Gamma_2 = 0, \tag{6}$$

where $\overline{u}_i$, $\overline{p}_i$ are boundary displacements and tractions given on Γ_1, Γ_2, respectively, $\Gamma = \Gamma_1 \cup \Gamma_2$ is the contour of the considered area and W is a weighting constant. When the integrals in (6) are calculated numerically using the

Gauss–Legendre procedure, this form is the particular case of the least-squares orthogonal collocation (in zeros of Legendre polynomials) [12]. The difference consists in the weights, which for the integral fit are equal to

$$A_i = \frac{1}{2}(t_2 - t_1)G_i, \tag{7}$$

where t_1, t_2 are the lower and upper limit of the integral and G_i are the Gaussian weights. For the orthogonal collocation, they can be chosen in an arbitrary way (in the numerical examples we applied $A_i = 1.0$).

The weighting constant W is necessary because of the difference in dimension of displacements and tractions. It can be avoided in the case of the indirect fit:

$$\int_{\Gamma_1} P_{ik}^{\nu}\left(u_i^g + u_i^{\mathrm{prt}} - \overline{u}_i\right)d\Gamma_1 - \int_{\Gamma_2} U_{ik}^{\nu}\left(p_i^g + p_i^{\mathrm{prt}} - \overline{p}_i\right)d\Gamma_2 = 0. \tag{8}$$

In this case, we also have symmetric matrices (the symmetry of the matrix in (6) is obvious), which can be proved analogically to the case of the Laplace equation [13].

In the proposed elastoplastic solution two steps can be observed. In the first step, we calculate stresses in the plastic zones from the incremental stress–strain relations; in the second one we correct the first-step solution by application of compensatory body forces to the investigated object. In the final algorithm, both steps are repeated inside the approximating loops.

For the first step solution, we begin with applying to the object the load p_Q, which causes yielding in the most highly stressed point. Then we increase the load:

$$p = \kappa\, p_Q = (1 + \Delta\kappa)p_Q \tag{9}$$

and we obtain zones in which in the purely elastic solution the reduced stress σ_0 (here in the sense of Huber–von Mises hypothesis) is larger than yield point σ_Q. In these zones, we apply the incremental Prandtl–Reuss plastic flow equations in the form:

$$\dot{\sigma}_{ij}^{\mathrm{ep}} = 2G\left[\dot{\varepsilon}_{ij} + \frac{\nu}{1-2\nu}\delta_{ij}\dot{\varepsilon}_{kk} - \frac{3S_{ij}S_{kl}}{2\sigma_Q^2}\dot{\varepsilon}_{kl}\right], \tag{10}$$

where $S_{ij} = \sigma_{ij} = 1/3\delta_{ij}\sigma_{kk}$ is the deviatoric stress. Here the dot over the stresses and strains can be defined as their increment caused by the increment of the load:

$$\dot{p} = \Delta\kappa\, p_Q. \tag{11}$$

Formula (10) describes the incremental stress–strain relations for the material without hardening (Prandtl model); however, generalization of the algorithm presented to materials with hardening does not introduce any basic difficulties.

The numerical tests suggested application of the load increment (inside a loop), in which it is divided into very small 'internal' increments ($\Delta\kappa_i$). In each cycle of this internal loop, the value of the reduced stress is checked in the particular internal points of the object and the relation $\sigma_0 > \sigma_Q$ involves application of (10). The current values of the deviatoric stress S_{ij} are also introduced in this level of calculations. The described procedure is very quick and seems to be easier than the introduction of the factor r [9] in calculation of the approximate beginning of the plastic flow.

In the second step, we apply compensatory forces inside the zones of plastic flow to correct the internal equilibrium of the object. The forces are here calculated in the algorithm called 'initial stress approach' [2]. In this approach, the initial stress rates are defined as

$$\dot{\sigma}_{ij}^{0} = \dot{\sigma}_{ij}^{\mathrm{e}} - \dot{\sigma}_{ij}^{\mathrm{ep}} \tag{12}$$

where

$$\dot{\sigma}_{ij}^{\mathrm{e}} = D_{ijkl}^{\mathrm{e}}\dot{\varepsilon}_{kl}, \quad \dot{\sigma}_{ij}^{\mathrm{ep}} = D_{ijkl}^{\mathrm{ep}}\dot{\varepsilon}_{kl} \tag{13}$$

and D_{ijkl}^{e}, D_{ijkl}^{ep} are the elastic and elastoplastic constitutive tensors, respectively.

The equilibrium equations take the form

$$\frac{\partial \dot{\sigma}_{ij}^{\mathrm{ep}}}{\partial x_j} = 0 \quad \text{or} \quad \frac{\partial \dot{\sigma}_{ij}^{\mathrm{e}}}{\partial x_j} - \frac{\partial \dot{\sigma}_{ij}^{0}}{\partial x_j} = 0. \tag{14}$$

It is visible that the term $\left(-\partial\dot{\sigma}_{ij}^{0}/\partial x_j\right)$ can be considered as a body force $\dot{f}_i$. Hence, in the second-step solution, we shall solve the elastic problem in which the particular integral can be written as

$$u_i^{\mathrm{prt}}(\mathbf{x}) = \int_{\Omega} U_{ij}(\mathbf{x},\xi)\dot{f}_j(\xi)d\xi \tag{15}$$

or

$$u_i^{\mathrm{prt}}(\mathbf{x}) = -\int_{\Omega} U_{ij}(\mathbf{x},\xi)\frac{\partial \dot{\sigma}_{jk}^{0}}{\partial \xi_k}d\xi. \tag{16}$$

The compensatory forces $\dot{f}_1$ change the distribution of stresses and the range of plastic zones inside the object. If the changes are bigger than acceptable norms, it is necessary to repeat the procedure applying the respective compensatory forces inside the corrected plastic zones [2]; the divergence of this procedure means 'collapse' of the object. The changes mentioned above can be diminished by choosing small increments $\Delta\kappa = \Delta\kappa_e$ in a so-called external loop developing the elastoplastic process.

The boundary tractions change their values according to the general formula:

$$\dot{p}_i^{\mathrm{ep}} = \dot{\sigma}_{ij}^{\mathrm{ep}} n_j = \dot{\sigma}_{ij}^{\mathrm{e}} n_j - \dot{\sigma}_{ij}^{0} n_j. \tag{17}$$

Hence, we model the elastoplastic process in which we increase the load step by step, reducing the solutions (also the tractions) in each step according to the given equations of plastic flow. The reduction of the load occurs if the plastic zones adjoin the place of its application. The complete algorithm is presented in detail in numerical examples in the next section. The presented approach was first proposed in [15].

2 Numerical illustration of the algorithm proposed

The numerical illustration of the method is presented in two examples shown in Figs. 1 and 14. In the first one, we used a square element loaded parabolically from all sides (compare also Chapter 9). This problem has the close elastic solution. The stress function can be written as [10].

$$\varphi = c\left(x^4 - y^4\right). \tag{18}$$

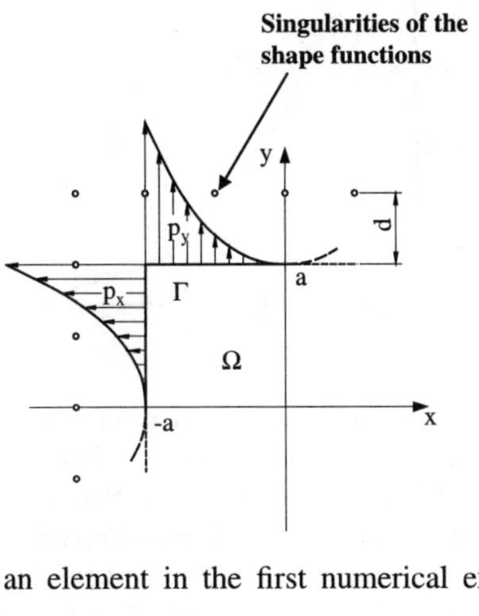

Figure 1: Scheme of an element in the first numerical example (elastoplastic square).

Hence, it is easy to observe convergence of the numerical elastic solutions and the influence of particular factors on accuracy of the results. These investigations were necessary before application of the Kupradze functions to elastic–plastic problems [15].

In Figs 2 and 3, we observe behaviour of the extreme error on the boundary (Collatz error measure [5, 12]) for increasing number of shape functions and control points on the boundary. In this example, as in many others, the considerable advantage of applying orthogonal collocation is visible. Optimality of this type of boundary collocation and its relations (similarities) to the numerical Gaussian integral fit were discussed in detail in paper [12].

Also, for varying distance of singularities from the given area (Fig. 4), this optimality was clearly visible. The behaviour of results in the case of the orthogonal collocation was very regular, with an extreme error in the corner (Fig. 5). On the contrary, in the equidistant collocation, the form of the error function was changing considerably and starting from $d \cong 30\ mm$, the extreme error was larger.

Figure 6 presents the result of the first-step elastoplastic solution and particularly the behaviour of the solution during application of the internal loop. In this loop, we applied a small increment $\Delta\kappa_i$, checked the plastic condition in a control point and introduced the current values of the deviatoric stress to (10). The total increment

$$\Delta\kappa_e = \kappa - 1 \tag{19}$$

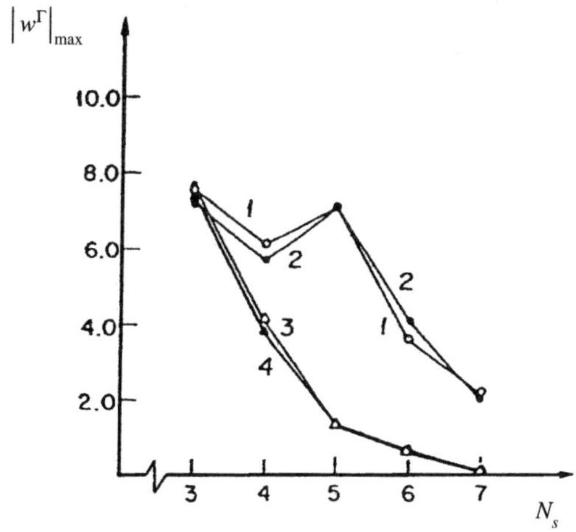

Figure 2: Extreme boundary error versus increasing number of shape functions; N_s is the number of singularities on one side of the square; $w^\Gamma = p_y - \overline{p}_y$, ($p_{xext} = p_{yext} = 100 MPa$, $a = 50 mm$, $d = 50 mm$ $G = 8 \cdot 10^4 MPa$, $\nu = 0.3$)1: Equidistant collocation – $N_c = 7$, $N_T = 24$ (N_c is the number of control points on each side of the square, N_T – is the total number of control points on the boundary). 2: Equidistant collocation – $N_c = 8$, $N_T = 28$. 3: Orthogonal collocation – $N_c = 6$, $N_T = 24$. 4: Orthogonal collocation – $N_c = 7$, $N_T = 28$.

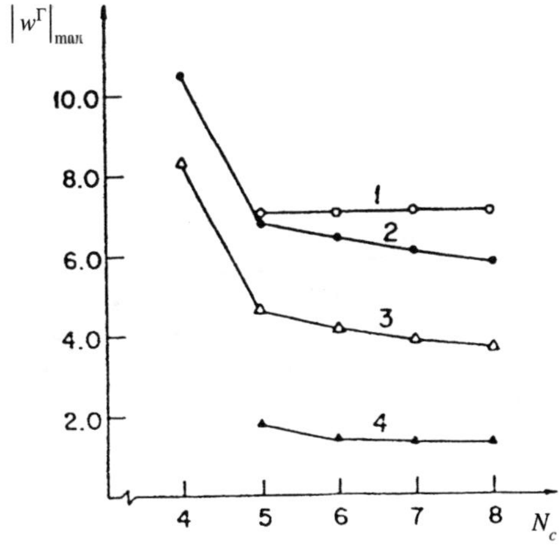

Figure 3: Extreme boundary error versus increasing number of the control points on the boundary (data as in Fig. 2). 1: Equidistant collocation – $N_s = 5$. 2: Equidistant collocation – $N_s = 4$. 3: Orthogonal collocation – $N_s = 4$. 4: Orthogonal collocation – $N_s = 5$.

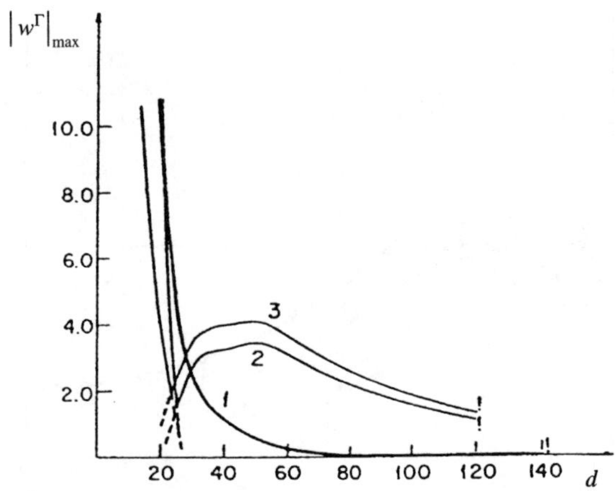

Figure 4: Extreme boundary error versus increasing distance d of singularities from the boundary, p_{xext}, p_{yext}, a, G, ν, as in Fig. 2; $N_s = 6$). 1: Orthogonal collocation – $N_c = 7$, $N_T = 24$; 2: Equidistant collocation – $N_c = 7$, $N_T = 24$; 3: Equidistant collocation – $N_c = 8$, $N_T = 28$. !: the last points which could be calculated because of ill-conditioned matrices.

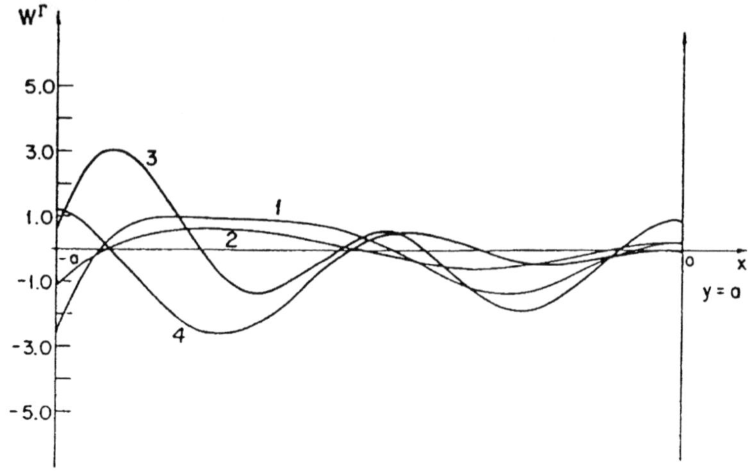

Figure 5: Boundary error function (data as in Fig. 4). 1: Orthogonal collocation – $d = 30$ mm, $N_s = 6$, $N_c = 7$. 2: Orthogonal collocation – $d = 40$ mm, $N_s = 6$, $N_c = 7$. 3: Equidistant collocation – $d = 22$ mm, $N_s = 6$, $N_c = 8$. 4: Equidistant collocation – $d = 26$ mm, $N_s = 6$, $N_c = 8$.

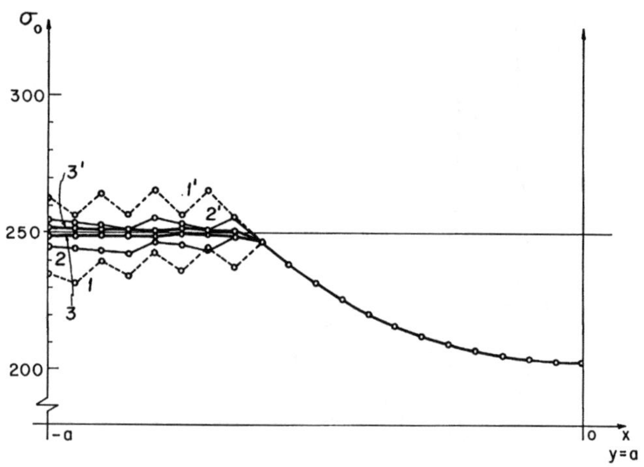

Figure 6: First-step elastoplastic solution, σ_0^{ep} – influence of number of cycles in the internal loop (load coefficient $\kappa = 1.4$, $a = 50$ mm, $d = 100$ mm, $G = 80\ 000$ MPa, v = 0.3, $N_s = 5$, $N_c = 7$; orthogonal collocation). Prime means purely elastic solution in the last cycle of the loop. 1: four cycles; 2: 10 cycles; 3: 40 cycles.

was equal to 0.4 in this example and the increment $\Delta\kappa_i$ varied. For the first curve, it was equal to 0.1 (four cycles), for the second –0.04 (10 cycles) and for the third –0.01 (40 cycles). The upper curves in which the last cycle was calculated for purely elastic formulas show the way of convergence of the solutions. Figs. 7 and 8 illustrate the final first-step results in this example and Fig. 9 explains the meaning of the quantities introduced in (12) and used in the second-step solution.

In our example, we applied on purpose a very large external increment $\Delta\kappa_e = 0.4$ and only one cycle in the external loop. This allowed us to clearly illustrate the quantities appearing in the second step of the method. For facility, we used the engineering notations:

$$\begin{gathered} u \equiv u_1, \quad v \equiv u_2 \\ \sigma_x \equiv \sigma_{11}, \quad \sigma_y \equiv \sigma_{22}, \quad \sigma_{xy} \equiv \sigma_{12} \end{gathered} \tag{20}$$

in the following figures and added the letter 'p' for the results obtained from (10) (and the internal loop) in the plastic zones.

Figure 10 shows the results of the second-step solution. In this step, we applied compensatory forces $\dot{f}_i$ to the object with zero boundary conditions. From (6) it is obvious that the general solution of this problem means in fact the solution with the boundary conditions

$$\overline{p}_i = -p_i^{\mathrm{prt}}. \tag{21}$$

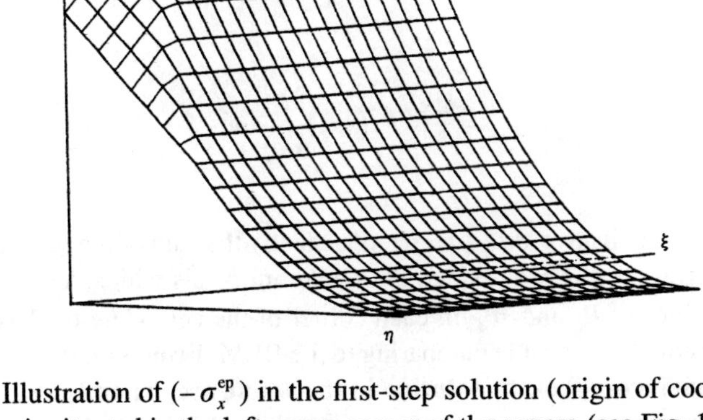

Figure 7: Illustration of $(-\sigma_x^{\mathrm{ep}})$ in the first-step solution (origin of coordinates ξ, η is situated in the left upper corner of the square (see Fig. 1)).

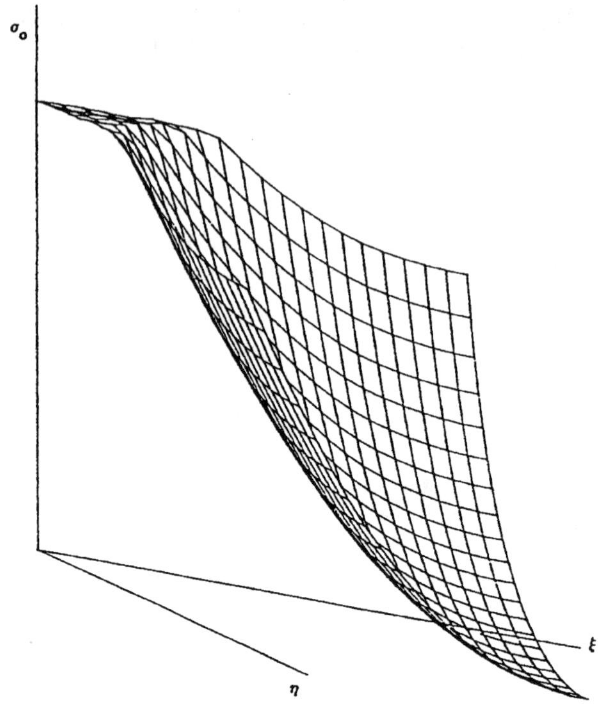

Figure 8: Illustration of σ_0^{ep} in the first-step solution (coordinate system as in Fig. 7).

Hence, the stresses σ_y^{prt} and σ_{xy}^{prt} in Fig. 10 illustrate these changed boundary conditions (but with opposite sign) and the stresses σ_y, σ_{xy} are the boundary error of the solution.

The particular traction solution used in this place has the form

$$p_i^{\text{prt}} = \int_\Omega P_{ij}(\mathbf{x},\xi)\dot{f}_j(\xi)d\xi \tag{22}$$

or

$$p_i^{\text{prt}} = -\int_\Omega P_{ij}(\mathbf{x},\xi)\frac{\partial\dot{\sigma}_{jk}^0}{\partial\xi_k}d\xi. \tag{23}$$

To calculate the derivative $\partial\dot{\sigma}_{jk}^0/\partial\xi_k$ (more exactly the sum of the derivatives) and then the integral over Ω, we divided the square into cells with sides $\delta = a/50$. We calculated values of P_{ij} and $\dot{\sigma}_{jk}^0$ in each corner of the cell. Here the basic advantage of the method was visible comparing to the BEM. **From eqn (6) we obtained the functional form (4) of the solution in the whole area Ω and the calculation of the functions mentioned above was very quick and simple**. Hence, finally we

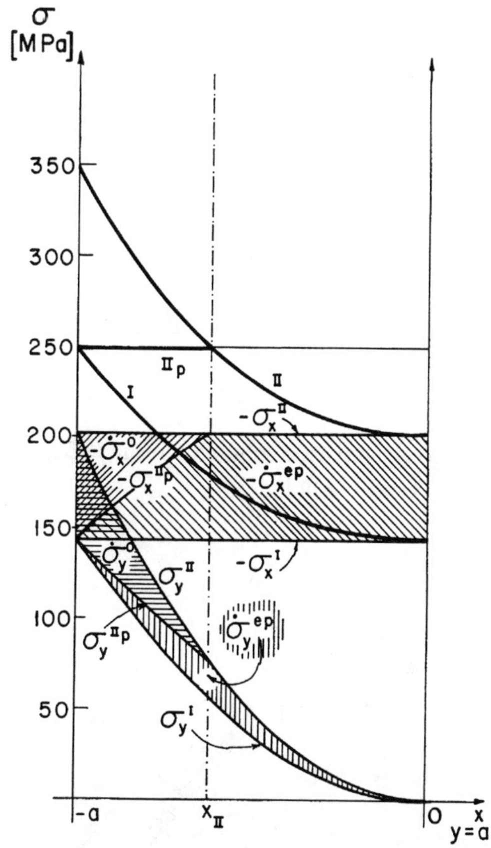

Figure 9: Illustration of quantities in the incremental equilibrium equations (data as in Fig. 12.6, except $d = 30$ mm, $N_s = 9$, $N_c = 11$). I: σ_0 for the elastic load capacity (also respective stresses) – $\kappa = 1.0$. II: σ_0 for load coefficient $\kappa = 1.4$. Letter p means elastoplastic solution in the plastic zone.

represented the integrals (23) as sums over the cells and the derivative $\partial\dot{\sigma}^0_{jk}/\partial\xi_k$ was calculated numerically in the centre of the cell.

The possibly exact calculation of integral (23) caused additional problem. The function p_i^{prt} must be calculated in the control points of integration (or collocation) used in eqn (6). Hence, during the integration of formula (23), we met local singularities of $P_{ij}(x,\xi)$ when the point (x) coincided with (ξ). In these cells, we applied local analytical integration inside the triangles adjoined to the singularity. This was done analogically to the method presented in monograph [2].

The system of coordinates used in this analytical integration is shown in Fig. 11 and the result in Table 1. The value of the body force was considered as locally constant inside the triangle. Also, to facilitate exchange of singular cell integrals,

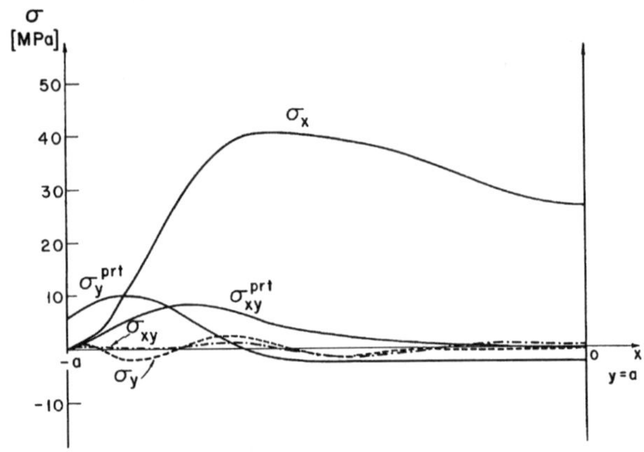

Figure 10: Second-step elastoplastic solution – results of application of compensatory forces (data as in Fig. 9).

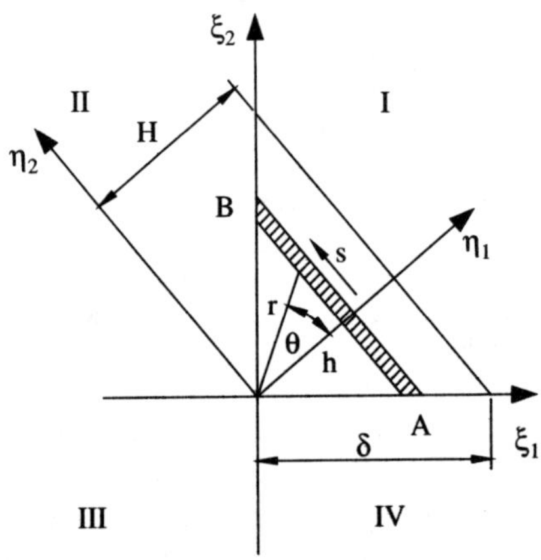

Figure 11: System of coordinates for analytical integration of singular functions.

the equidistant collocation was used in this second-step solution. However, *the numerical tests showed that the simple cut of the singularity* (assuming $r = 10^{-2} \div 10^{-4}$ for its smaller values) *did not change visibly the final results comparing to the integration described above.*

Figure 12 presents results of the second-step solution in which the stresses resulting from application of compensatory forces $\dot{f}_i$ (Fig. 10) were added to the first-step solution. The tractions in the example were changed in the first-step solution according to formula (17) (this is illustrated by the curve σ_y^{IIp}). Hence, it was necessary to introduce the notion of the reduced load increment $\Delta\kappa_{er}$ in which the integral load in the purely elastic solution was equal to the integral load calculated

Table 1: Integrals of the singular functions ($x_1 = x_2 = 0$) over the triangle (Fig. 11).

		$i=1, j=1$	$i=2, j=2$	$i=1, j=2$	$i=2, j=1$
$\int_\Delta U_{ij} d\xi$	I	$+C_1 \frac{\delta^2}{4}\left[C_2\left(2\ln\delta+\frac{\pi}{2}-3\right)-1\right]$		$-C_1 \frac{\delta^2}{4}\left(\frac{\pi}{2}-1\right)$	
	II	$+C_1 \frac{\delta^2}{4}\left[C_2\left(2\ln\delta+\frac{\pi}{2}-3\right)-1\right]$		$+C_1 \frac{\delta^2}{4}\left(\frac{\pi}{2}-1\right)$	
	III	$+C_1 \frac{\delta^2}{4}\left[C_2\left(2\ln\delta+\frac{\pi}{2}-3\right)-1\right]$		$-C_1 \frac{\delta^2}{4}\left(\frac{\pi}{2}-1\right)$	
	IV	$+C_1 \frac{\delta^2}{4}\left[C_2\left(2\ln\delta+\frac{\pi}{2}-3\right)-1\right]$		$+C_1 \frac{\delta^2}{4}\left(\frac{\pi}{2}-1\right)$	
$\int_\Delta P_{ij} d\xi \quad n_1=0, \quad n_2=1$	I	$-C_3 \frac{\delta}{2}\left(C_4 \frac{\pi}{2}+1\right)$	$-C_3 \frac{\delta}{2}\left(C_4 \frac{\pi}{2}+\pi-1\right)$	$-C_3 \frac{\delta}{2}\left(C_4 \frac{\pi}{2}+1\right)$	$+C_3 \frac{\delta}{2}\left(C_4 \frac{\pi}{2}-1\right)$
	II	$-C_3 \frac{\delta}{2}\left(C_4 \frac{\pi}{2}+1\right)$	$-C_3 \frac{\delta}{2}\left(C_4 \frac{\pi}{2}+\pi-1\right)$	$+C_3 \frac{\delta}{2}\left(C_4 \frac{\pi}{2}+1\right)$	$-C_3 \frac{\delta}{2}\left(C_4 \frac{\pi}{2}-1\right)$
	III	$+C_3 \frac{\delta}{2}\left(C_4 \frac{\pi}{2}+1\right)$	$+C_3 \frac{\delta}{2}\left(C_4 \frac{\pi}{2}+\pi-1\right)$	$+C_3 \frac{\delta}{2}\left(C_4 \frac{\pi}{2}+1\right)$	$-C_3 \frac{\delta}{2}\left(C_4 \frac{\pi}{2}-1\right)$
	IV	$+C_3 \frac{\delta}{2}\left(C_4 \frac{\pi}{2}+1\right)$	$+C_3 \frac{\delta}{2}\left(C_4 \frac{\pi}{2}+\pi-1\right)$	$-C_3 \frac{\delta}{2}\left(C_4 \frac{\pi}{2}+1\right)$	$+C_3 \frac{\delta}{2}\left(C_4 \frac{\pi}{2}-1\right)$

(Continued)

Table 1: Integrals of the singular functions ($x_1 = x_2 = 0$) over the triangle (Fig. 11).

		$i=1, j=1$	$i=2, j=2$	$i=1, j=2$	$i=2, j=1$
$\int_\Delta P_{ij} d\xi \quad n_1 = 1, \quad n_2 = 0$	I	$+C_3 \frac{\delta}{2}\left(C_4 \frac{\pi}{2} - \pi + 1\right)$	$+C_3 \frac{\delta}{2}\left(C_4 \frac{\pi}{2} - 1\right)$	$+C_3 \frac{\delta}{2}\left(C_4 \frac{\pi}{2} - 1\right)$	$-C_3 \frac{\delta}{2}\left(C_4 \frac{\pi}{2} + 1\right)$
	II	$-C_3 \frac{\delta}{2}\left(C_4 \frac{\pi}{2} - \pi + 1\right)$	$-C_3 \frac{\delta}{2}\left(C_4 \frac{\pi}{2} - 1\right)$	$+C_3 \frac{\delta}{2}\left(C_4 \frac{\pi}{2} - 1\right)$	$-C_3 \frac{\delta}{2}\left(C_4 \frac{\pi}{2} + 1\right)$
	III	$-C_3 \frac{\delta}{2}\left(C_4 \frac{\pi}{2} - \pi + 1\right)$	$-C_3 \frac{\delta}{2}\left(C_4 \frac{\pi}{2} - 1\right)$	$-C_3 \frac{\delta}{2}\left(C_4 \frac{\pi}{2} - 1\right)$	$+C_3 \frac{\delta}{2}\left(C_4 \frac{\pi}{2} + 1\right)$
	IV	$+C_3 \frac{\delta}{2}\left(C_4 \frac{\pi}{2} - \pi + 1\right)$	$+C_3 \frac{\delta}{2}\left(C_4 \frac{\pi}{2} - 1\right)$	$-C_3 \frac{\delta}{2}\left(C_4 \frac{\pi}{2} - 1\right)$	$+C_3 \frac{\delta}{2}\left(C_4 \frac{\pi}{2} + 1\right)$

for the elastoplastic tractions (the equivalence of the shaded areas in Fig. 12). The increment with which we started the calculations is called 'nominal' – $\Delta\kappa_{en}$. The results for $\Delta\kappa_{er}$ have notations III and for the $\Delta\kappa_{en}$ notations II.

After application of the compensatory body forces, the plastic zone obviously increases in comparison to the first-step zone for $\Delta\kappa_{er}$, but in this example decreases comparing to the zone calculated for $\Delta\kappa_{en}$, as the reduced load

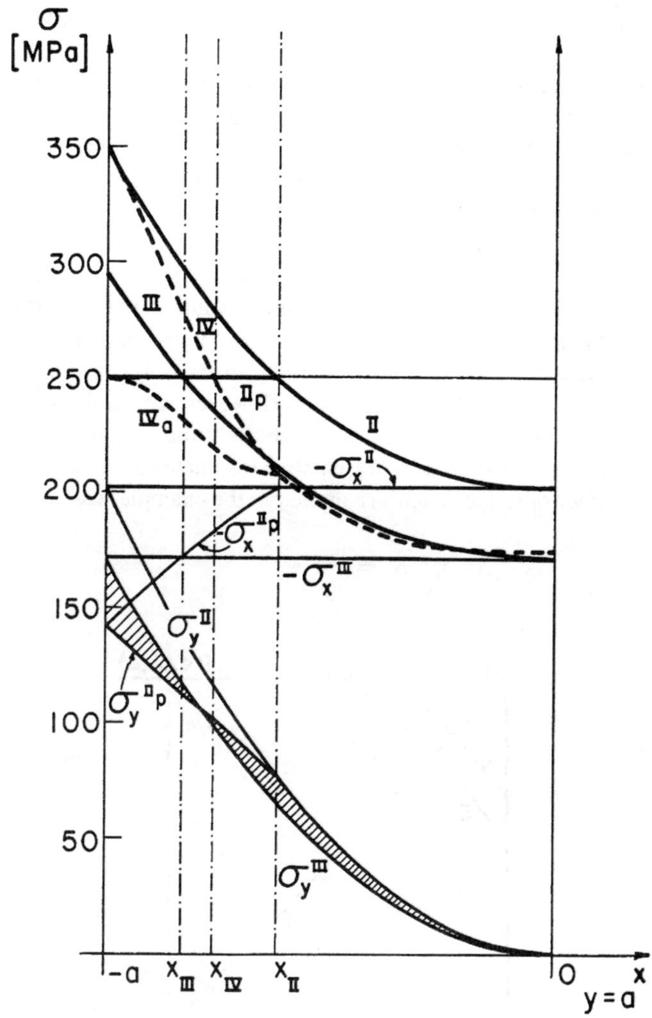

Figure 12: Results of the second-step elastoplastic solution on boundary of the square (data as in Fig. 9). II: σ_0 for load coefficient $\kappa_n = 1.4$ (also respective stresses). III: σ_0 for load coefficient $\kappa_r = 1.18$. IV, IVa: σ_0 after addition of stresses resulting from application of compensatory forces (Fig. 10) to stresses σ_{ij}^{II}, $\sigma_{ij}^{\text{II}p}$, respectively.

(elastoplastic tractions) was considerably smaller than the nominal one (purely elastic tractions). This situation does not occur always, as in many cases the appearance of plastic zones does not change boundary conditions. In such cases, $\Delta\kappa_{er} = \Delta\kappa_{en}$ and the body forces $\dot{f}_i$ cause increase of the plastic zones.

The choice of the large $\Delta\kappa_{en}$ in our example demonstrated certain difficulties in the algorithm when the active boundary tractions are changed inside the loops. In particular, it appeared more convenient to add the stresses caused by the compensatory forces to the purely elastic solution II (curve IV in Figs. 12 and 13) reducing their sum according to relation (10) in the zones where $\sigma_0 > \sigma_Q$. Addition of the compensatory stresses to the elastoplastic first-step solution (curve IV a) gave unacceptable results in this case; however, for much smaller $\Delta\kappa_e$ and many cycles in the external loop, this variant can also be accepted.

The problem signalized above and further corrections of the compensatory forces need more detailed investigations; however, they are not directly connected with the boundary collocation method presented.

The purpose of the second example (Fig. 14) was to present behaviour of the method proposed in case of a solution of an atypical, singular problem. It again started from preliminary investigations of elastic formulation. The elastic solution can in this case be treated as a certain approximation for particular errors of boundary conditions (it will be changed considerably in the singular zones (corners) while changing accuracy of these conditions). However, the whole procedure will not be changed significantly and the elastoplastic solution (cutting singularities of the stresses) can be considered as representative results for the case.

The total number of shape functions applied in the example was

$$N_n = 2N_s + 1 \tag{24}$$

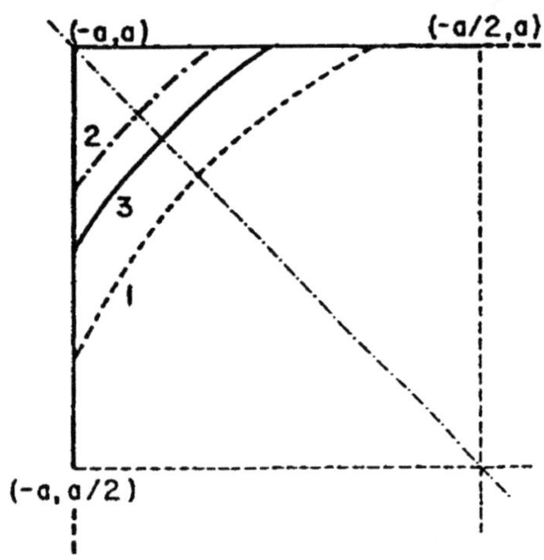

Figure 13: Ranges of plastic zones (data as in Fig. 9). 1: for $\kappa_n = 1.4$; 2: $\kappa_r = 1.18$; 3: after application of compensatory forces.

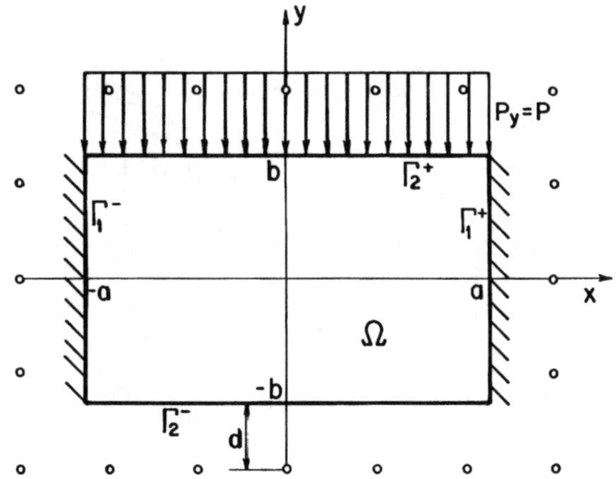

Figure 14: Scheme of element in second example.

because it was necessary to use the function representing the vertical rigid body motion in this case (otherwise we obtained false results for displacements). To calculate this additional coefficient, we added the condition

$$\int_{\Gamma_1} u_2 d\Gamma_1 = 0 \tag{25}$$

to the system resulting from (6):

$$\sum_{\mu=1}^{M}\sum_{j=1}^{2} a_{j\mu}\left[\int_{\Gamma_1} U_{ik}^{\nu} U_{ij}^{\mu} d\Gamma_1 + W\int_{\Gamma_2} P_{ik}^{\nu} P_{ij}^{\mu} d\Gamma_2\right] = W\int_{\Gamma_2^+} P_{2k}^{\nu} p d\Gamma_2^+ \tag{26}$$
$$k = 1,2, \quad \nu = 1,2,\ldots,N, \quad N = M.$$

The relations between the formulas (2) and (3) suggested the choice of the value of weighting constant W as

$$W = \frac{R_a^2}{4G^2} \tag{27}$$

where R_a is the average distance of singularities from the origin of coordinates (centre of the element).

For this value of W, the extreme relative errors of the boundary conditions on the clamped and free edges were similar (the traction error referred to the load p and the displacement error referred to extreme deflection of the object).

Figures 15 and 16 show increasing accuracy of the boundary conditions in this second example, with the increase of the number of shape functions applied.

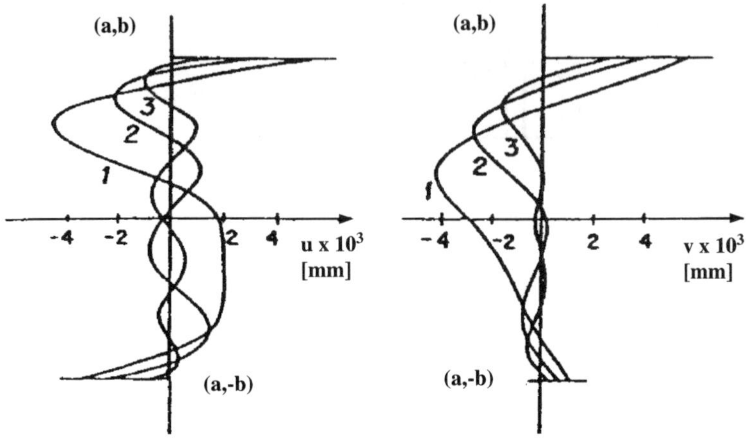

Figure 15: Convergence of the boundary conditions – displacements along clamped edges ($a = 50$ mm, $b = 30$ mm, $d = 15$ mm, $p = 100$ MPa; $N_x = 11$, number of control points on free edge; $N_y = 11$, number of control points on clamped edge; $G = 8{\cdot}10^4$ MPa, $\nu = 0.3$, orthogonal collocation), 1: $N_{sx} = 6$, number of singularities in horizontal row; $N_{sy} = 5$, number of singularities in vertical row; $N_s = 18$; 2: $N_{sx} = 8$, $N_{sy} = 6$, $N_s = 24$; 3: $N_{sx} = 10$, $N_{sy} = 7$, $N_s = 30$.

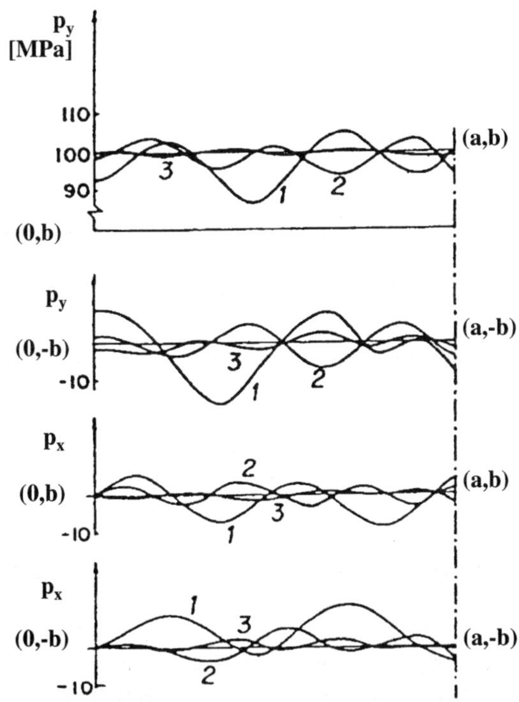

Figure 16: Convergence of the boundary conditions – tractions on free edges (data as in Fig. 15), 1: $N_{sx} = 6$, $N_{sy} = 5$, $N_s = 18$; 2: $N_{sx} = 8$, $N_{sy} = 6$, $N_s = 24$; 3: $N_{sx} = 10$, $N_{sy} = 7$, $N_s = 30$.

All the further calculations were carried out for $N_s = 30$ shape function singularities (61 shape functions). For this case, the extreme relative errors were equal to

$$\begin{aligned} \delta_u &= 5.60\% \quad \text{for the displacements} \\ \delta_p &= 3.19\% \quad \text{for the tractions.} \end{aligned} \tag{28}$$

In this example, only the first-step solutions were calculated because of local character of the plastic zones (the compensatory forces applied very near to the corners of the object did not change considerably the final solution).

Figures 17 and 18 show elastoplastic results near the corner. The singularities of the stresses in the purely elastic solution have been softened by inaccuracies in the boundary conditions. As we can also see the difference between results $\Delta\kappa_{en} = 0.40$ and $\Delta\kappa_{er} = 0.38$ is very small in this case. Figure 19 presents the range of plastic zones for various κ and Figs. 20 and 21 illustrate the final shape of the reduced stress function in the purely elastic and elastoplastic solutions.

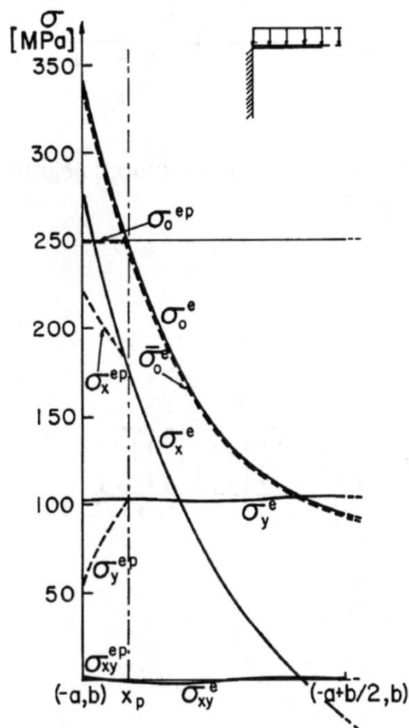

Figure 17: Elastoplastic results (first-step solution) near the corner. Free edge (data as in Fig. 15, except for p which depends on load coefficient κ_n), $\kappa_n = 1.4$; $\kappa_r = 1.38$ – results with the bar $^{-}$; $N_{sx} = 10$, $N_{sy} = 7$, $N_s = 30$.

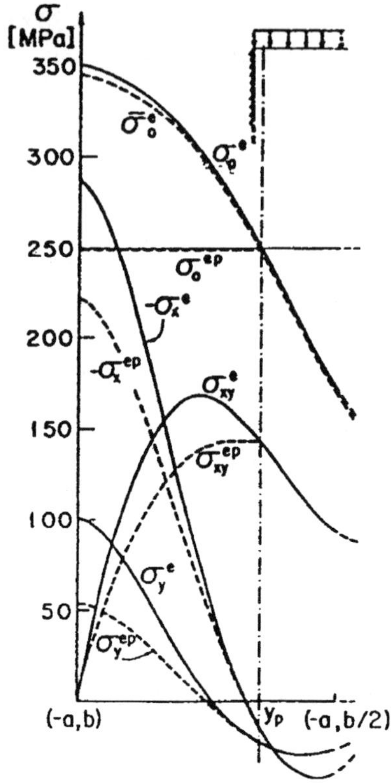

Figure 18: Elastoplastic results (first-step solution) near the corner, clamped edge (data as in Fig. 17).

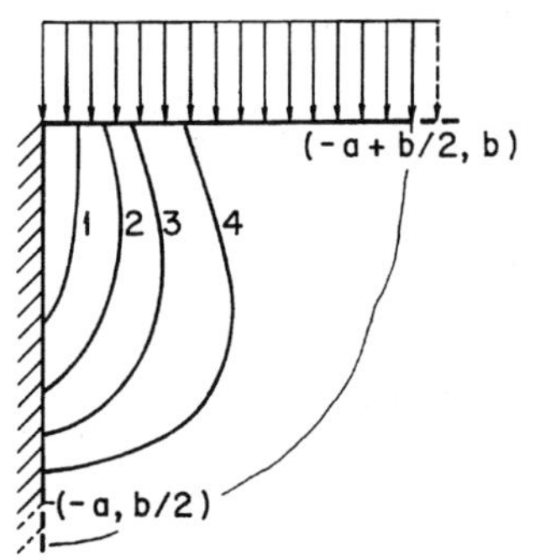

Figure 19: Range of plastic zones (data as in Fig. 17). 1: $\kappa_n = 1.2$; 2: $\kappa_n = 1.4$; 3: $\kappa_n = 1.6$; 4: $\kappa_n = 1.8$.

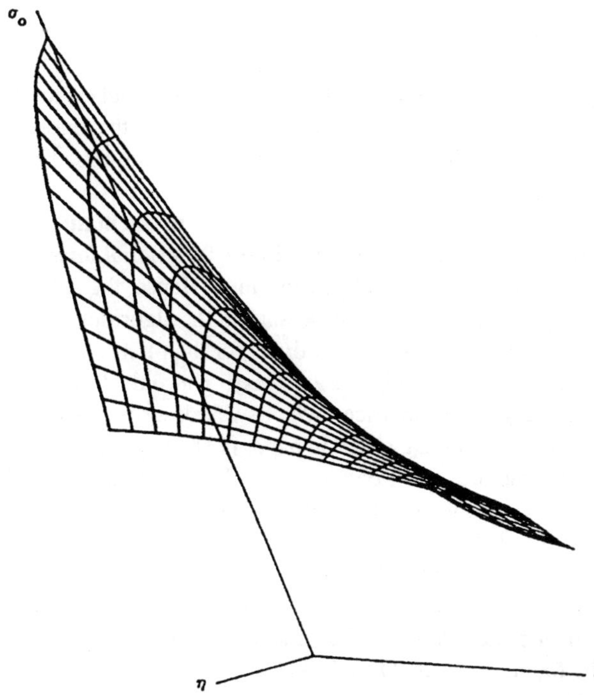

Figure 20: Illustration of purely elastic solution in the corner – σ_x^{ep} (data as in Fig. 17). Origin of coordinates ξ, η is situated in left upper corner of the element.

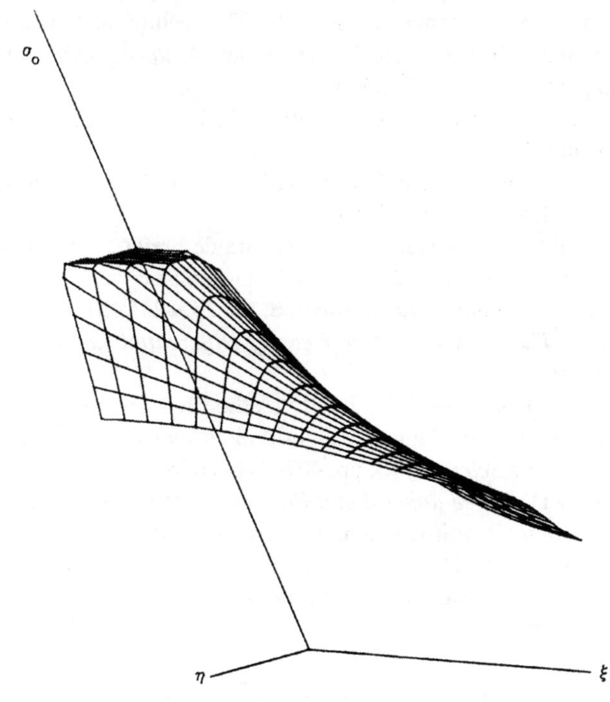

Figure 21: Illustration of elastoplastic first-step solution in the corner – σ_x^{ep} (data as in Fig.17).

3 Final remarks

The chapter presents the boundary collocation method applied to elastic–plastic problems. The elastoplastic solution proposed is based on the algorithm presented in the monograph [2] and it is essentially similar to that described by Zienkiewicz and others [16, 17].

The advantage of proposal consists in the fact that **in the boundary collocation method the linear solutions are obtained in the functional form – we calculate directly the coefficients of the Trefftz functions**. Hence, the calculation of even a very dense mesh of internal points, necessary in the algorithm, is very quick and simple. This results in surprisingly short times for the final solutions.

The presented algorithm is logical and easy. The only limitation consists in conditioning of the solution matrices which was described and investigated in detail in [12]. Hence, as was suggested in paper [13], in applications of the method to more complex structures, their division into elements with T-complete solution functions should be considered. This path was followed by Bussamra, Pimenta & Teixeira de Freitas in [4].

Application of the boundary collocation method for solution of other non-linear boundary value problems, for example, connected with linear diffusion reaction, large deflection of plates, functionally graded materials and gas flow in porous medium can be found in papers [1, 6, 8, 11].

References

[1] Balakrishnan, K. & Ramachadran, P.A., The method of fundamental solutions for linear diffusion-reaction equations. *Mathematical and Computer Modeling*, **31**, pp. 221–237, 2000.

[2] Banerjee, P.K., *Boundary Element Methods in Engineering Science*, McGraw-Hill: London, 1981.

[3] Brebbia, C.A., Telles, J.C.F. & Wrobel, L.C., *Boundary Element Techniques*, Springer-Verlag: New York, 1984.

[4] Bussamra, F.L.S., Pimenta, P.M. & Teixeira de Freitas, J.A., Hybrid-Trefftz stress elements for three-dimensional elastoplasticity. *Computer Assisted Mechanics and Engineering Sciences*, **8**, pp. 235–246, 2001.

[5] Collatz, L., *The Numerical Treatment of Differential Equations*, Springer: Berlin, 1960.

[6] Klekiel, T. & Kołodziej, J.A., Trefftz method for large deflection of plates with application of evolutionary algorithms. *Computer Assisted Mechanics and Engineering Sciences*, **13**, pp. 407–416, 2006.

[7] Kupradze, V.D., *Three-dimensional Problems of the Theory of Elasticity and Thermoelasticity*, North-Holland: Amsterdam, 1979.

[8] Martin, L. & Lesnic, D., The method of fundamental solutions for nonlinear functionally graded materials. *International Journal of Solids and Structures*, **44**, pp. 6878–6890, 2007.

[9] Nayak, G.C. & Zienkiewicz, O.C., Elasto-plastic stress analysis. A generalization for various constitutive relations including strain softening. *International Journal for Numerical Methods in Engineering*, **5**, pp. 113–135, 1972.

[10] Timoshenko, S., *Theory of Elasticity*, 3rd ed., Mc-Graw-Hill: New York, 1982.

[11] Uscıłowska, A. & Kołodziej, J.A., Solution of the nonlinear equation for isothermal gas flows in porous medium by Trefftz method. *Computer Assisted Mechanics and Engineering Sciences*, **13**, pp. 445–456, 2006.

[12] Zieliński, A.P. & Herrera, I., Trefftz method: Fitting boundary conditions. *International Journal for Numerical Methods in Engineering*, **24**, pp. 871–891, 1987.

[13] Zieliński, A.P. & Zienkiewicz, O.C., Generalized finite element analysis with T-complete boundary solution functions. *International Journal for Numerical Methods in Engineering*, **21**, pp. 509–528, 1985.

[14] Zieliński, A.P., Correct form of the influence functions in elasticity. *Communications in Applied Numerical Methods*, **3**, pp. 453–455, 1987.

[15] Zieliński, A.P., Trefftz method: elastic and elastoplastic problems. *Computer Methods in Applied Mechanics and Engineering*, **69**, pp. 185–204, 1988.

[16] Zienkiewicz, O.C. & Taylor, R.L., *The Finite Element Method*, 5th ed., Butterworth-Heinemann: Oxford, 2000.

[17] Zienkiewicz, O.C., Valliapan, S. & King, I.P., Elasto-plastic solutions of engineering problems, 'initial stress', finite element approach. *International Journal for Numerical Methods in Engineering*, **1**, pp. 75–100, 1969.

CHAPTER 13

Problems described by parabolic equations

1 The boundary collocation method in boundary–initial problems

1.1 Introduction

The number of papers in which the boundary collocation approach has been applied to solve boundary–initial problems, that is, problems in which differential equations are of parabolic type, is limited. Since almost all present parabolic equation applications refer to non-stationary temperature field in flat two-dimensional areas (exception is paper [15] where non-steady Stokes problem is considered), thus the boundary collocation method is presented here for that particular case.

Let us consider a boundary–initial problem as:

$$\nabla_1^2 T(\mathbf{x},t) = a\frac{\partial T(\mathbf{x},t)}{\partial t} \quad \text{in } \Omega \tag{1a}$$

$$T(\mathbf{x},t) = F(\mathbf{x}) \quad \text{on } \Gamma \tag{1b}$$

$$T(\mathbf{x},0) = G(\mathbf{x}) \quad \text{in } \Omega \tag{1c}$$

where $T(\mathbf{x},t)$ is temperature, t denotes time, $F(\mathbf{x})$, $G(\mathbf{x})$ are known functions, a is a constant depending on the material and ∇_1^2 is Laplace operator in a polar coordinate system.

Known procedures applied to solve boundary–initial problems by means of the boundary collocation method are based on:

- application of the boundary collocation to determine eigenvalues in solving the conductive equation [10, 12];
- application of the Laplace transformation to the time variable [2, 3, 7, 9, 14];

- substituting time derivative in eqn (1a) by difference quotient [2, 3];
- application of fundamental solutions to the conductive equation [1, 5, 7, 13].

1.2 The Ojalvo–Linzer method

The Ojalvo and Linzer method proposed by the authors in 1965 [10] and applied by Sparrow and Haji-Sheikh in 1968 [12] depends on substituting in the first stage:

$$T_\tau(\mathbf{x},t) = T(\mathbf{x},t) - T_z(\mathbf{x}), \tag{2}$$

where $T_z(\mathbf{x})$ is a solution of the boundary problem

$$\nabla_1^2 T_z(\mathbf{x}) = 0 \quad \text{in } \Omega \tag{3a}$$

$$T_z(\mathbf{x}) = F(\mathbf{x}) \quad \text{on } \Gamma \tag{3b}$$

The problem formulated in this way can be solved by the boundary collocation method described in Chapter 11.

Function $T_\tau(\mathbf{x},t)$ is a solution of the following initial–boundary problem:

$$\nabla_1^2 T_\tau(\mathbf{x},t) = a \frac{\partial T_\tau(\mathbf{x},t)}{\partial t} \quad \text{in } \Omega \tag{4a}$$

$$T_\tau(\mathbf{x},t) = 0 \quad \text{on } \Gamma \tag{4b}$$

$$T_\tau(\mathbf{x},0) = G(\mathbf{x}) - T_z(\mathbf{x}) \quad \text{in } \Omega \tag{4c}$$

Due to substitution defined by formula (2), the algorithm of initial–boundary solution is as follows:

1. Boundary problem (3) is solved by the boundary collocation method described in Chapter 6.
2. Initial–boundary problem solution is assumed in the form:

$$T_\tau(\mathbf{x},t) = \hat{T}(\mathbf{x}) \exp(-\gamma t), \tag{5}$$

where γ is a temporarily unknown constant. Substituting solution (5) into (4a) and (4b) leads to the following problem on eigenvalues:

$$\nabla_1^2 \hat{T}(\mathbf{x}) + a\gamma \hat{T}(\mathbf{x}) = 0 \quad \text{in } \Omega \tag{6a}$$

$$\hat{T}(\mathbf{x}) = 0 \quad \text{on } \Gamma \tag{6b}$$

3. Solutions of the problem (6) is assumed in the form:

$$\hat{T}(\mathbf{x}) = \sum_{i=1}^{M} b_i \psi_i(\mathbf{x}, a\gamma), \tag{7}$$

where ψ_i ($\mathbf{x}$,$a\gamma$) are trial functions, which exactly fulfil eqn (6a) and b_i are temporarily unknown constants. After substituting solution (7) to boundary condition (6b) and then fulfilling it in M points on the boundary Γ, we obtain a set of homogenous linear equations. The condition to get non-trivial solutions of the set is zeroing of its determinant:

$$\begin{vmatrix} \psi_1(\mathbf{x}_1, a\gamma) & \psi_2(\mathbf{x}_1, a\gamma) & \dots & \psi_M(\mathbf{x}_1, a\gamma) \\ \psi_1(\mathbf{x}_2, a\gamma) & \psi_2(\mathbf{x}_2, a\gamma) & \dots & \psi_M(\mathbf{x}_2, a\gamma) \\ \dots & \dots & \dots & \dots \\ \psi_1(\mathbf{x}_M, a\gamma) & \psi_2(\mathbf{x}_M, a\gamma) & \dots & \psi_M(\mathbf{x}_M, a\gamma) \end{vmatrix} = 0, \tag{8}$$

where $\mathbf{x}_i = (r_i, \theta_i)$, $i = 1, 2, ..., M$ are coordinates of the collocation points on the boundary Γ.

4. After determining a sequence of constants $\gamma_1, \gamma_2, ..., \gamma_M$ from non-linear equation solution (8), a solution of initial–boundary problem (4) is assumed in the form:

$$T_\tau(\mathbf{x}, t) = \sum_{j=1}^{Q} \sum_{i=1}^{M} b_{ij} \psi_i(\mathbf{x}, a\gamma_j) \exp(-\gamma_j t) \tag{9}$$

Constants b_{ij} are defined after fulfilling initial condition (4c) in Q. M points in area Ω. It leads to a linear equation set in the form:

$$\sum_{j=1}^{Q} \sum_{i=1}^{M} b_{ij} \psi_i(\tilde{\mathbf{x}}_k, a\gamma_j) = I(\tilde{\mathbf{x}}_k) - T_z(\tilde{\mathbf{x}}_k), \tag{10}$$

where $(\tilde{\mathbf{x}}_k)$, $k = 1, 2, .., Q \cdot M$ are collocation points in Ω.

Finally, it should be noticed that the described approach is not the pure BCM; the whole algorithm consists of the boundary and inner collocation.

1.3 Method based on the Laplace transformation

Another solution of initial–boundary problem (1) using the boundary collocation method, which has been applied in papers by Lo [9], Takeuti & Noda [14], Kołodziej Stefaniak & Kleiber [7], Chen, Golberg & Hon [2], Chen, Golberg & Muleshkov [3], depends on application of the Laplace transformation with regard to time variable. Limiting the case to zero initial conditions, though it is not necessary, we obtain the following algorithm:

1. Application of the Laplace transformation in eqn (1a) and boundary condition (1b) obtaining:

$$\nabla_1^2 \overline{T}(\mathbf{x}, p) = a p \overline{T}(\mathbf{x}, p) \tag{11a}$$

$$\overline{T}(\mathbf{x}, p) = \overline{F}(\mathbf{x}, p), \tag{11b}$$

where

$$\overline{T}(r,\theta,p) = \int_0^\infty \exp(-pt) T(r,\theta,p)\, dt. \tag{11c}$$

2. Search for the boundary problem solution (11a and b) in the form:

$$\overline{T}(\mathbf{x}, p) = \sum_{i=1}^{M} a_i(p) \phi_i(\mathbf{x}, p), \tag{12}$$

where $\phi_i(r,\theta,p)$ are trial functions which exactly fulfil eqn (11a), and $a_i(p)$ are temporarily unknown functions of transformation p.

3. Choice of a certain discrete value set of transformation parameters $(p_1, p_2, \ldots, p_N)$ [11] and fulfilling boundary condition (11b) in M points on the boundary Γ. Thus, N linear equation sets in the form:

$$\sum_{i=1}^{M} a_i(p_k) \phi_i(\mathbf{x}_j, p_k) = \overline{F}(\mathbf{x}_j, p_k) \qquad j = 1,\ 2,\ldots,\ M,\quad k = 1,\ 2,\ldots,\ N \tag{13}$$

are obtained.

4. Determination of functions $a_i(p_k)$ allows to calculate the initial–boundary problem solution by numerical inversion of the Laplace transform.

1.4 The boundary collocation method with use of time discretization

Chen, Golberg & Hon [2] and Chen, Golberg & Muleshkov [3] used the finite difference method for discretizing time to solve the initial–boundary problem with help of the BCM. They introduced a discretized time variable in form $t_n = n\Delta t,\ \ n \geq 0$, where $\Delta t > 0$. For $t_n \leq t \leq t_{n+1}$, temperature and its derivative approximation with regard to time was defined by:

$$T(\mathbf{x},t) \approx \chi T(\mathbf{x},t_{n+1}) + (1-\chi) T(\mathbf{x},t_n) \tag{14}$$

and

$$\frac{\partial T(\mathbf{x},t)}{\partial t} \approx \frac{T(\mathbf{x},t_{n+1}) - T(\mathbf{x},t_n)}{\Delta t}, \tag{15}$$

where $(0 \leq \chi \leq 1)$. After substituting formulas (14) and (15) into eqn (1a) and introducing designation $T(\mathbf{x},t_n) = T_n$, they obtained:

$$\chi \nabla_1^2 T_{n+1} + (1-\chi)\nabla_1^2 T_n = \frac{a(T_{n+1} - T_n)}{\Delta t} \tag{16}$$

and after some transformations

$$\nabla_1^2 T_{n+1} - \frac{a}{\chi \Delta t} T_{n+1} = -\frac{a}{\chi \Delta t} T_n - \frac{(1-\chi)\nabla_1^2 T_n}{\chi}. \tag{17}$$

The last equation may be regarded as a non-homogenous, modified Helmholtz's equation.

In this way, the initial–boundary problem has been transformed to a solution, in each time step, of a non-homogenous boundary value problem. The non-homogenous part is a solution function from the previous time step; for the first step the initial condition is used. Hence, applying the procedure described in Chapter 11, the problem step by step can be solved.

1.5 Method based on fundamental solutions of heat transfer equation

In this method, for the first time presented in Stefaniak's paper [13], the author assumes that concentrated heat sources with intensity W_{ik} are active in $\mathbf{p}_i$ points situated outside the area Ω and that they change jump-way in time. The temperature distribution resulted from those sources is following:

$$T(\mathbf{x},t)=\sum_{i=1}^{N}\sum_{k=0}^{M}W_{ik}G(\mathbf{x},\mathbf{p}_i)\eta(t-t_k), \tag{18}$$

where W_{ik} are temporarily unknown constants and functions G are fundamental solutions of eqn (1a) defined as

$$G(\mathbf{x},\mathbf{p}_i,t,t_k)=\frac{1}{4\pi}\mathrm{Ei}\left(-\frac{(x-\xi_i)^2+(y-\zeta_i)^2}{\sqrt{t-t_k}}\right), \tag{19}$$

where (see [8], p. 39):

$$\mathrm{Ei}(-z)=-\int_z^{\infty}\frac{1}{u}\exp(-u)du. \tag{20}$$

In eqn (19) (x,y) are coordinates of point $\mathbf{x}$ and (ξ_i,ζ_i) are coordinates of the source points $\mathbf{p}_i$.

Temperature defined in expression (18) fulfils identically eqn (1a) and initial condition (1b). After choosing M time moments and fulfilling boundary condition in N points $\mathbf{x}_j$ on boundary Γ (for those moments), $N{\cdot}M$ linear equations on unknown constants W_{ik} in the form:

$$\sum_{i=1}^{N}\sum_{k=0}^{M}W_{ik}G(\mathbf{x}_j,\mathbf{p}_i,t_l,t_k)\,\eta\,(t_l-t_k)=F(\mathbf{x}_j) \qquad l=0,1,2,\ldots M-1 \quad j=1,2,\ldots N \tag{21}$$

can be obtained. In this case, we have the boundary collocation combined with the time collocation procedure.

2 The boundary collocation for non-stationary temperature field; comparison with the FEM

2.1 Formulation of the example

Let us consider temperature field within a cooled, rectangular cylinder. Determination of transient temperature field leads to solution of an initial–boundary problem in the form of conductivity equation:

$$\frac{\partial^2\phi}{\partial X^2}+\frac{\partial^2\phi}{\partial Y^2}=\frac{\partial\phi}{\partial\tau} \quad \text{in } 0<X<1,\ \ 0<Y<D \tag{22a}$$

with initial and boundary conditions

$$\phi = 0 \text{ and } \tau = 0 \quad \text{for } 0 \le X \le 1,\ \ 0 \le Y \le D \tag{22b}$$

$$\phi = -1 \text{ and } \tau > 0 \quad \text{for X} = 1,\ \ 0 < Y < D \tag{22c}$$

$$\phi = -1 \text{ and } \tau > 0 \quad \text{for } Y = D,\ \ 0 < X < 1 \tag{22d}$$

$$\frac{\partial \phi}{\partial Y} = 0 \text{ and } \tau > 0 \quad \text{for } Y = 0,\ \ 0 < X < 1 \tag{22e}$$

$$\frac{\partial \phi}{\partial X} = 0 \text{ and } \tau > 0 \quad \text{for } X = 0,\ \ 0 < Y < D \tag{22f}$$

where ϕ is a non-dimensional temperature, $\tau = \lambda t / \rho c_p l^2$ is a non-dimensional time (Fourier's number), λ is a thermal conductivity coefficient, t is dimensional time, l is a characteristic length (half of rectangle side), ρ is material density, c_p – specific heat and X, Y are non-dimensional coordinates. This initial–boundary problem has exact the analytical solution in form:

$$\phi = -1 + \frac{16}{\pi^2} \sum_{m=1}^{\infty} \sum_{n=1}^{\infty} \frac{(-1)^{m+n+2} \cos\left[(2m-1)\pi X/2\right] \cos\left[(2n-1)\pi Y/2\right]}{(2m-1)(2n-1)} \times \exp\left\{ -\left[(2m-1)^2 + (2n-1)^2 D^2\right] \frac{\pi^2 \tau}{4} \right\}. \tag{23}$$

In non-dimensional coordinates, the cylinder rectangle cross-section has dimensions $2 \times 2D$; because of symmetry we consider only its one quarter. Problems I, II and III are defined according to values D ($D = 1$, $D = 0.5$ or $D = 0.25$, respectively). We present the solution of this problem with application of two boundary collocation variants, namely, the method based on Laplace's transformation and another – with the heat conductivity fundamental solutions. These results are compared below with the exact solution (23) and with one obtained using the finite element method (FEM).

2.2 The BCM with Laplace's transformation (BCL)

After application of the Laplace transformation in eqn (22a) and in boundary condition (22c–f), applying initial condition (22b), we obtain a boundary problem expressed by the equation:

$$\left(\frac{\partial^2}{\partial X^2} + \frac{\partial^2}{\partial Y^2} \right) \tilde{\phi}(X, Y, p) = p \tilde{\phi}(X, Y, p) \quad \text{in } 0 < X < 1,\ \ 0 < Y < D \tag{24a}$$

with boundary conditions

$$\tilde{\phi} = -\frac{1}{p} \quad \text{for } X = 1,\ 0 < Y < D \tag{24b}$$

$$\tilde{\phi} = -\frac{1}{p} \quad \text{for } Y = D,\ 0 < X < 1 \tag{24c}$$

$$\frac{\partial \tilde{\phi}}{\partial X} = 0 \quad \text{for } X = 0,\ 0 < Y < D \tag{24d}$$

$$\frac{\partial \tilde{\phi}}{\partial X} = 0 \quad \text{for } X = 0,\ 0 < Y < D \tag{24e}$$

where

$$\tilde{\phi}(X,Y,p) = \int_0^\infty \exp(-p\tau)\tilde{\phi}(X,Y,\tau)d\tau. \tag{24f}$$

We assume a solution of the boundary value problem (24) in the form:

$$\tilde{\phi} = \frac{1}{2\pi}\sum_{j=1}^{N} a_j(p)\varphi\left(X,Y,\xi_j,\zeta_j,p\right), \tag{25}$$

where $a_j(p)$ are temporarily unknown functions of transformation parameter p and the trial functions ϕ are as follows:

$$\begin{aligned}\varphi\left(X,Y,\xi_j,\zeta_j,p\right) = {} & K_0\left(\left[p\left(\left(X-\xi_j\right)^2+\left(Y-\zeta_j\right)^2\right)\right]^{\frac{1}{2}}\right) \\ & + K_0\left(\left[p\left(\left(X+\xi_j\right)^2+\left(Y-\zeta_j\right)^2\right)\right]^{\frac{1}{2}}\right) \\ & + K_0\left(\left[p\left(\left(X-\xi_j\right)^2+\left(Y+\zeta_j\right)^2\right)\right]^{\frac{1}{2}}\right) \\ & + K_0\left(\left[p\left(\left(X+\xi_j\right)^2+\left(Y+\zeta_j\right)^2\right)\right]^{\frac{1}{2}}\right),\end{aligned} \tag{26}$$

where K_0 are the modified Bessel functions of the second kind.

Points with coordinates (ξ_j,ζ_j) are situated outside the considered area. The arrangement of the source and collocation points is presented in Fig. 1.

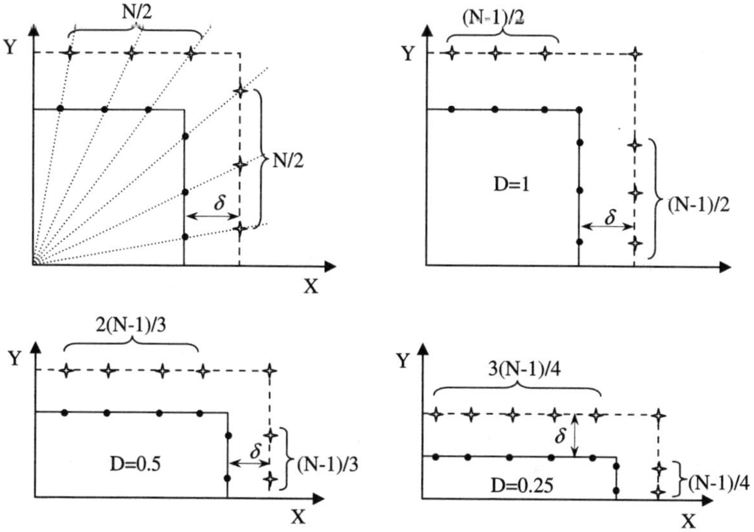

Figure 1 : Positions of source and collocation points.

Solution (25) exactly fulfils differential equation (24a) and boundary conditions (24d and e) resulting from symmetry of the problem.

Now, the solution of the problem means determination of functions $a_j(p)$ and then the inverse Laplace transformation. Analytically, the transform can be inverted only in case of particular, simple transform functions. In this example, the searched Laplace transform functions are given in discrete form, so the only possible method is their numerical inversion. According to the paper by Davies & Martin [4], there is no universal method of the numerical inversion suitable for any given function. Thus, we should consider character of the inverted function while choosing the method.

Since we consider the problem of transient temperature field, we have chosen the method suggested by Shapery [11] for similar problems of determination of stresses in material with visco-elastic properties. This method is based on choice of finite sequence of real values of transformation parameters $P_1, P_2, \ldots, P_M$. After substituting those values to eqn (25) and then using boundary conditions (24b and c), we obtain:

$$\sum_{j=1}^{N} a_{jk}\varphi\left(X^{(0)}, Y^{(0)}, \xi_j, \zeta_j, p_k\right) = -\frac{2\pi}{p_k}, \quad k = 1,2,\ldots,M \tag{27}$$

where $(X^{(0)}, Y^{(0)})$ are coordinates of points on the boundary $X = 1$, $0 < Y < D$ and $Y = D$, $0 < X < 1$, whereas $a_{jk} = a_j(p_k)$.

One of the methods to determine unknown constants a_{jk} is collocational fulfilling of boundary conditions (24b and c). After selecting N collocation points on the boundary $X = 1$ and $Y = D$, we get:

$$\sum_{j=1}^{N} a_{jk}\,\varphi\left(X_i^{(0)}, Y_i^{(0)}, \xi_j, \zeta_j, p_k\right) = -\frac{2\pi}{p_k}, \quad k = 1, 2, \ldots, M \;\; i = 1, 2, \ldots, N. \tag{28}$$

Having discrete functions $a_j\,(P)$, we assume temperature field in the form:

$$\varphi(X, Y, \tau) = -1 + \sum_{l=1}^{M} c_l\,(X, Y)\exp(-b_l \tau) \tag{29}$$

where $c_l\,(X, Y)$ are unknown functions and b_l are unknown constants.

After applying the Laplace transformation in eqn (29), we obtain:

$$\tilde{\varphi}(X, Y, p) = -\frac{1}{p} + \sum_{l=1}^{M} \frac{c_l\,(X, Y)}{p + b_l}. \tag{30}$$

Now comparing right-hand sides of eqns (30) and (25), we get:

$$-\frac{1}{p} + \sum_{l=1}^{M} \frac{c_l\,(X, Y)}{p + b_l} = \frac{1}{2\pi} \sum_{j=1}^{N} a_j\,(p)\varphi\left(X, Y, \xi_j, \zeta_j, p\right). \tag{31}$$

After assuming b_l equals P_l (as it was stated in Shapery's paper [11], where encouraging results had been achieved) and after assigning coordinates of the spatial points (X_k, Y_k), we obtain a set of linear equations on unknown constants $c_1^{(k)}, c_2^{(k)}, \ldots, c_M^{(k)}$ in the form:

$$\mathbf{Ac} = \mathbf{b}, \tag{32a}$$

where

$$A_{ij} = \frac{1}{p_i + p_j}, \quad i, j = 1, 2, \ldots, M \tag{32b}$$

$$b_i = \frac{1}{p_i} + \frac{1}{2\pi} \sum_{j=1}^{N} a_{ji}\,\varphi\left(X_k, Y_k, \xi_j, \zeta_j, p_i\right). \tag{32c}$$

The algorithm presented above shows that in order to determine temperature in any selected point, it is necessary each time to solve a different equation set (32a).

2.3 The BCM with fundamental solutions as trial functions (BCF)

Let us consider a solution of the problem (22) in form of linear superposition of fundamental solutions (18):

$$\phi(X,Y,\tau)=\sum_{i=1}^{N}\sum_{k=0}^{M}W_{ik}G(X,Y,\xi_i,\zeta_i,\tau,\tau_k)\eta(\tau-\tau_k), \tag{33}$$

where W_{ik} are temporarily unknown constants, and functions G are the equation fundamental solutions expressed in the form:

$$\begin{aligned} G(X,Y,\xi_i,\zeta_i,\tau,\tau_k) = {} & \frac{1}{4\pi}\mathrm{Ei}\left(-\frac{(X-\xi_i)^2+(Y-\zeta_i)^2}{\sqrt{\tau-\tau_k}}\right) \\ & +\frac{1}{4\pi}\mathrm{Ei}\left(-\frac{(X+\xi_i)^2+(Y-\zeta_i)^2}{\sqrt{\tau-\tau_k}}\right) \\ & +\frac{1}{4\pi}\mathrm{Ei}\left(-\frac{(X-\xi_i)^2+(Y+\zeta_i)^2}{\sqrt{\tau-\tau_k}}\right) \\ & +\frac{1}{4\pi}\mathrm{Ei}\left(-\frac{(X+\xi_i)^2+(Y+\zeta_i)^2}{\sqrt{\tau-\tau_k}}\right), \end{aligned} \tag{34}$$

where Ei(...) has been defined by (20). This form of the solution exactly fulfils conditions (22e and f) resulting from symmetry of the problem. After having fulfilled boundary conditions (22c and d) within a finite number of points on the boundary Γ (for a finite number of moments), we obtain:

$$\begin{gathered}\sum_{i=1}^{N}\sum_{k=0}^{M}W_{ik}G(X_j,Y_j,\xi_i,\zeta_i,\tau_l,\tau_k)\eta(\tau_l-\tau_k)=-1 \\ l=0,1,2,\ldots M-1 \quad j=1,2,\ldots N.\end{gathered} \tag{35}$$

Now, after having solved the above linear system, we get the solution in the closed form (33).

In order to present accuracy of the boundary collocation method, the above problem has been solved by means of the FEM, using a variant of this approach described by Kleiber & Służalec [6]. We apply division of the considered area into rectangular elements with four or eight nodes (as shown in Fig. 2). The matrix equation for transient heat transfer can be written as:

$$\mathbf{C}\frac{d}{d\tau}[\phi]+\mathbf{K}[\phi]=\mathbf{q}, \tag{36}$$

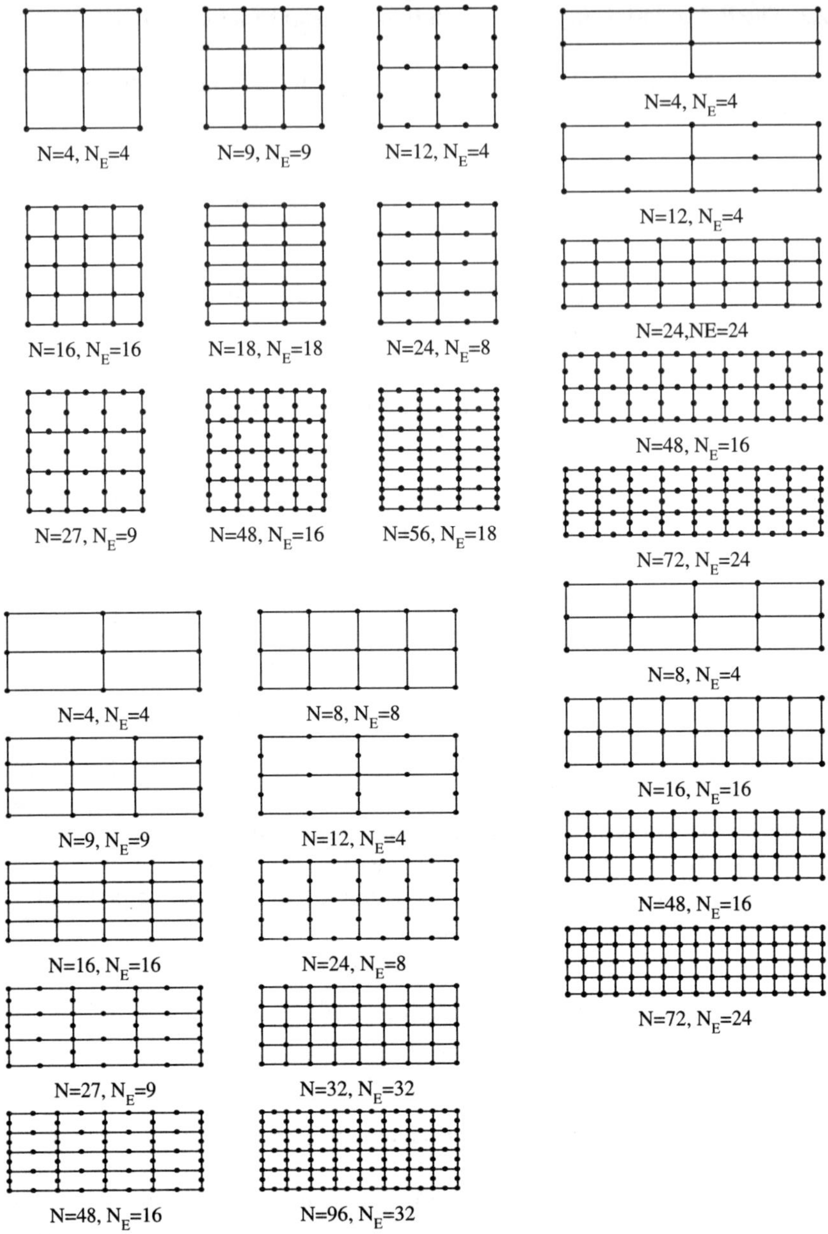

Figure 2 : Division of investigated area into finite elements; N – number of active degrees of freedom, N_E – number of elements.

where **C** and **K** are global volume and heat transfer matrices, **q** is a heat load vector and $[\phi]$ is a vector of non-dimensional temperature nodal values.

In order to compare the results obtained by the described boundary collocation method with the exact solution (23) and with the results from the FEM, two different error measures have been assumed. The first one, global measure, has been defined as:

$$\varepsilon_g = \frac{1}{N_T \cdot N_P} \sum_{i=1}^{N_T} \sum_{j=1}^{N_P} \frac{\left|\phi_{ex}\left(X_j, Y_j, \tau_i\right) - \phi_a\left(X_j, Y_j, \tau_i, N\right)\right|}{\left|\phi_{ex}\right|_{\max}}, \tag{37}$$

where index *ex* means a value of the exact solution (23) and index *a* means solution obtained by one of the approximate methods. Points (X_j,Y_j) with error calculated for the boundary collocation method are uniformly located on the area Ω. For the finite element solution, the error is calculated in the nodes.

The second measure is the maximum local error value:

$$\varepsilon_l = \max \frac{\left|\phi_{ex}\left(X_j, Y_j, \tau_i\right) - \phi_a\left(X_j, Y_j, \tau_i, N\right)\right|}{\left|\phi_{ex}\right|_{\max}}. \tag{38}$$

As we can observe, the following conclusions come from the comparison in Tables 1–5:

- Errors in the BCF for given number of collocation points depend on distance δ from boundary Γ to the line with source points (Table 1). The choice of optimal distance δ is a non-linear problem and it is not discussed in this example (see Section 3.2 in Chapter 11).

Table 1: Errors ε_g (37) and ε_l (38) in BCF versus distance δ of sources from boundary Γ (problem II – D = 0.5, $N = 17$).

δ	ε_g	ε_l
0.005	0.0656	0.272
0.01	0.0483	0.247
0.02	0.0331	0.226
0.03	0.0261	0.219
0.04	0.0225	0.219
0.05	0.0205	0.221
0.06	0.0194	0.225
0.07	0.0189	0.229
0.09	0.0186	0.240
0.12	0.0189	0.257

Table 2: Comparison of FEM, BCL and BCF errors – problem I – D = 1.0, (time step for FEM $\Delta\tau = 0.05$, number of time steps $N_T = 20$; in case of FEM N is a number of active degrees of freedom – see Fig. 2).

	FEM		BCL		BCF	
			$\delta = 0.2$		$\delta = 0.04$	
N	ε_g	ε_l	ε_g	ε_l	ε_g	ε_l
4	0.0184	0.0716	–	–	–	–
7	–	–	0.00208	0.00941	0.0668	0.385
9	0.0175	0.112	0.000743	0.00756	0.0507	0.215
11	–	–	0.000244	0.000536	0.0335	0.247
12	0.0144	0.0977	–	–	–	–
13	–	–	0.000139	0.00222	0.0269	0.234
15	–	–	0.0000990	0.00121	0.0236	0.225
16	0.176E–1	0.165	–	–	–	–
17	–	–	0.0000927	0.000189	0.0225	0.219
18	0.175E–1	0.0163	–	–	–	–
19	–	–	0.0000865	0.00121	0.0198	0.215
23	–	–	0.0000852	0.00121	0.0182	0.211
24	0.158E–1	0.170	–	–	–	–
25	–	–	0.0000857	0.00121	0.0180	0.210
27	0.162E–1	0.166	0.0000854	0.00121	0.0174	0.209
31	–	–	–	–	0.170E–1	0.209
48	0.172E–1	0.195	–	–	–	–
56	0.172E–1	0.192	–	–	–	–
64	0.184E–1	0.169	–	–	–	–

- In both methods (BCL and BCF), the global error and the maximum local error decrease monotonously with the growth of number of collocation points. In the FEM, the decrease of errors is not so regular. However, it should be noted that the number of finite elements is relatively small.
- In all the compared methods, the maximal local error can be noticed in the first time step (Table 5). The basic reason of this phenomenon is the temperature jump in moment $t = 0$. Particularly sensitive to that jump is the boundary collocation approach with fundamental solutions as trial functions (BCF).

Table 3: Comparison of FEM, BCL and BCF errors – problem II – D = 0.5, (time step for FEM $\Delta\tau$ = 0.05, number of time steps N_T = 20; in case of FEM N is a number of active degrees of freedom – see Fig. 2).

	FEM		BCL		BCF	
N	ε_g	ε_1	ε_g	ε_1	ε_g	ε_1
4	0.0178	0.0865	–	–	–	–
7	–	–	0.00160	0.0316	0.0517	0.271
8	0.0168	0.0698	–	–	–	–
10	–	–	0.000427	0.00645	0.0485	0.263
12	0.0164	0.125	–	–	–	–
13	–	–	0.000275	0.00362	0.0236	0.224
16	0.0174	0.144	0.000216	0.00362	0.0218	0.223
19	–	–	0.000255	0.00362	0.0189	0.223
22	–	–	0.000265	0.00362	0.0182	0.222
27	0.0171	0.163	–	–	0.0173	0.222
32	0.0178	0.174	–	–	–	–
48	0.0154	0.197	–	–	–	–

Table 4: Comparison of FEM, BCL and BCF errors – problem III – D = 0.25, (time step for FEM $\Delta\tau$ = 0.05, number of time steps N_T = 20; in case of FEM N is a number of active degrees of freedom – see Fig. 2).

	FEM		BCL		BCF	
N	ε_g	ε_l	ε_g	ε_l	ε_g	ε_l
4	0.0170	0.0570	–	–	–	–
8	0.0167	0.0805	–	–	–	–
11	–	–	0.00367	0.111	0.0201	0.300
12	0.0286	0.0221	–	–	–	–
16	0.0169	0.104	0.00215	0.0640	0.0192	0.245
21	–	–	0.00154	0.04006	0.0182	0.232
24	0.0175	0.141	–	–	–	–
26	–	–	0.00121	0.0327	0.0177	0.229
31	–	–	0.00102	0.0275	0.0175	0.227
36	0.0177	0.170	0.000889	0.0237	–	–
41	–	–	0.000784	0.0209	–	–

Table 5: Local maximum error ε_l versus non-dimensional time τ; problem I – D = 1.0, $N = 9$.

τ	BCF	BCL	FEM
0.05	0.271	0.00795	0.112
0.10	0.170	0.00759	0.107
0.15	0.118	0.00130	0.0735
0.20	0.104	0.00224	0.0512
0.25	0.101	0.00148	0.0562
0.30	0.0929	0.00193	0.0637
0.35	0.0834	0.00249	0.0672
0.40	0.0739	0.00225	0.0657
0.45	0.0652	0.00163	0.0615
0.50	0.0574	0.009088	0.0561
0.55	0.0505	0.000436	0.0502
0.60	0.0446	0.000411	0.044
0.65	0.0396	0.000504	0.0388
0.70	0.0353	0.000568	0.0336
0.75	0.0316	0.000600	0.0290
0.80	0.0285	0.000604	0.0248
0.85	0.0268	0.000583	0.0212
0.90	0.0261	0.000543	0.0179
0.95	0.0255	0.000491	0.0152
1.00	0.0249	0.000431	0.0128

References

[1] Chantasiriwan, S., Method of fundamental solutions for time-dependent heat conduction problems. *International Journal for Numerical Methods in Engineering*, **66**, pp. 147–165, 2006.

[2] Chen, C.S., Golberg, M.A. & Hon, Y.C., The method of fundamental solutions and quasi-Monte-Carlo method for diffusion equations. *International Journal for Numerical Methods in Engineering*, **43**, pp. 1421–1435, 1998.

[3] Chen, C.S., Golberg, M.A. & Muleshkov, A.S., The method of fundamental solutions for time dependent problems. In: *Proceedings of Thirteenth International Conference on Boundary Element Technology: 'Boundary Element Technology XII'*, eds. C.S. Chen, C.A. Brebbia, D.W. Pepper, WITpress: Southampton, pp. 377–386, 1999.

[4] Davies, B. & Martin, B., Numerical inversion of the Laplace transform: a survey and comparison of methods. *Journal of Computational Physics*, **33**, pp. 1–32, 1979.

[5] Dong, C.F., Sun, F.Y. & Meng, B.Q., A method of fundamental solutions for inverse heat conduction problems in anisotropic medium equations. *Engineering Analysis with Boundary Elements*, **31**, pp. 75–82, 2007.

[6] Kleiber, M. & Śluzalec, A.S., Numerical analysis of heat flow in flash welding. *Archives of Mechanics*, **35**, pp. 687–699, 1983.

[7] Kołodziej, J.A., Stefaniak, J. & Kleiber, M., Transient heat conduction by boundary collocation methods and FEM – a comparison study. In: *Numerical Techniques for Boundary Element Methods. Proceedings of the Seventh GAMM-Seminar Kiel, January 25–27, 1991, Notes on Numerical Fluid Mechanics*, **33**, pp. 104–115, 1992.

[8] Lebiediew, N.N., *Special Functions and their Application* (in Polish), PWN: Warszawa, 1957.

[9] Lo, C.F., Numerical solution of the unsteady heat equation. *AIAA Journal*, **7**, pp. 973–975, 1969.

[10] Ojalvo, L.U. & Linzer, F.D., Improved point-matching techniques. *Quarterly Journal of Mechanics & Applied Mathematics*, **18**, pp. 41–56, 1965.

[11] Shapery, R.A., Approximate methods of transform inversion for viscoelastic stress analysis. In: *Proceedings 3rd U.S. National Congress on Applied Mechanics*, **2,** pp. 1075–1085, 1962.

[12] Sparrow, E.M. & Haji-Sheikh, A., Transient and steady heat conduction in arbitrary bodies with arbitrary boundary and initial conditions. *Journal of Heat Transfer*, **90**, pp. 103–108, 1968.

[13] Stefaniak, J., Controlling the concentrated sources in some problems of heat conduction. *Journal of Technical Physics*, **26**, pp. 349–358, 1985.

[14] Takeuti, Y. & Noda, N., Transient thermoelastic problem in a polygonal cylinder with a circular hole. *Journal of Applied Mechanics*, **40**, pp. 935–940, 1973.

[15] Tsai, C.C., Young, D.L., Fan, C.M. & Chen, C.W., MFS with time-dependent fundamental solutions for unsteady Stokes equations. *Engineering Analysis with Boundary Elements*, **30**, pp. 897–908, 2006.

CHAPTER 14

Eigenvalue problems

1 Problem formulation and general description of the method

Eigenvalue problem leads to search of non-zero solutions λ and function $u(\mathbf{x})$, which fulfil the equation:

$$Du(\mathbf{x}) + \lambda p(\mathbf{x})u(\mathbf{x}) = 0 \quad \text{in } \Omega, \tag{1a}$$

with boundary conditions

$$Bu(\mathbf{x}) = 0 \quad \text{on} \Gamma, \tag{1b}$$

where D is an elliptical operator and B is a known linear operator.

Applying the boundary collocation method to solve that problem means assuming the solution in the form:

$$u(\mathbf{x}) = \sum_{i=1}^{N} a_i \phi_i(\mathbf{x}, \lambda), \tag{2}$$

where $\phi_i(\mathrm{x},\lambda)$ are trial functions, which exactly fulfil eqn (1a) and a_i are unknown constants.

After choosing N collocation points on the boundary Γ and fulfilling boundary condition (1b) in those points, we obtain homogenous equation set:

$$\sum_{i=1}^{N} a_i B\phi_i(\mathbf{x}_j, \lambda) = 0 \quad j = 1, 2, \ldots, N, \tag{3}$$

where $\mathbf{x}_1, \ldots, \mathbf{x}_j, \ldots, \mathbf{x}_N$ are collocation points.

This homogenous, linear equation set has a non-trivial solution, if its determinant equals zero:

$$\begin{vmatrix} \phi_1(\mathbf{x}_1,\lambda) & \phi_2(\mathbf{x}_1,\lambda) & \dots & \phi_N(\mathbf{x}_1,\lambda) \\ \phi_1(\mathbf{x}_2,\lambda) & \phi_2(\mathbf{x}_2,\lambda) & \dots & \phi_N(\mathbf{x}_2,\lambda) \\ \dots & \dots & \dots & \dots \\ \phi_1(\mathbf{x}_N,\lambda) & \phi_2(\mathbf{x}_N,\lambda) & \dots & \phi_N(\mathbf{x}_N,\lambda) \end{vmatrix} = 0. \tag{4}$$

After solution of non-linear equation (4). we obtain a sequence of constants λ_1, $\lambda_2,\dots,\lambda_N$. After determination of those constants, eigenfunctions are calculated in the next stage.

2 Example. Determination of eigenvalues for a silencer

The boundary collocation method was applied for determination of eigenvalues regarding various mechanical and electrical problems, for example frequencies of free vibrations of membranes, plates and waveguides. Table 1 presents related papers. Below, as an example, determination of eigenvalues in case of a combustion engine silencer has been presented [24].

Let us consider an infinite cylinder channel with an arbitrary cross-section, divided into two sub-areas of various acoustic properties. Since we model the engine silencer, we assume that the internal area contains exhaust fumes, whereas the external area is an absorber, which is porous and filled with gases. Because of geometry, it is easier to apply a cylindrical coordinate system (r,θ,z) with channel axis z, as presented in Fig. 1.

We assume that the acoustic problem is stationary and the area is homogenous and acoustically isotropic. Hence, from linear acoustics, we obtain:

$$\nabla^2\Psi_1 + k_1^2\Psi_1 = 0 \quad \text{in interior 3D domain } \Omega_1 \tag{5a}$$

Table 1: List of works, in which the boundary collocation method has been applied to determine eigenvalues.

Eigenvalue problems	References
Eigenfrequencies of membrane vibrations	[6, 7, 10, 13, 16, 19, 20, 22, 30, 38, 40, 45–47]
Eigenfrequencies of thin plate vibrations	[1, 11, 12, 14, 15, 17, 21, 26, 32, 34, 37, 42, 44]
Eigenfrequencies of waveguides	[4, 5, 18, 35, 36, 39, 41, 43, 48–53]
Others	[2, 3, 8, 9, 15–17, 23–25, 27–29, 31, 33]

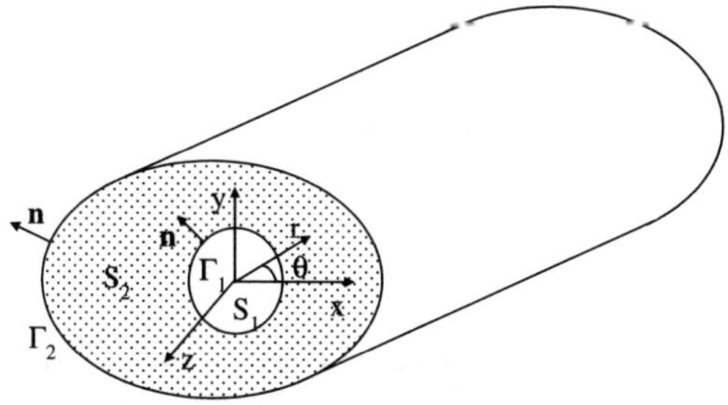

Figure 1: Scheme of silencer.

$$\nabla^2\Psi_2 + k_2^2\Psi_2 = 0 \quad \text{in exterior 3D domain } \Omega_2, \tag{5b}$$

where $\Psi_{1,2}(r,\theta,z,\omega)$ is the Fourier transform with regard to time, that is $\Psi(\omega)=\int_{\infty}^{\infty}\phi(t)e^{i\omega t}dt$, (1b) and acoustic velocity $\mathbf{v}$ is related to the potential ϕ as: $\mathbf{v} = -\nabla\phi$ Besides, ∇ is a three-dimensional nabla operator and $k_{1,2} = \omega/c_{1,2}$ are wavenumbers, which can be real or complex. External area is treated as dissipative liquid, hence, it is characterized by complex density ρ_2 and complex velocity c_2. The internal area is assumed to be non-dissipative, thus ρ_1 and c_1 are real.

The Fourier spatial transform $\psi(k_z) = \int_{-\infty}^{\infty}\Psi(z)\exp(-ik_z z)dz$ leads to two-dimensional (2D) problem on eigenvalues for the velocity potential defined in the whole cross-section:

$$\psi(r,\theta,k_z,\omega)=\begin{cases}\psi_1(r,\theta,k_z,\omega), & (r,\theta)\in S_1\\ \psi_2(r,\theta,k_z,\omega), & (r,\theta)\in S_2\end{cases}, \tag{6}$$

where S_1 and S_2 mean internal and external part of the cross-section. In each section, the potential is governed by the equations:

$$\nabla_\perp^2\psi_1 + k_{\perp 1}^2\psi_1 = 0 \tag{7a}$$

$$\nabla_\perp^2\psi_2 + k_{\perp 2}^2\psi_2 = 0, \tag{7b}$$

where $\nabla_\perp$ is a 2D del operator in polar coordinate system $(\mathbf{e}_r,\mathbf{e}_\theta)$:

$$\nabla_\perp = \mathbf{e}_r\frac{\partial}{\partial r} + \mathbf{e}_\theta\frac{1}{r}\frac{\partial}{\partial\theta}. \tag{7c}$$

Transversal and longitudinal wavenumbers are connected by following relations:

$$k_{\perp 1}^2 = k_1^2 - k_z^2 \tag{8a}$$

$$k_{\perp 2}^2 = k_2^2 - k_z^2. \tag{8b}$$

The internal area is limited by the curve Γ_1 given in the form of a parametrical equation as $r = a(\theta)$, and the whole cross-section area is bounded by contour Γ_2 presented in the form $r = b(\theta)$. If we assume that external surface of the silencer is rigid, the boundary condition on that surface will be defined as

$$\nabla_{\perp} \psi_2 \cdot \mathbf{n}\big|_{r=b} = 0, \tag{9}$$

where **n** is an externally perpendicular vector (Fig. 1). On the interface surface between the two areas, continuity of acoustic pressure and the normal velocity component is assumed:

$$\rho_1 \psi_1 (a,\theta) = \rho_2 \psi_2 (a,\theta) \tag{10}$$

$$\nabla_{\perp} \psi_1 \cdot \mathbf{n}\big|_{r=a} = \nabla_{\perp} \psi_2 \cdot \mathbf{n}\big|_{r=a}. \tag{11}$$

The functions $\psi_{1,2}$ and their derivatives must be periodical in a polar coordinate system, namely:

$$\left.\frac{\partial^n \psi_{1,2}}{\partial \theta^n}\right|_{\theta} = \left.\frac{\partial^n \psi_{1,2}}{\partial \theta^n}\right|_{\theta \pm m2\pi}, \quad m = 1,\ 2, \ldots, \quad n = 0,\ 1. \tag{12}$$

Generally, the eigenvalue problem presented in eqns (7–12) cannot be solved in an exact way. In the considered case, the boundary collocation method is applied and the following forms of solutions are assumed:

$$\psi_1 = A_0 J_0 (k_{\perp 1} r) + \sum_{n=1}^{N} J_n (k_{\perp 1} r)(A_n \cos n\theta + B_n \sin n\theta) \tag{13a}$$

$$\begin{aligned} \psi_2 = {} & C_0 J_0 (k_{\perp 2} r) + E_0 N_0 (k_{\perp 2} r) \\ & + \sum_{n=1}^{M} J_n (k_{\perp 2} r)(C_n \cos n\theta + D_n \sin n\theta) \\ & + \sum_{n=1}^{M} N_n (k_{\perp 2} r)(E_n \cos n\theta + F_n \sin n\theta), \end{aligned} \tag{13b}$$

where J_n is the Bessel function of order n, N_n is Neuman's function of the same order and $A_n \div F_n$ are unknown coefficients.

It should be noticed that wavenumbers $k_{\perp 1}$ and $k_{\perp 2}$ in solutions (13) are unknown. For any contour shapes Γ_1 and Γ_2, we determine those numbers substituting the solutions to boundary conditions (9–11). When internal contour Γ_1 is a circle, boundary conditions (10 and 11) can be fulfilled in an exact way. Then we assume in solutions (13) that $N = M$. After situating P collocation points on contour Γ_2 by defining $b_j = b(\theta_j)$, $j = 1,2,\ldots,P$ and then fulfilling boundary condition (9) in those points, we obtain a homogenous linear equation system in the form:

$$\begin{aligned} & A_0 k_{\perp 2}\left[K_{1,0} J_0'(k_{\perp 2} b_j) + K_{2,0} N_0'(k_{\perp 2} b_j)\right] \\ & + \sum_{n=1}^{N} A_n \left[k_{\perp 2}[K_{1,n} J_n'(k_{\perp 2} b_j) + K_{2,n} N_n'(k_{\perp 2} b_j)\right] \cos(n\theta_j) \\ & \qquad + \frac{n b_j'}{b_j^2}\left[K_{1,n} J_n(k_{\perp 2} b_j) + K_{2,n} N_n(k_{\perp 2} b_j)\right] \sin(n\theta_j) \\ & + \sum_{n=1}^{N} B_n \left[k_{\perp 2}[K_{1,n} J_n'(k_{\perp 2} b_j) + K_{2,n} N_n'(k_{\perp 2} b_j)\right] \sin(n\theta_j) \\ & \qquad - \frac{n b_j'}{b_j^2}\left[K_{1,n} J_n(k_{\perp 2} b_j) + K_{2,n} N_n(k_{\perp 2} b_j)\right] \cos(n\theta_j) = 0, \end{aligned} \tag{14a}$$

where

$$K_{1,n} = \frac{\pi a}{2\rho_2}\left[\rho_1 k_{\perp 2} J_n(k_{\perp 1} a) N_n'(k_{\perp 2} a) - \rho_2 k_{\perp 1} J_n'(k_{\perp 1} a) N_n(k_{\perp 2} a)\right] \tag{14b}$$

$$K_{2,n} = \frac{\pi a}{2\rho_2}\left[\rho_2 k_{\perp 1} J_n(k_{\perp 1} a) N_n'(k_{\perp 2} a) - \rho_1 k_{\perp 2} J_n'(k_{\perp 1} a) J_n'(k_{\perp 2} a)\right]. \tag{14c}$$

The number of unknowns A_0, A_n, B_n is equal to the number of equations, if $P = 2N + 1$.

The above system of equations can be written in the matrix form:

$$\mathbf{S}\mathbf{a} = 0 \tag{15a}$$

where

$$\mathbf{a} = (A_0, A_1, B_1, \ldots, A_N, B_N)^T. \tag{15b}$$

Table 2: Normalized fourth eigenvalue $k_{\perp 1}^{(4)} b_m$ for elliptic external contour with $k_1 b_m = 2$, for SAAB silencer [24].

N	$k_{\perp 1}^{(4)} b_m$
2	2.3871 – 1.7639i
4	2.3983 – 1.7534i
6	2.3986 – 1.7525i
8	2.3947 – 1.7491i
10	2.4010 – 1.7540i
12	2.3919 – 1.7494i
14	2.4053 – 1.7528i
16	2.3854 – 1.7515i

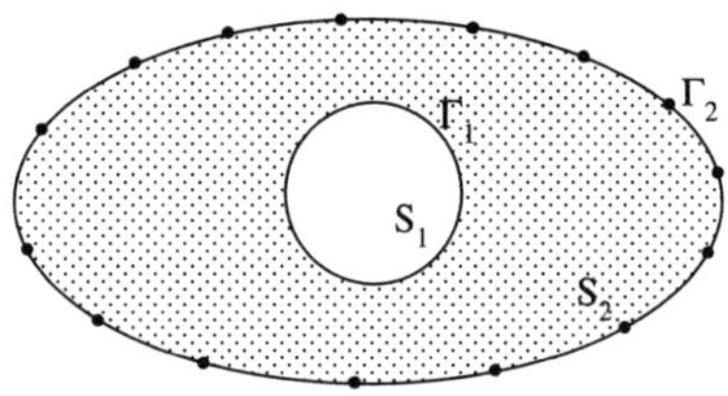

Figure 2: Location of collocation points.

The condition for existing non-zero solutions is equating the determinant of matrix **S** to zero:

$$\det(\mathbf{S}) = 0. \tag{16}$$

As a result of solution of this non-linear algebraic equation, we obtain the sought wavenumbers. It has been found that with very few collocation points, the calculated solutions are stable and do not significantly change after increase of number of the collocation points, which has been illustrated in Table 2. This table presents fourth eigenvalue $k_{\perp 1}^{(4)} b_m$ for the SAAB silencer, in which the external contour is an ellipse (Fig. 2).

References

[1] Akkari, R.F. & Hutchinson, J.R., An improved boundary method for plate vibrations. In: *Boundary Element Techniques: Applications in Steress Analysis and Heat Transfer*, eds. C.A. Brebbia, W.S. Venturini, Computational Mechanics Publications: Southampton, pp. 89–103, 1987.

[2] Akkari, R.F. & Hutchinson, J.R., Boundary point collocation method for vibration of the Mindlin thick plate. In: *Boundary Elements X*, ed. C.A. Brebbia, Springer: Berlin, **3**, pp. 465–476, 1988.
[3] Baltrukonis, J.H., Axial shear vibrations of star shaped bars by the collocation method. *Hercules Powder Company*, Salt Lake City, Utah, Rept. 4, 1963.
[4] Bates, R.H.T. & Ng, F.L., Point matching computation of transverse resonance. *International Journal of Numerical Methods in Engineering*, **6**, pp. 155–168, 1973.
[5] Bates, R.H.T., Rayleigh hypothesis the extend-boundary-condition and point matching. *Electronic Lectures*, **5**, pp. 654–655, 1969.
[6] Bates, R.H.T., The theory of the point matching method for perfectly conducting waveguides and transmission lines. *IEEE Transactions on Microwave Theory and Techniques*, MTT-**17**, pp. 294–301, 1969.
[7] Cap, F.F., Boundary value problems of unseparable linear partial differential equations. *Zeitschrift für Angewandte Mathematik und Mechanik (ZAMM)*, **66**, pp. 248–250, 1986.
[8] Chen, J.T., Chang, M.H., Chen, K.H. & Chen, I.L., The boundary collocation method for acoustics eigenanalysis of three-dimensional cavities using radial basis function. *Computational Mechanics*, **29**, pp. 392–408, 2002.
[9] Chen, J.T., Chang, M.H., Chen, K.H. & Lin, S.R., The boundary collocation method with meshless concept for acoustic eingenanalysis of two-dimensional cavities using radial basis function. *Journal of Sound and Vibrations*, **257**, pp. 667–711, 2002.
[10] Chen, J.T., Chen, I.L. & Lee, Y.T., Eigensolutions of multiply connected membranes using the method of fundamental solutions. *Engineering Analysis with Boundary Elements*, **29**, pp. 166–174, 2005.
[11] Chen, J.T., Chen, I.L., Chen, K.H. & Lee, Y.T., Comments on 'Free vibration analysis of arbitrary shaped plates with clampede edges using wave-type function'. *Journal of Sound and Vibrations*, **262**, pp. 370–378, 2003.
[12] Chen, J.T., Chen, I.L., Chen, K.H., Lee, Y.T. & Yeh, Y.T., A meshless method for free analysis of circular and rectangular clamped plates using radial basis function. *Engineering Analysis with Boundary Elements*, **28**, pp. 535–545, 2004.
[13] Christiansen, S. & Kleinman, R.E., On a misconception involving point collocation and the Rayleigh hypothesis. *IEEE Transactions on Antennas and Propagation*, **44**, pp. 1309–1316, 1996.
[14] Conway, H.D. & Farnham, K.A., The free flexural vibrations of triangular rhombic parallelogram plates and some analogies. *International Journal of Mechanical Sciences*, **7**, pp. 811–819, 1965.
[15] Conway, H.D. & Leissa, A.W., A method for investigating eigenvalue and vibration plates. *Journal of Applied Mechanics*, **27**, pp. 557–558, 1960.
[16] Conway, H.D., Analogies between the buckling and vibration of polygonal plates and membranes. *Canadian Aeronautical Journal*, **1**, pp. 93–97, 1960.
[17] Conway, H.D., The bending, buckling, and flexural vibration of simply supported polygonal plates by point matching. *Journal of Applied Mechanics*, **28**, pp. 288–291, 1961.

[18] Davies, J.B. & Nagenthiram, P., Irregular fields, nonconvex shapes and the point matching method for hollow waveguides. *Electronic Letters*, **7**, pp. 401–404, 1971.

[19] Descloux, J. & Tolley, M.D., An accurate algorithm for computing the eigenvalues of polygonal membrane. *Computer Methods in Applied Mechanics and Engineering*, **39**, pp. 37–53, 1983.

[20] Descloux, J. & Tolley, M.D., Approximation of the Poisson problem and of the eigenvalue problem for the Laplace operator by the method of large singular elements. *Research Report* 81-01, *Seminar für Angewandte Mathematic,* Eidgenossiche Technical Hochschule, 1980.

[21] Deverall, L.I., Modified method of collocation for vibrating plate problems. In: *International Conference on Partial Differential Equations and Continuum Mechanics*, Mathematics Research Centre U.S. Army, University of Wisconsin, Madison, Wisconsin, 1960.

[22] Donnelly, J.D.P., Eigenvalues of membranes with reentrants corners. *SIAM Journal of Numerical Analysis*, **6**, pp. 47–61, 1969.

[23] Fleming, D.P. & Sparrow, E.M., Flow in the hydrodynamic entrance region of ducts of arbitrary cross section. *Journal of Heat Transfer*, **91**, pp. 345–354, 1969.

[24] Glav, R., The point-matching method on dissipative silencers of arbitrary cross-section. *Journal of Sound and Vibration*, **189**, pp. 123–135, 1996.

[25] Glav, R., The transfer matrix for a dissipative silencer of arbitrary cross-section. *Journal of Sound and Vibration*, **236**, pp. 575–594, 2000.

[26] Hegerty, R.F. & Ariman, T., Elasto-dynamic analysis of rectangular plates with circular holes. *International Journal of Solids and Structures*, **11**, pp. 895–906, 1975.

[27] Houmat, A. & Hutchinson, J.R., A boundary method for free vibration of solids of revolution. In: *Advances in Boundary Elements Methods in Japan and USA*, eds. M. Tanaka, C.A. Brebbia, R. Shaw, Computational Mechanics Publikations, pp. 121–135, 1990.

[28] Houmat, A. & Hutchinson, J.R., Free vibration of bodies of revolution by boundary collocation. *Journal of Sound and Vibration*, **171**, pp. 35–48, 1994.

[29] Houmat, A., Flexural and longitudinal vibrations of deep beams by boundary collocation. *European Journal of Mechanics, A/Solids*, **13**, pp. 267–276, 1994.

[30] Kang, S.W. & Lee, J.M., Application of free vibration analysis of membranes using the non-dimensional dynamic influence function. *Journal of Sound and Vibration*, **234**, pp. 455–470.

[31] Kang, S.W. & Lee, J.M., Eigenmode analysis of arbitrary shaped two-dimensional cavities by the method of point-matching. *Journal of Acoustical Society of America*, **107**, pp. 1153–1160, 2000.

[32] Kang, S.W., Free vibration analysis of arbitrary shaped plates with a mixed boundary condition using non-dimensional dynamic influence functions. *Journal of Sound and Vibrations*, **256**, pp. 533–549, 2002.

[33] Karageorghis, A., The method of fundamental solutions for the calculation of the eigenvalues of the Helmholtz equation. *Applied Mathematics Letters*, **14**, pp. 837–842, 2001.

[34] Klein, A., Vibration of simply supported isosceles trapezoidal plates. *Journal of Acoustical Society of America*, **27**, pp. 1059–1060, 1955.

[35] Laura, P.A.A., Application of the point-matching method in waveguide problems. *IEEE Transactions MTT (Correspondence)*, MTT-**14**, pp. 251, 1966.

[36] Laura, P.A.A., Determination of cut-off frequencies of wave-guides with arbitrary cross sections by point matching. *Proceedings IEEE* (Correspondence), **53**, pp. 1660–1661, 1965.

[37] Laura, P.A.A. & Rodriguez, K., Comments on 'The point-matching method on dissipative silencers of arbitrary cross-sections'. *Journal of Sound and Vibration*, **201**, pp. 127–128, 1997.

[38] Laura, P.A.A., Gutierrez, R.H., Nagaya, K. & Sanchez Sarmieto, G., Vibration of rectangular membrane with an eccentric inner circular boundary: a comparison of approximate methods. *Journal of Sound and Vibration*, **75**, pp. 109–115, 1981.

[39] Laura, P.A.A., Nagaya, K. & Sanchez Sarmieto, G., Numerical experiments on the determination of cutoff frequencies of waveguides of arbitrary cross section. *IEEE Transactions MTT*, MTT-**28**, pp. 568–572, 1980.

[40] Moler, C.B., Accurate bounds for the eigenvalues of the Laplacian and application to rhombical domains. *Stanford Technical Report* CS69-121, Stanford University, February, 1969.

[41] Ng, F.L. & Bates, R.H.T., Bull field method for waveguides of arbitrary cross section. *IEEE Transactions MTT*, MTT-**20**, pp. 658–662, 1972.

[42] Quinlan, P.M., The edge function method (EFM) for 2D regions with arbitrary boundaries in fracture, elastodynamics, thin plates and Laplace/Poisson problems. In: *Boundary Elements X*, **1**, ed. C.A. Brebbia, Springer: Berlin, pp. 381–396, 1988.

[43] Raciti, F. & Venturino, E., Eigenfrequency determination for arbitrary cross-section waveguides. *Computers and Mathematics with Applications*, **41**, pp. 319–326, 2001.

[44] Subrahmanyan, M.B. & Wah, T., Vibration of quadrilateral plates. *Computer Methods in Applied Mechanics and Engineering*, **43**, pp. 315–323, 1984.

[45] Torvik, P.J. & Eastep, F.E., A method for improving the estimation of membrane frequencies. *Journal Sound and Vibrations*, **21**, pp. 285–294, 1972.

[46] Wang, C.Y., Fundamental frequencies of a membrane strip with periodic boundary constraints. *Journal of Sound and Vibration*, **214**, pp. 389–393, 1998.

[47] Wang, C.Y., On the polygonal membrane with a circular core. *Journal of Sound and Vibration*, **215**, pp. 195–199, 1998.

[48] Yee, H.Y. & Audeh, N.F., Attenuation constants of waveguides with general cross sections. *IEEE Transactions on Microwave Theory and Techniques*, MTT-**14**, pp. 252–253, 1966.

[49] Yee, H.Y. & Audeh, N.F., Cutoff frequencies of eccentric waveguides. *IEEE Transactions on Microwave Theory and Techniques*, MTT-**14**, pp. 487–493, 1966.

[50] Yee, H.Y. & Audeh, N.F., Cutoff frequencies of waveguides with arbitrary cross section. *Proceedings IEEE*, **53**, pp.637–638, 1965.

[51] Yee, H.Y. & Audeh, N.F., Uniform waveguides with arbitrary cross section considered by point matching method. *IEEE Transactions on Microwave Theory and Techniques*, MTT-**13**, pp. 847–851, 1965.
[52] Yee, H.Y., On determination of cutoff frequencies of waveguides with arbitrary cross sections. *Proceedings IEEE (Correspondence)*, **54**, pp. 64, 1966.
[53] Young, D.L., Hu, S.P., Chen, C.W., Fan, C.M. & Murugesan, K., Analysis of elliptical waveguides by the method of fundamental solutions. *Microwave and Optical Technology Letters*, **44**, pp. 552–558, 2005.

CHAPTER 15

Final remarks, conclusions and perspectives

The main idea as well as mathematical basis of the presented boundary collocation method (BCM) are very simple. A solution of any boundary value problem is searched in a form of a finite functional series, the terms of which identically fulfill a governing partial differential equation of the discussed problem. The unknown coefficients of this series can be calculated from algebraic systems of equations resulting from boundary and connectivity conditions.

Because of this form of solution, the most convenient for application of the boundary collocation approach are linear problems defined in simple-shape regions. Sometimes even in such simple cases, the mesh-depended methods can meet difficulties, for example, in solutions with unbounded domains or in case of wave equations for larger eigenfrequencies. Obviously, the boundary collocation can be used not only in the problems mentioned above; however, for more complex regions and non-linear formulations, it should be used more carefully. Conditioning of the resulting algebraic equations should then be observed and convergence of iterations should be investigated.

The ways of proceeding is clearly presented in Chapter 5 on easy but illustrative numerical examples. Also in further chapters, the particular examples elucidate behaviour of results while solving different engineering problems. Hence, this book exposes more practical, applicability side of the Trefftz method in comparison to recent monograph [1] broader presenting its mathematical background.

The analytical–numerical boundary collocation approach is the nearest relative of the regular boundary element method (RBEM). In case of the Kupradze T-functions and numerical formulation of the boundary integrals, the RBEM can be treated as a specific form of the BCM. Therefore, their properties are similar. In comparison to the classical, singular BEM, the boundary collocation approach has two different positive features: (a) fully regular formulation in any investigated area Ω including its boundary Γ and (b) solution in whole Ω defined in form of a functional series. They result in easy calculation of searched results in the interesting domain and enable easier application to non-linear problems (see Chapter 12). On the other hand, the difficulties with conditioning are more serious in the BCM than in the singular BEM. Certain conditioning improvement procedures must sometimes be used (see Chapter 5) and substructuring more often applied (Chapter 4).

A strong side of the BCM is the possibility of application of the special T-complete functions (Section 5 in Chapter 3), which enable obtaining very accurate results in vicinity of edges or corners, where solution gradients are large. Also introduction of large Trefftz finite elements (Chapter 4) makes this form of mathematical modelling more attractive.

As the main direction of development of the BCM, the authors consider the following areas:

- Optimization procedures, in which certain solutions should be repeated very many times with variable design parameters (see Section 4 and also [23, 24, 61, 63] in Chapter 8)
- Modelling of transient, non-stationary effects (see Chapter 13)
- Inverse problems (see Chapter 9 and also [2])
- Certain non-linear models [3]
- Structural high-cycle fatigue problems, in which equivalent stresses can be calculated from linear elastic solutions (plastic effects are not allowed)

At the end it should be noted that the recent quick development of computational basis makes the boundary collocation approach more effective and reliable.

References

[1] Li, Z.C., Lu T.T., Hu, H.Y. & Cheng, A.H.D., *Trefftz and Collocation Methods,* WITpress: Southampton and Boston, 2008.

[2] Maciąg, A., *The Trefftz Functions for Selected Direct and Inverse Problems of Mechanics* (in Polish), Habilitation Thesis, Kielce University of Technology Press: Kielce, 2009.

[3] Uściłowska, A., *Solution of Selected Problems of Nonlinear Mechanics by the Method of Fundamental Solutions* (in Polish), Habilitation Thesis, Poznan University of Technology Press: Poznan, 2008.

Index